S0-BLS-185

Laser-Induced Chemical Processes

Laser-Induced Chemical Processes

Edited by

Jeffrey I. Steinfeld

Massachusetts Institute of Technology
Cambridge, Massachusetts

PLENUM PRESS • NEW YORK AND LONDON

Library of Congress Cataloging in Publication Data

Main entry under title:

Laser-induced chemical processes.

Includes index.
1. Lasers in chemistry. I. Steinfeld, Jeffrey I.
QD63.L3L37 541.3 80-20478
ISBN 0-306-40587-3

A Division of Plenum Publishing Corporation
227 West 17th Street, New York, N.Y. 10011

Printed in the United States of America

Contributors

Jay R. Ackerhalt, Theoretical Division, Los Alamos Scientific Laboratory, Los Alamos, New Mexico

W. Roger Cannon, Energy Laboratory, Massachusetts Institute of Technology, Cambridge, Massachusetts

Wayne C. Danen, Department of Chemistry, Kansas State University, Manhattan, Kansas

Harold W. Galbraith, Theoretical Division, Los Alamos Scientific Laboratory, Los Alamos, New Mexico

John S. Haggerty, Energy Laboratory, Massachusetts Institute of Technology, Cambridge, Massachusetts

J. C. Jang, Department of Chemistry, Kansas State University, Manhattan, Kansas

J. I. Steinfeld, Department of Chemistry, Massachusetts Institute of Technology, Cambridge, Massachusetts

Preface

The possibility of initiating chemical reactions by high-intensity laser excitation has captured the imagination of chemists and physicists as well as of industrial scientists and the scientifically informed public in general ever since the laser first became available. Initially, great hopes were held that laser-induced chemistry would revolutionize synthetic chemistry, making possible "bond-specific" or "mode-specific" reactions that were impossible to achieve under thermal equilibrium conditions. Indeed, some of the early work in this area, typically employing high-power continuous-wave sources, was interpreted in just this way. With further investigation, however, a more conservative picture has emerged, with the laser taking its place as one of a number of available methods for initiation of high-energy chemical transformations. Unlike a number of these methods, such as flash photolysis, shock tubes, and electron-beam radiolysis, the laser is capable of a high degree of spatial and molecular localization of deposited energy, which in turn is reflected in such applications as isotope enrichment or localized surface treatments.

The use of lasers to initiate chemical processes has led to the discovery of several distinctly new molecular phenomena, foremost among which is that of multiple-photon excitation and dissociation of polyatomic molecules. This research area has received the greatest attention thus far and forms the focus of the present volume. The intent here is to provide a comprehensive picture of the chemical effects of high-power infrared radiation, including theoretical analysis, large-molecule behavior, and thermal effects at the gas–solid interface, along with a survey of the literature.

The first chapter, by Drs. H. W. Galbraith and J. R. Ackerhalt of the Theoretical Division of the Los Alamos Scientific Laboratory, presents a detailed derivation of their general model for the multiple-photon excitation process in light of the most recent experimental data. Calculations based on this model are presented for the specific examples of SF_6 and S_2F_{10}, and a comparison is developed with other theoretical approaches, namely that of Stone, Goodman, and Dows, and the "thermal quasicontinuum" model of Black and Yablonovitch.

In Chapter 2, multiple-photon excitation, isomerization, and dissociation of "large" organic molecules are discussed from the perspective of physical–organic chemistry. The authors are Drs. W. C. Danen and J. C. Jang of the Department of Chemistry of Kansas State University. They review data for a variety of experimental systems, with particular emphasis on ethyl acetate, a molecule that has been thoroughly explored as a model system in the authors' laboratory. The use of a "chemical thermometer" as a monitor of the thermal component of the laser-induced reaction is discussed. The authors also present a phenomenological, incoherent-pumping model with which they analyze their results.

In the third chapter, Drs. J. S. Haggerty and W. R. Cannon describe a new application of high-power laser radiation to the synthesis of finely divided powders of refractory ceramic materials in a homogeneous gas-phase medium. These powders appear to be ideally suited for production of high-temperature components by a diffusional sintering process. Although the laser synthesis is established to be essentially thermal in nature, the unique way in which the energy is deposited leads to particle shapes, sizes, and compositions that are superior to those produced in nonlaser processes. Included in this chapter are measurements of the optical properties of the process gases, comparison with collisionless multiple-photon excitation, and a model for the thermal process. Drs. Haggerty and Cannon are with the Energy Laboratory at the Massachusetts Institute of Technology.

Chapter 4 is an attempt, by the editor of the present volume, to survey the literature in laser-induced chemistry. Tables of reported reaction data are presented, arranged according to the laser-excited principal reacting species, and giving such information as added reagents, products observed, identity, wavelength, and intensity of the laser line employed, and other pertinent reaction conditions. References for these tables are then given, along with a selection of published reviews and monographs. It is our hope that this chapter will serve as an entry to the rapidly growing body of work in this field for those who wish to begin research in this area, as well as a convenient tool for checking known laser effects on a specific compound.

The possibility of applying laser-induced chemical processes to industrial operations has received at least as much attention as the basic physics and chemistry, frequently anticipating the actual result from the laboratory. Thus, a few words about this aspect may be appropriate here. The primary original impetus was provided by the isotopic-specific nature of a number of laser-induced processes, especially multiple-photon dissociation. Indeed, laser isotope separation schemes have now been demonstrated, at least on a laboratory scale, for many species ranging from deuterium to uranium, frequently with high single-step enrichment ratios. The use of lasers for general chemical synthesis has not progressed nearly so far, in part as a result of the high specific cost of the laser photons as an energy source and in part because of the lack of discovery, to date, of a truly "mode-specific" reaction. These limitations have been discussed in a number of reports, the most recent a "Jason" summer study.‡ The authors of that study concluded that, indeed, at its present stage of development, laser-induced photochemistry was not in a position to make a major impact on industry over the next few years. A somewhat more optimistic conclusion was reached by the members of a panel subsequently convened by the National Science Foundation and the Department of Energy in the summer of 1979. Their report,§ which takes a somewhat broader view, concludes that the use of lasers for *diagnostic* purposes has already had a major impact in the chemical process industries, and that actual laser-induced chemical processes, such as those discussed in this volume, are likely to be significant in a variety of specialized industrial applications, such as semiconductor device processing.

The authors of these chapters would like collectively to thank their colleagues, secretaries, and family members who aided in the production of their individual contributions. A special acknowledgment is due Ellis Rosenberg, Senior Editor at Plenum Publishing Corporation, for his encouragement of this project.

Cambridge, Massachusetts J. I. Steinfeld

‡ "Laser Induced Photochemistry," W. Happer, J. Chamberlain, H. Foley, N. Fortson, J. Katz, R. Novick, M. Ruderman, and K. Watson, SRI International Report No. JSR-78-11 (February 1978).

§ "Laser Photochemistry and Diagnostics: Recent Advances and Future Prospects," J. Davis, M. Feld, C. P. Robinson, J. I. Steinfeld, N. Turro, W. S. Watt, and J. T. Yardley (November 1979).

Contents

1. Vibrational Excitation in Polyatomic Molecules

Harold W. Galbraith and Jay R. Ackerhalt

2. Multiphoton Infrared Excitation and Reaction of Organic Compounds

Wayne C. Danen and J. C. Jang

3. Sinterable Powders from Laser-Driven Reactions

John S. Haggerty and W. Roger Cannon

4. Laser-Induced Chemical Reactions: Survey of the Literature, 1965–1979

J. I. Steinfeld

1

Vibrational Excitation in Polyatomic Molecules

Harold W. Galbraith and Jay R. Ackerhalt

1.1. Introduction

In this article we will discuss what happens to a polyatomic molecule that is driven by a high-power, resonant, infrared laser. It is assumed that the molecules are in the gas phase, and our discussion centers around systems at low pressure so that we may focus our attention on one individual molecule immersed in the photon field of the laser. We will present detailed calculations simulating the response of two molecular species, SF_6 and S_2F_{10}. In this introduction we will discuss briefly our (admittedly subjective) view of the current status of the multiphoton excitation–dissociation problem (mpe, mpd) with primary emphasis on SF_6, the most studied molecule.

In this paper we will draw upon the results of many experiments related to the mpe and mpd of polyatomic molecules. We divide these experiments into four general classes: (1) molecular structure experiments, (2) multiple-photon absorption experiments, (3) dissociation experiments, and (4) two frequency experiments.

The molecular structure experiments provide the basic building blocks of all further discussion. Here we include Fourier transform, Raman, and laser diode work which gives information as to the frequencies of the normal modes, absorption band intensities from which the mode-dipole moment is derived, rotational constants and even some high-order Coriolis

Harold W. Galbraith and Jay R. Ackerhalt • Theoretical Division, Los Alamos Scientific Laboratory, Los Alamos, New Mexico

effects. Also fundamental are the electron diffraction studies from which the symmetry and bond lengths are determined. From these data we obtain the basic molecular constants of the normal mode picture of the molecule, a picture that is strictly correct at low levels of excitation and a picture to which our model (Ackerhalt and Galbraith, 1979) *must limit* as the laser flux is reduced. Of the molecules studied in mpe and mpd, SF_6 is the one that has received the most detailed analysis, and we will draw upon these results (at least to first order) (Ackerhalt and Galbraith, 1978). S_2F_{10}, the other molecule that we study here, is much less well understood spectroscopically although significant advances have recently been reported (Jones and Ekberg, 1979).

The second class of experiments are the multiple-photon absorption measurements. Here the fundamental dependent variable is the average number of photons absorbed, $\langle n \rangle$ [or absorption cross section, σ, given by $\langle n \rangle$ multiplied by the laser fluence in appropriate units (Lyman *et al.*, 1980)]. The experimental parameters that are varied consist of the laser fluence, ϕ, pulse length, τ_p, frequency, ω_L, the gas temperature, and gas pressure. A critical analysis of the absorption data in SF_6 was recently carried out by Judd (1979). Although there are some major inconsistencies in some of the data, consistent trends can be observed. For example in SF_6 absorption at CO_2–$P(20)$, one finds quite generally that $\langle n \rangle$ increases with ϕ at roughly the two-thirds power for all fluences from 10^{-4} to 3 J/cm^2. This means a steady falloff in the absorption cross section. For S_2F_{10} the absorption at low fluence is even stronger (Lyman and Leary, 1978). Quite generally the absorption increases steadily with fluence, and no threshold for mpe is observed in the data (Bagratashvili *et al.*, 1976 and 1979; Ambartzumian *et al.*, 1976a; Deutsch, 1977; Letokhov, 1977; Ham and Rothchild, 1977; Lyman *et al.*, 1978b; Letokhov, 1978; Akmanov *et al.*, 1978). Furthermore, Lyman *et al.* (1980) has pointed out that the deviation from linear absorption can occur at very low fluences (μJ/cm^2 in SF_6). Evidently, most polyatomic molecules having CO_2 laser resonances (even combination bands with their corresponding weak dipole are included) (Ambartzumian *et. al.*, 1975; Akhmanov *et al.*, 1977) can absorb many infrared photons quite easily. The difficulty arises from our analysis of the class 1 experiments. There we have a picture of the molecule as a collection of anharmonic oscillators. As the laser fluence increases, the absorption advances to the point where the molecular anharmonicities should, in principle, prevent further absorption. Clearly this picture of standard spectroscopy is incomplete if directly applied to mpe. Our model is an attempt at correcting this picture.

The pulselength dependence of $\langle n \rangle$ has been shown to be very weak, much weaker than is expected in a standard spectroscopic theory. For $P(20)$ pumping of SF_6 at room temperature, one finds a change of laser intensity by a factor of 100 gives only a factor of 2 change in the absorption (Lyman *et al.*, 1980; Smith *et al.*, 1979). A comparison of high-fluence sub-nanosecond pulse excitation with standard TEA laser pulse data are even more convincing evidence of the weak dependence of $\langle n \rangle$ on the laser intensity (Kolodner *et al.*, 1977). While increased laser intensity may increase the absorption slightly (factors of 2 or 3 at best), the functional form of $\langle n \rangle$ versus ϕ remains unchanged. [This effect is mirrored in comparison of mode-locked pulse and smooth pulse data (Lyman *et al.*, 1978a).]

As the laser energy is increased, a scan of $\langle n \rangle$ versus laser frequency shows that the absorption peak is slightly redshifted and that the spectra are generally smoothed (Letokhov, 1977; Deutsch, 1977; Stafast *et al.*, 1977; Knyazev *et al.*, 1978; Alimpiev *et al.*, 1978a,b; Letokhov, 1978; Fuss and Hartmann, 1979). This behavior is quite easily understood in terms of simple pumping of anharmonically redshifted multiple-photon resonances of the normal mode picture (Larsen, 1976; Sazonov and Finkelshtein, 1977; Galbraith and Ackerhalt, 1978a,b; Ackerhalt and Galbraith, 1978; Knyazev *et al.*, 1978; Alimpiev *et al.*, 1978a,b; Akulin *et al.*, 1978). The results are also found in other molecules (Kolomiiskii and Ryabov, 1978; Rabinowitz *et al.*, 1978; Ambartsumian *et al.*, 1978a,b,c; Karl and Lyman, 1978; Proch and Schröder, 1979; Tiee and Wittig, 1978b; Horsley *et al.*, 1980; Lyman *et al.*, 1979; Ambartzumian *et al.*, 1979; Freund and Lyman, 1978).

The last two parameters, the temperature and pressure, refer to the condition of the gas rather than the laser beam. It is found that increasing the temperature increases the absorption very strongly (Tsay *et al.*, 1979) especially on the low-frequency side of the resonant band center. Similarly, increasing the pressure increases the absorption, but only markedly below about 0.4 J/cm^2 in SF_6 (Quigley, 1978).

Contrary to the absorption data, the dissociation rate versus fluence does show a steep threshold (Gower and Billman, 1977; Grant *et al.*, 1977a,b,c; Coggiola *et al.*, 1977; Kolodner *et al.*, 1977; Rothchild *et al.*, 1978; Brunner and Proch, 1978; Black *et al.*, 1979). This is physically reasonable since no dissociation can occur below the activation energy, and once this level of excitation is reached, the normal mode density of states is enormously large so that a statistical rate model appears to be in order. The work of the Berkeley group (Grant *et al.*, 1977a,b,c; Coggiola *et al.*, 1977) indicates that the laser-excited molecules can be fairly high in the true continuum before dissociation sets in. This would be in accordance with standard RRKM theory (Robinson and Holbrook, 1972) although the issue is still not settled (Thiele *et al.*, 1980).

As in mpe, the intensity dependence of the dissociation rate is weak (Kolodner *et al.*, 1977; Woodin *et al.*, 1978), and there is a similar red shift in the frequency dependence (Gower and Billman, 1977). Surprisingly as the gas temperature is increased, the dissociation drops strongly at all but the very highest fluences. Again SF_6 is the molecule studied (Duperrex and Van den Bergh, 1979b) for temperatures above 300°K. With increased gas pressure one finds an overall increase in the dissociation, but the effect has not been studied to any great extent (Rothchild, 1979; Duperrex and Van den Bergh, 1979a; Gordinko *et al.*, 1978).

From all of the above mpe and mpd studies we obtain only general trends defining a so-called new phenomenon. As we will see in later sections, however, these data are insufficient for distinguishing some very different theoretical models. A much more sensitive probe of the molecular dynamics has appeared on the scene in the form of a combination of mpe and spectroscopy. These are the two frequency pump and probe experiments (Steinfeld and Jensen, 1976; Moulton *et al.*, 1977; Steinfeld *et al.*, 1979; Jensen *et al.*, 1979) and the two frequency mpe and mpd experiments (Ambartzumian *et al.*, 1976b; Gower and Gustafson, 1977; Ambartzumian *et al.*, 1977; Quigley, 1979; Tiee and Wittig, 1978a), where both lasers are high-powered pulse lasers. The pump and probe experiments on SF_6 can give us information regarding excited state spectroscopy (hence yielding more and more accurate molecular constants), decay rates for excited states (Steel and Lam, 1979), as well as values for higher-order terms in the expansion of the molecular dipole moments (Jensen *et al.*, 1979), Convoluted in such data will also be information on specific V–V (intramolecular) rates and the actual temporal flow of laser-induced populations. It is a virtual certainty that the untangling of such data will mean the downfall of many of the current theoretical models. (Perhaps all!) Unfortunately, we will have little more to say about the two frequency experiments since to date we have not completed any such model calculations.

Let us now briefly review the theoretical situation. We first divide all models into quantum versus classical. The classical models have certain advantages as well as some shortcomings relative to the quantum models (Lamb, 1977 and 1979; Narducci *et al.*, 1977; Walker and Preston, 1977). Quite generally, however, the strengths of the classical models correspond exactly to the weaknesses in the quantum models and vice versa. The first requirement of the classical models is a potential energy surface. This is a difficult problem but not an unsurmountable one as shown by Walker (1979). Once the potential surface is constructed, the classical equations of motion are solved as a function of time over the laser pulse. Here is where the great difficulty arises: The laser pulses correspond to a very large number of oscillations ($\sim 10^6$) of the normal modes. Hence, the time steps must be very small in the calculations leading to inaccuracies as the integration proceeds. Furthermore, each trajectory must be run for several sets of initial phase for accurate statistics. These problems of high computer cost and inaccuracy are sidestepped by turning up the laser power and integrating for short times. As we will see below, this leads to an artificially large amount of coherence in the excitation that may be misleading if taken out of context. This is not a shortcoming at all if comparisons are made with strictly short-pulse, high-intensity data (Kolodner *et al.*, 1977; Black *et al.*, 1979). As we will see, the quantum models have a particularly bad time under these conditions. One of the great difficulties of the quantum calculations is the correct incorporation of intramolecular V–V relaxation and the determination of exactly how the energy is distributed over the normal modes. This is automatic in the classical calculations and follows directly from the form of the potential surface. It is our opinion that many of the results of classical calculations will prove useful for the quantum models. For example, Walker (1979) finds that the spectroscopic ΔB varies linearly with the number of quanta absorbed in SF_6, and this appears to be confirmed in recent work on $3\nu_3$ (Pine and Robiette, 1979).

Most quantum models are either a direct application of the normal mode picture driven by a powerful laser (fully coherent Schrödinger equation dynamics) (Larsen, 1976; Jensen *et al.*, 1977; Cantrell and Galbraith, 1976 and 1977; Cantrell *et al.*, 1977; Galbraith and Cantrell, 1977; Galbraith and Ackerhalt, 1978a,b; Ackerhalt and Galbraith, 1978) or make no attempt at being consistent with spectroscopy (usually incoherent rate equation dynamics) (Hodgkinson and Briggs, 1976; Bloembergen and Yablonovitch, 1977; Black *et al.*, 1977; Kuzmin, 1978; Quack, 1978; Taylor *et al.*, 1977; Steverding *et al.*, 1978; Shuryak, 1978; Jensen *et al.*, 1978; Platonenko, 1978; Fuss, 1979; Mukamel, 1979; Emanuel, 1979; Black *et al.*, 1979). The fundamental conceptual breakthrough came with the work of Isenor *et al.* (1973) and the introduction of the molecular quasicontinuum. This general model is now well accepted among quantum modellers, though some insist that they can still do a part of the problem and have that part describe the essential physics. The picture is as follows: Initial states are represented in accordance with spectroscopy, discrete states in the normal mode representation (Cantrell *et al.*, 1979). As the excitation advances, anharmonic mixing begins to occur, gradually blending together normal mode oscillator states. The blending of states leads to a flow of population out of the overtone levels of the pumped mode and into certain background states characterized by a density of states at the given energy (Deutsch and Brueck, 1978; Frankel, 1976; Frankel and Manuccia, 1978; Steel and Lam, 1979; Kay, 1974; Kwok and Yablonovitch, 1978; Mukamel, 1978; Yablonovitch, 1977). The excitation advances still further through this quasicontinuum, QC, until the activation region is reached and dissociation occurs. The level structure in the QC is found to be quite dense so that the anharmonic bottleneck of spectroscopy is replaced with the (slow) pumping of very broad resonant levels.

Our model (Ackerhalt and Galbraith, 1979) incorporates all of the above structure into one unified picture. By assuming weak, selection rule limited, anharmonic mixing of the normal mode states, we find that the pumped mode ladder naturally blends into the molecular QC for SF_6 and S_2F_{10}. We are able to derive from first principles the onset of this QC and relate all of our QC parameters to basic spectroscopic quantities. We always work with the normal mode basis set, which allows us to make connections with low-power spectroscopy easily, and from which the molecular–laser coupling is straightforward. As the calculations show, it is quite easy for our model (in the cases of SF_6 and S_2F_{10}) to accommodate all of the data of types 1–3.

Ours is not the only "complete" model describing ground state through dissociation (Letokhov and Makarov, 1976; Mukamel and Jortner, 1976; Goodman *et al.*, 1976; Stone *et al.*, 1976; Tamir and Levine, 1977; Letokhov and Makarov, 1978; Zi-Zhao *et al.*, 1978; Horsley *et al.*, 1979), but we do incorporate a true first-order spectroscopic model with the broader QC in a natural manner.

1.2. Complete Model

Almost every day we learn of another molecule that has been dissociated by stepwise absorption of infrared laser radiation up through its vibrational manifold. Most of these molecules have certain general features in common enabling their mpe and mpd to be characterized in a single unified picture. We are specifically referring to molecules that are excited and dissociated in a statistically random fashion; i.e., energy is easily distributed among all the vibrational motions of similar energy, and dissociation is described by RRKM rates (Robinson and Holbrook, 1972). Molecules of this type can be characterized simply by their density of states. Molecules that may be excited via a bond specific local mode that is weakly coupled to other molecular vibrations are considered anomalous and are not describable in the framework of a general model. We do not consider such molecules here (Heller and Mukamel, 1979).

In general, mpe is characterized by three regions of excitation: low-level discrete excitation, a transition region to the molecular quasicontinuum, and the high QC from which dissociation eventually occurs. Depending on the temperature, density of states, and anharmonicity of the pumped mode, these regions can play a greater or lesser role in the mpe process. For example, in S_2F_{10} the data at 300°K combined with the very large density of states reduces the mpe entirely to the QC and dissociative regions. We have found in SF_6, however, that all three regions play an important role.

The low-level discrete excitation region is characterized by the molecule's rotation–vibration spectrum and is therefore extremely sensitive, not only to properties of the specific molecule, but also to the laser power, laser frequency, and laser bandwidth.

From an analysis of the fundamental absorption band, we obtain the rotational B value as well as the transition dipole moment, μ. In addition, we must have a knowledge of the molecular symmetry in conjunction with spectra of all the IR-active modes including possibly a force field model for a complete assignment of the fundamental modes and their respective degeneracies. In order to have some insight into the higher-order vibration and Coriolis terms, in particular the anharmonicity, spectra of at least one overtone is necessary. As we can see, this region of mpe is very complex requiring a large amount of *a priori* information. Modeling molecules whose mpe physics depends heavily on this region can be very nearly impossible in many cases. In the two molecules we study, this part of the problem is tractable. In SF_6 there exists spectroscopy of both the fundamental (Aldridge *et al.*, 1975; Loëte *et al.*, 1977; McDowell *et al.*, 1976a, 1977, and 1978) and $3\nu_3$ (Pine and Robiette, 1979). In S_2F_{10} the discrete levels play no role since most of the population is initially thermally in the QC (Lyman and Leary, 1978). In practice, however, it should only be necessary to know the mode frequencies and degeneracies, the anharmonicity, and the B value such that the center of gravity and the spread of the band is reasonable well approximated (Brock *et al.*, 1979). In SF_6 we use a complete first-order Hamiltonian for modeling the excitation (Ackerhalt and Galbraith, 1978; Galbraith and Ackerhalt, 1978a and b).

Using this molecular Hamiltonian obtained from spectroscopy and the appropriate electric dipole selection rules, the mpe Hamiltonian,

$$H = \mathscr{H}_{\mathrm{MOL}} - \boldsymbol{\mu} \cdot \mathbf{E} \tag{1.1}$$

can be used to solve the time-dependent Schrödinger equation (Steinfeld, 1974; Allen and Eberly, 1975; Larsen and Bloembergen, 1976). Since the rotational ground states can be chosen to be independent, the excitation modeling is done individually for each rotational ground state and the resulting populations are averaged with the appropriate Boltzmann factors at the end of the calculation. The number of vibrational steps necessary for a complete description of the low-level excitation ladder is determined by the maximum available laser power and the pumped mode anharmonicity, i.e., these two parameters determine the location of the pump mode ladder bottleneck. In principle, some molecules may be pumped almost completely to dissociation while remaining in the initially pumped mode.

If a molecule dissociates and the pump mode ladder bottlenecks at some level below the dissociation limit, then anharmonic coupling must play a role in the mpe physics (Deutsch and Brueck, 1978; Frankel, 1976;

Frankel and Manuccia, 1978; Steel and Lam, 1979; Kay, 1974; Kwok and Yablonovitch, 1978; Mukamel, 1978; Yablonovitch, 1977).

We will now briefly describe the mathematical framework upon which our model (Ackerhalt and Galbraith, 1979) is built. At low energies in the molecule it is well known that the molecular vibrational motion is almost harmonic. Hence it is reasonable (and always done in IR spectroscopy work) to begin in a basis set that consists of a direct product of harmonic oscillators (some of which are degenerate), one for each normal mode of vibration. In zeroth order the molecule's rotational motion is described (for nonlinear molecules) by the rotor wave function, D^J_{MK} (Hecht, 1960). To first order in SF_6 there is a vector coupling of the rotational and vibrational motions but no anharmonic coupling (i.e., coupling of oscillator wave functions) occurs. Anharmonic couplings can be of two types: Coriolis and pure vibrational. The Coriolis interaction between ν_3 and ν_4 of SF_6 has been considered in the analysis of ν_4 by Kim *et al.* (1979). Pure vibrational couplings have been written out by Shaffer *et al.* (1939) for tetrahedral molecules and by Fox *et al.* (1979) for octahedral XY_6 molecules. In the usual treatment of the molecular Hamiltonian by spectroscopists, these off-diagonal terms are removed by performing a Van Vleck contact transformation (Van Vleck, 1929; Chedin and Cihla, 1974). This procedure is legitimate only if the detunings of the interacting levels are large, i.e., if a perturbation expansion is rapidly converging. The contact transformation is usually a good technique for low-lying levels where the density of vibrational states is very sparse. A famous exception is ν_2–ν_4 of CH_4 as treated by Jahn (1938a,b and 1939); Childs and Jahn (1939). In our case, we are interested in very high levels of molecular excitation where the density of nearby vibrational states is large and perturbation theory is *not* applicable. Our approach is based upon the assumption that individual level-to-level couplings are weak (they are at least second order in the molecular Hamiltonian) but that the overall interaction leading to a flow of population out of the excited level is large because of the large number of states involved in the coupling. Hence, we treat the problem wholly in the harmonic oscillator basis. This has two distinct advantages: (1) the connection with low-level spectroscopy is immediate and automatic; (2) it is easy for us to keep track of all dipole-allowed transitions and their strengths.

In the expansion of the molecular Hamiltonian, one writes down all operators that are polynomials of some degree k in the normal coordinates (for pure V–V interaction) and that are invariant with respect to the molecular point group (Shaffer *et al.*, 1939; Amat *et al.*, 1971; Fox *et al.*, 1979). This implies that only those levels with the same symmetry interact—a strong selection rule. Since the application of this selection rule in the case of a molecule like SF_6 (or S_2F_{10}) is practically unfeasible, we consider implementing one based more upon the ordering of the molecular Hamiltonian. The net effect is essentially the same: to limit the density of background states in the V–V interactions to some subdensity. According to the fundamental principles, those combination states interact most strongly which are most similar in their oscillator content. If we are at some energy ε in the molecule, then states having n quanta of the pumped mode (called ν_p) will interact most strongly with states of $n \pm 1$ quanta and less strongly with states of $n \pm 2$, all other things being equal. (Since we do not keep track of any but the ν_p content of the states at energy ε, we assume that all other things are in fact equal.)

We assume that no group of state-to-state couplings are forbidden, only that some interactions are stronger, and that Fermi's golden rule (FGR) describes the anharmonic interaction once the subdensity of states is sufficiently large,

$$\Gamma_{n,\,n\pm 1}(m) = 2\pi g^2 \chi_{n\pm 1}(m) \tag{1.2}$$

where $\Gamma_{n,\,n\pm 1}(m)$ is the FGR rate for population flow from states with $n\nu_p$ character to states with $(n \pm 1)\nu_p$ character at an energy equal to $m\nu_p$ quanta. The pumped mode is defined by ν_p; i.e., in SF_6 $\nu_p = \nu_3$. The density of states $\chi_{n\pm 1}(m)$ with character $(n \pm 1)\nu_p$ at energy $m\nu_p$ quanta is computed using standard techniques (Whitten and Rabinovitch, 1963; Thiele, 1963; Stone *et al.*, 1976). The coupling strength g^2 in Eq. (1.2) represents the statistical squared average of the state-to-state coupling between states that change ν_p character by $1\nu_p$ quantum. Because it is the average, it is constant throughout the molecule. Since couplings that change ν_p by more than one quantum are assumed to be of higher order in the molecular Hamiltonian, we neglect those terms in the model.

As the density of states increases with increasing energy in the molecule, the ν_p ladder becomes more and more strongly coupled to other modes because of this anharmonic interaction. At each level in the ν_p ladder, the rates Eq. (1.2) can be calculated and compared with the Rabi frequencies and the pulse duration. Once these rates become dynamically important, they must be included as imaginary terms on the diagonal in the Hamiltonian Eq. (1.1). While in principle, coupling back to the ν_p ladder exists, we would expect it to be small because of the small number of states in the ν_p ladder as compared with the rapidly increasing density of background states. From a practical point of view, we cannot include homogeneous rates back into the ν_p ladder since this would require the use of Bloch equations making the problem an $N^2 \times N^2$ matrix diagonalization. (A molecule like SF_6 would require a 256 × 256 matrix diagonalization.) Since the ν_p ladder and the leakage rates into the QC make the low-level excitation problem independent of the QC dynamics, this piece of the problem can be solved exactly, giving the population at different locations in the QC as a function of time. This aspect of the problem is shown pictorially in Figure 1.1 where the pump mode is labeled ν_3 as in SF_6. As previously discussed, the number of levels in the ν_p ladder that need to be considered is limited because of (1) increasing molecular anharmonicity preventing further excitation up the ladder and (2) large leakage rates that can also inhibit the pumping. For example, in SF_6 we treat levels only up to $3\nu_3$. We do not distinguish coupling to the QC from different rotational levels, but only take into account their vibrational character (pure vibrational anharmonic coupling), i.e., all the rotational levels at $2\nu_p$ in the ν_p ladder are coupled at the same rate with the QC level with $1\nu_p$ character.

In Figure 1.1 the QC is shown as a function of its separate ν_p components. The anharmonic coupling between these levels at an energy of $m\nu_p$ quanta is given by Eq. (1.2) where the rates from $n\nu_p \rightarrow (n+1)\nu_p$ and from $(n+1)\nu_p \rightarrow n\nu_p$ are both taken into account. There are no unidirectional rates in the QC as we required with the ν_p ladder leakage into the QC.

The transition dipole strength must be computed for each step in the QC. For example, let us consider the transition from states $\chi_n(m)$ to states $\chi_{n+1}(m+1)$. If the ν_p mode has degeneracy d_p, then the squared dipole

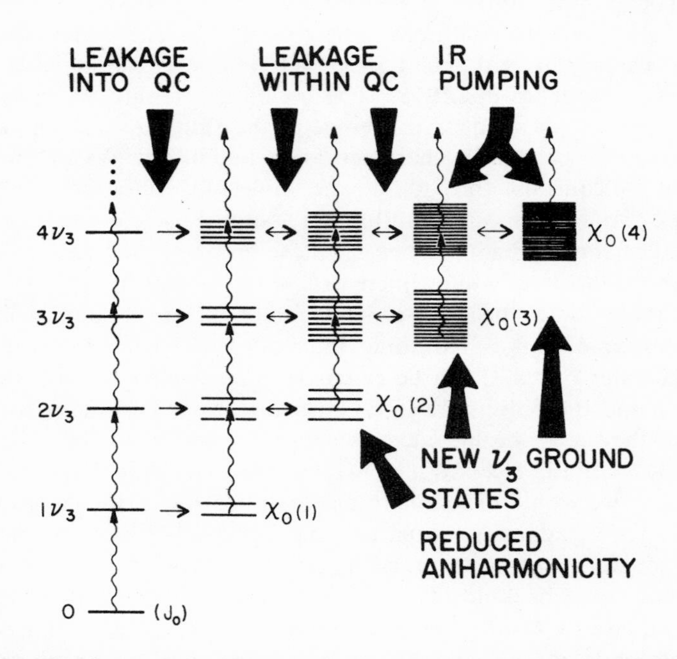

Figure 1.1. Molecular energy-level diagram. Regions where primary characteristics play a role are indicated by large arrows. Vertical (horizontal) arrows indicate laser excitation (intramolecular V–V relaxation). Pump mode labeled as in SF_6, $\nu_p = \nu_3$.

strength for each level at $\chi_n(m)$ is

$$D_n^2(m) = N_n(m)(n + d_p)\alpha_{01}^2/2 \tag{1.3}$$

where α_{01} is the bare Rabi frequency and $N_n(m)$ is the number of effective states in the sense of Stone *et al.* (1976),

$$N_n(m) = \bar{g}\chi_n(m) \tag{1.4}$$

The anharmonic coupling parameter $\bar{g}$ represents the coupling between states of the same ν_p character. Since the number of quanta in all modes but ν_p can change for these states, we expect this parameter to be smaller than g (Amat *et al.*, 1971). In SF_6 where ν_3 is the highest frequency mode, we expect many states of very different character for each $\chi_n(m)$ such that $\bar{g}$ should be smaller than g by roughly two orders in the molecular vibrational Hamiltonian

$$\bar{g} \sim (X_{33}/\omega_3)g \tag{1.5}$$

where X_{33} and ω_3 are the ν_3 anharmonicity and mode frequency, respectively. In S_2F_{10}, however, there are many modes of similar frequency to the pumped mode ν_9, so that many states will have quantum changes similar to that represented by the coupling g. We would expect, therefore, that $\bar{g}$ should be only slightly smaller than g. We take one order in the molecular Hamiltonian,

$$\bar{g} \sim (X_{99}/\omega_9)^{1/2}g \tag{1.6}$$

where X_{99} and ω_9 are the anharmonicity and mode frequency of the ν_9 mode, the pumped mode in the CO_2-laser excitation experiments (Lyman and Leary, 1978).

In general, the population dynamics in the QC as shown in Figure 1.1 must be solved using Bloch equations. If the molecule absorbs many photons in the QC, then the problem becomes intractable because of the large size of the matrix involved. For SF_6 and S_2F_{10}, however, a major simplification occurs such that the QC can be reduced to rates at every step and in these two cases to a single ladder. For these molecules, we have found that every laser transition in the QC is dominated by the anharmonic coupling rates such that each transition is governed by a rate (Ackerhalt and Eberly, 1976; Ackerhalt and Shore, 1977),

$$R_n(m) = \frac{D_n^2(m)\gamma_n(m)/4}{\Delta_n^2(m) + \gamma_n^2(m)/4} \tag{1.7}$$

where

$$\gamma_n(m) \equiv \Gamma_{n,n+1}(m) + \Gamma_{n,n-1}(m) + \Gamma_{n+1,n}(m+1) + \Gamma_{n+1,n+2}(m+1) \tag{1.8}$$

$$\Delta_n(m) = \omega_p - \omega_L - 2nX_{pp} - 2mX_H \tag{1.9}$$

The frequency of the pump mode and of the laser are given by ω_p and ω_L, respectively. The pump mode anharmonicity and effective "hotband" anharmonicity are given by X_{pp} and X_H, respectively. The "hotband" anharmonicity can, in principle, be calculated from a least-squares fit of the hotband band origins as a function of energy in the molecule. The mean-square deviation of the fit can be used as a minimum frequency condition on $\Delta_n(m)$ so that spurious resonances in the QC are avoided. We have been able to perform this calculation for SF_6 using the anharmonicity parameters of McDowell *et al.* (1976b) where $X_H = 1\ \text{cm}^{-1}$ and the frequency minimum is $\sim 2\ \text{cm}^{-1}$.

Some points should be emphasized before moving too far from Eq. (1.7). Equation (1.7) follows in our model for two reasons: First, we consider the problem in the harmonic oscillator basis where each state $|\alpha\rangle$ having $(m)v_p$ quanta is pumped to a state $|\alpha + v_p\rangle$ having $(m+1)v_p$ content. Second, after the laser pumping, the state rapidly decays because of the anharmonic V–V relaxation. If $\gamma_n(m)$ is fast enough, the laser coherence between states $|\alpha\rangle$ and $|\alpha + 1v_p\rangle$ is lost, and the rate equation approach is justified. For high enough laser powers or in molecules with very small mixing, Eq. (1.7) would not follow. Our approach is to assume Eq. (1.7), to obtain our anharmonic parameters g and $\bar{g}$ by fitting absorption data, and then to go back to Eq. (1.7) and show the consistency of the model. This approach has proven itself for SF_6 and S_2F_{10}. Note also that we *cannot* obtain FGR pumping rates for the laser excitation since each initial state is coupled to only d_p final states (d_p is degeneracy of the pumped mode). We do, however, obtain FGR rates for the V–V relaxation since each pumped state is coupled to all background states of one less (greater) v_p quanta (our selection rule). *We have a "2-level-like" incoherent laser pumping and subsequent FGR relaxation.*

In Figure 1.2 we show the states $\chi_n(m)$, with a population given by $P_n(m)$, and all the remaining states to which they are directly coupled. The rate equations governing these subpopulations are

$$\begin{aligned}\dot{P}_n(m) = &-[R_n(m) + R_{n-1}(m-1) + \Gamma_{n,n+1}(m) + \Gamma_{n,n-1}(m)]P_n(m) \\ &+ \Gamma_{n+1,n}(m)P_{n+1}(m) + \Gamma_{n-1,n}(m)P_{n-1}(m) \\ &+ R_n(m)P_{n+1}(m+1) + R_{n-1}(m-1)P_{n-1}(m-1) \end{aligned} \tag{1.10}$$

Comparing the laser rates in the QC Eq. (1.7) with the V–V relaxation rates [Eq. (1.2)] at every step, we find that the V–V rates dominate. The population at energy mv_p is in *statistical equilibrium* with respect to the subdensities $\chi_n(m)$; i.e. the subpopulations $P_n(m)$ can be related to the total population $P(m)$ at energy mv_p (Stone and Goodman, 1979),

$$P_n(m) = \frac{\chi_n(m)P(m)}{\sum_n \chi_n(m)} \equiv \frac{\chi_n(m)P(m)}{\chi(m)} \tag{1.11}$$

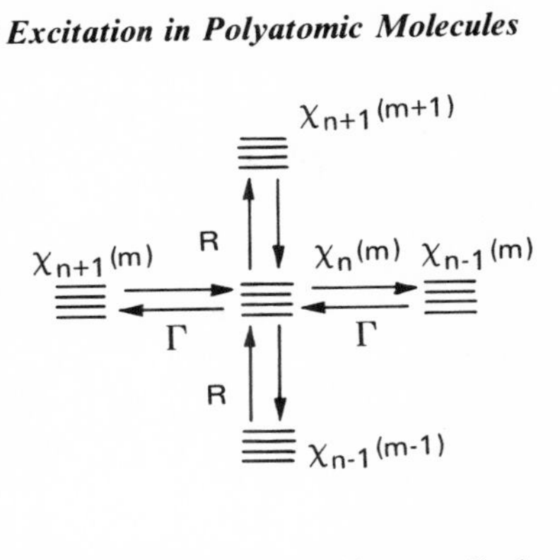

Figure 1.2. Dynamics of population $P_n(m)$. Vertical and horizontal arrows indicate laser excitation rates and intramolecular V–V relaxation rates, respectively.

After substituting Eq. (1.11) into Eq. (1.10) and summing over all n, we find a set of single-ladder rate equations:

$$\dot{P}(m) = -\left(\sum_{n=0}^{m} \frac{R_n(m)\chi_n(m)}{\chi(m)} + \sum_{n\neq 0}^{m} \frac{R_{n-1}(m-1)\chi_n(m)}{\chi(m)} \right) P(m)$$
$$+ \sum_{n=0}^{m-1} \frac{R_n(m-1)\chi_n(m-1)}{\chi(m-1)} P(m-1)$$
$$+ \sum_{n\neq 0}^{m+1} \frac{R_{n-1}(m)\chi_n(m+1)}{\chi(m+1)} P(m+1) \tag{1.12}$$

Equations (1.3) and (1.7) show that the QC rates in Eq. (1.12) depend only on laser intensity while the ν_p ladder excitation is coherent and depends on the Rabi frequency. If dissociation requires many QC steps as compared with the ν_p ladder, as for example in SF_6, then we would expect dissociation data to show primarily a laser fluence dependence (Kolodner *et al.*, 1977). Excitation data for low $\langle n \rangle$ would be flux dependent (Lyman *et al.*, 1978a; Smith *et al.*, 1979).

Since our model of the QC is inherently statistical, we find it consistent to use RRKM theory rates to describe the dissociation process (Robinson and Holbrook, 1972). Therefore, above the dissociation threshold we include RRKM rates in Eq. (1.12) (Grant *et al.*, 1977b; Mukamel and Jortner, 1976; Stone and Goodman, 1979).

At this point our model (Ackerhalt and Galbraith, 1979) is complete. In order to make comparisons with experiment, however, we must describe how to use the model in case the gas temperature is greater than 0°K. If the data were taken for a gas at very low temperature, then the ground vibrational state could have most of the population. Only the rotational Boltzmann distribution is important, and we account for it as discussed earlier. Our comparison of theory and experiment for SF_6 is for a gas at 140°K, which means over 90% of the population is thermally in the ground state. If the experimental gas temperature is high enough that hotbands must be included in the calculation, then the calculation should be done for each hotband and averaged over the vibrational Boltzmann distribution. Since hotbands may be treated as effective "ground states" with some energy above the actual ground state, it is necessary to recompute the hotband anharmonicity and the density of states for each hotband calculation at the shifted energy. We are assuming the spectroscopic structure of the absorption to be independent of the initial hotband state. (While this approximation is dubious, there will usually be insufficient information for making any more intelligent estimates of the detailed structure.) As the energy of the hotbands increases, the size of the corresponding ν_p ladder portion of the calculation decreases until eventually the hotband pumping takes place entirely in the QC. Since the QC absorption is usually greater than that in the ν_p ladder because of detuning and anharmonicity factors, we would expect greater excitation with increasing temperature (Tsay *et al.*, 1979; Duperrex and van den Bergh, 1979). If the thermal excitation is truly large (as in S_2F_{10} at 300°K), then a further simplification occurs in that thermal populations can be put into bins according to the numbers of quanta of the pumped mode so that a single ladder can represent the dynamics.

1.3. Application to SF_6

Of the molecules studied in mpe and mpd, SF_6 is the one about which the most is known. Spectra of ν_3 (Shimizu, 1969; Rabinowitz *et al.*, 1969; Goldberg and Yusek, 1970; Houston and Steinfeld, 1975; Aldridge *et al.*, 1975; McDowell *et al.*, 1976a, 1977, and 1978; Loëte *et al.*, 1977; Ouhayoun and Bordé, 1977; Bordé *et al.*, 1978; Bordé *et al.*, 1979) have been analyzed in such detail that the band origin for ν_3 is known to the same degree of accuracy as are the speed of light and the transitions pumped by the CO_2 laser. As far as excited states are concerned, double resonance data have been obtained by Steinfeld and Jensen (1976), by Moulton *et al.* (1977), and more recently by Steinfeld *et al.* (1979) and Jensen *et al.* (1979). The overtone $3\nu_3$ at 2829 cm^{-1} is also infrared active and has been measured in both FTIR (Kildal, 1977; Fox, 1978; Marcott *et al.*, 1978; Ackerhalt *et al.*, 1978) and at higher resolution by Pine and Robiette (1979). Analysis of these data is currently under way.

The basic spectroscopic parameters, which are needed to give an adequate picture of the molecule are (1) the fundamental vibrational band frequencies, from which the molecular harmonic densities of states can be computed, (2) the rotational B values (inverse moments of inertia) and Coriolis constants giving the rotational spread of the pumped mode, (3) the transition dipole moment, μ_0, giving the strength of the laser molecule coupling (Fox and Person, 1976; Fox, 1977; Galbraith, 1978), (4) the anharmonicities (diagonal in the oscillator picture), X_{ij}, which give the overtone anharmonicities as well as the locations of the hotband origins (Bertsev *et al.*, 1974; McDowell *et al.*, 1976b). More exotic quantities such as the anharmonic splitting constants G_{33} and T_{33} of Hecht (1960), must await the final analysis of the data of $3\nu_3$. The above parameters are essentially "spectroscopic," and the model that is thereby inferred has been solved for high-power laser fields (Ackerhalt and Galbraith, 1978). In these calculations a complete first-order molecule model was pumped by a CO_2 laser in the ν_3 band at several frequencies. The solutions were obtained from the Schrödinger equation in the rotating wave approximation (Allen and Eberly, 1975) for each initial angular momentum state J_0 coupled to all final states J (as allowed by dipole selection rules). The final populations were then summed over J_0 weighted by the rotational Boltzmann factor for that J_0. In this way we obtained the laser frequency dependent and power dependent $\langle n \rangle$ curves shown in Figure 1.3. As can be seen, there is the mpe red shift evident in the results. We find sharp resonances, which are especially strong at the multiple photon Q branches, but the experimental data exist only on the CO_2-laser frequencies, which are spaced by about 2 cm^{-1} so that detailed comparison is not yet possible. In these calculations, vibrational tensor splitting has not been included, and the model is purely first-order spectroscopy. With all of the above ingredients included, this represents the most straightforward solution to the mpe problem based on a simple spectroscopic model. That this model is incomplete follows under more detailed examination of the results of the calculations: (1) We find that the excitation is strongly laser intensity dependent because of the multiple photon resonances, and (2) the excitation is too small, and, in fact, population never reaches $4\nu_3$ even at high powers because of the strong anharmonicity of the pumped ν_3 mode. The new data on $3\nu_3$ indicates that the earlier assignments of that band may be in error (Pine and Robiette, 1979) and that anharmonic splitting may be large. Inclusions of the splitting operators, however, only means a change of basis from the first-order theory above, and even though there may be more positions of multiple photon Q branches, the overall absorption calculated cannot change very much. (This follows directly from the fact that including splitting merely redistributes the dipole strength but does not increase its absolute value.) Because of these results, the "spectroscopy model" fails to account for the observed new effect. It is the quasicontinuum hypothesis that is needed to "complete" the spectroscopy. More fundamentally the answer lies in the concepts of anharmonic coupling and the gradual loss of the strict normal mode picture as the excitation advances.

As discussed in Section 1.2, the excitation proceeds through three steps in SF_6: low-level intensity dependent, coherent pumping in the ν_3 ladder; leakage and fluence dependent pumping in the QC; and finally dissociation. Our model (Ackerhalt and Galbraith, 1979) calculations as described

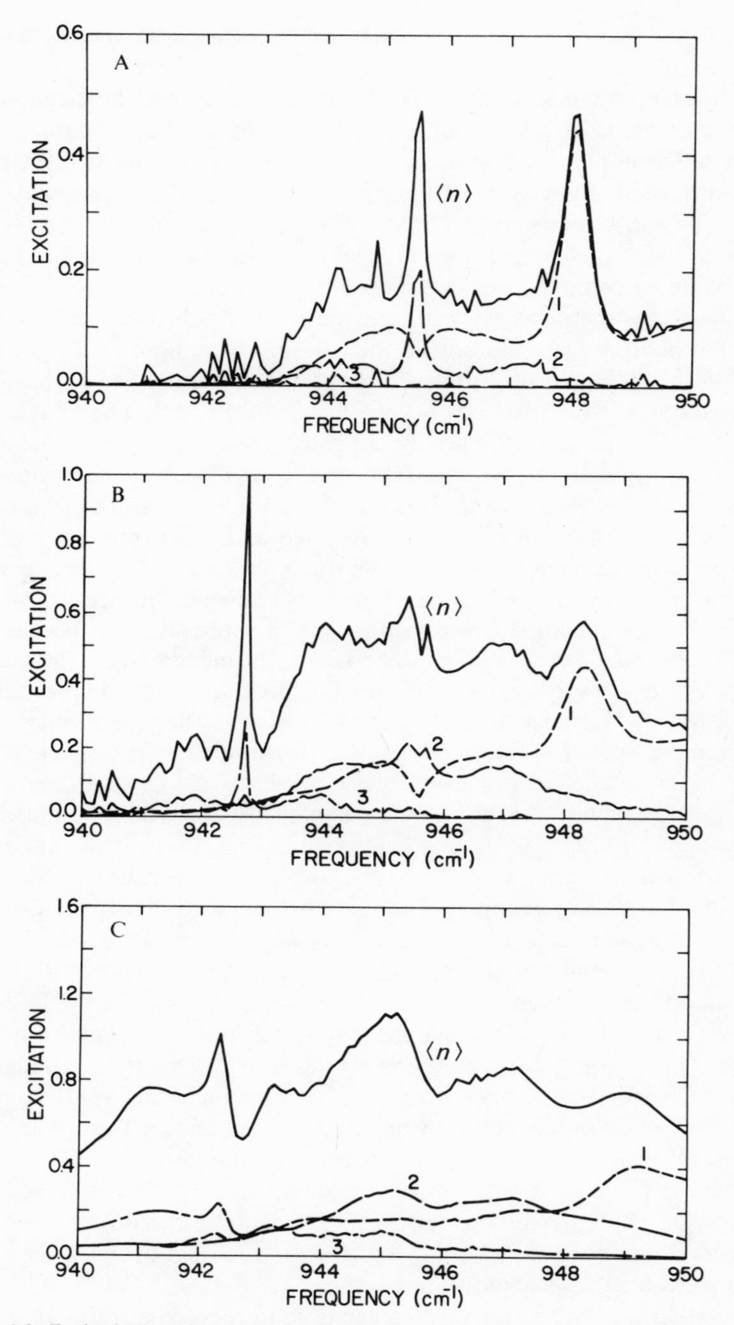

Figure 1.3. Excitation vs. frequency. Excitation in levels 1, 2, 3, ν_3, and average number of photons absorbed, $\langle n \rangle$, for case where QC coupling is zero. Figures 1.3a, 1.3b, 1.3c refer to laser powers 4 MW/cm^2, 23 MW/cm^2, 100 MW/cm^2, respectively. Rotational temperature is 300°K with vibrational temperature 0°K. Pulselengths are assumed to be 100 nsec.

in Section 1.2 applied to SF_6 are strictly at low temperature. The up and down rates are shown for convenience in Figure 1.4 in the form of cross sections. The steep rise in rate at the low levels is because of the off-resonant pumping before the turn on of effective states

$$R_k(n) \sim \frac{\Gamma_k(n)}{\Delta_k{}^2(n)} \frac{k+3}{2} \alpha_{01}^2 \tag{1.13}$$

The increase is with $\Gamma_k(n)$ or the subdensities of states. At the turnover point $\Gamma_k(n) \gtrsim \Delta_k(n)$, we obtain resonant pumping but still no self-mixing enhancement of the dipole strength,

$$R_k(n) \sim \frac{1}{\Gamma_k(n)} \frac{k+3}{2} \alpha_{01}^2 \tag{1.14}$$

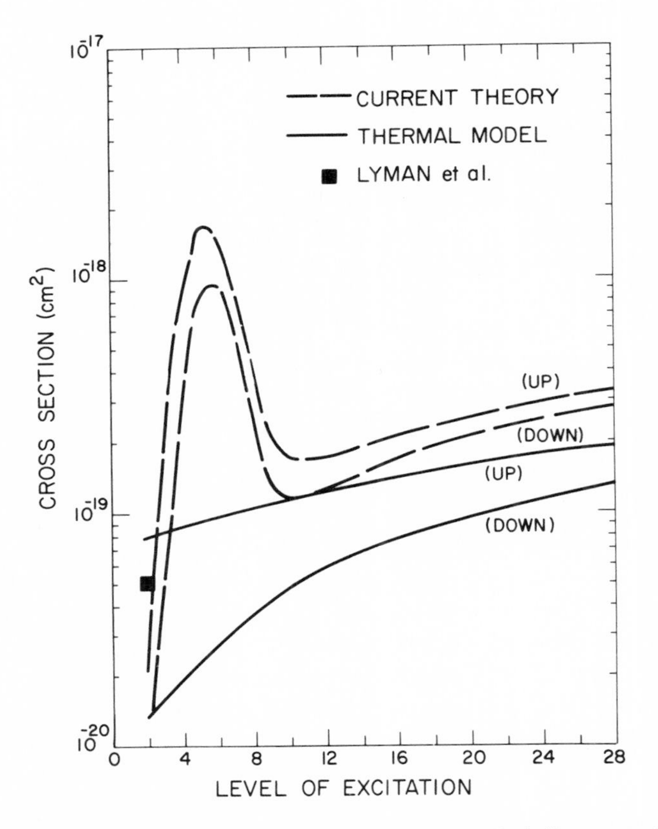

Figure 1.4. QC cross sections for absorption and emission including present theory and thermal model of Black *et al.* (1979) vs. level of excitation in molecule; measured in ν_3 quanta.

Finally we obtain in the asymptotic region,

$$R_k(n) \sim \frac{\bar{g}\chi_k(n)}{\Gamma_k(n)} \frac{k+3}{2} \alpha_{01}^2 \tag{1.15}$$

and the dipole is enhanced by $\chi_k(n)$. For SF_6 where the pumped mode is the highest energy mode, we expect that $\bar{g}$ will be much smaller than g. We write

$$\bar{g} = \left(\frac{X_{33}}{\omega_3}\right)^{\gamma} g \tag{1.16}$$

where X_{33} and ω_3 are the ν_3 anharmonicity and mode frequencies, respectively, and γ is a vibrational ordering (Amat *et al.*, 1971). In our fitting of the data we chose $\gamma = 1$ corresponding to two vibrational orders. The value of g itself is found by fitting the absorption data. We get a rough estimate from the absorption cross section at high fluence [Eq. (1.15) valid] and make any fine tuning from there. We find for SF_6

$$g = 0.035 \text{ cm}^{-1} \tag{1.17}$$

This value gives a leakage time of 0.83 nsec at $3\nu_3$ and 1.6 nsec at $2\nu_3$ in excellent agreement with recent data of Steel and Lam (1979).

Some points need to be clarified at this juncture since homogeneous leakage rates generally imply linewidths. Let us consider $3\nu_3$ for the moment. Generally speaking, a leakage rate of 1 nsec corresponds to a homogeneous width of ~ 0.01 cm^{-1}, however, Pine and Robiette (1979) claim that their $3\nu_3$ data are Doppler limited. This apparent contradiction is resolved by noting that in reality we are modeling the coupling of one level to a *finite number* of background levels. If the laser intensity of the transition is enough that the powerwidth overlaps the line spacing $1/[6\chi_0(1)] \sim 0.01$ cm^{-1} ~ 10 kW/cm^2 then the "line" appears broadened and FGR reasonably represents the population dynamics. At low powers, as in Pine and Robiette's (1979) case, however, these same levels appear Doppler limited since the level spacing is greater than the powerwidth. Our time of 0.83 nsec is a time scale at which the population flows out of $3\nu_3$ and appears as a line broadening only at high laser powers. (This has been checked in numerical model calculations.) Steel and Lam (1979) actually measure the $1/T_2$ decay of a four-wave process that we attribute to population flow out of $2\nu_3$. We will see that the agreement of our model with these measurements and the mpe and mpd experiments is quite satisfactory.

Our calculation of the mpe and mpd for SF_6 are as shown in Figure 1.5. We used RRKM rates based on dissociation at 33 CO_2 photons absorbed (Lyman, 1977; Bott and Jacobs, 1969; Benson, 1978; Kiang *et al.*, 1979).

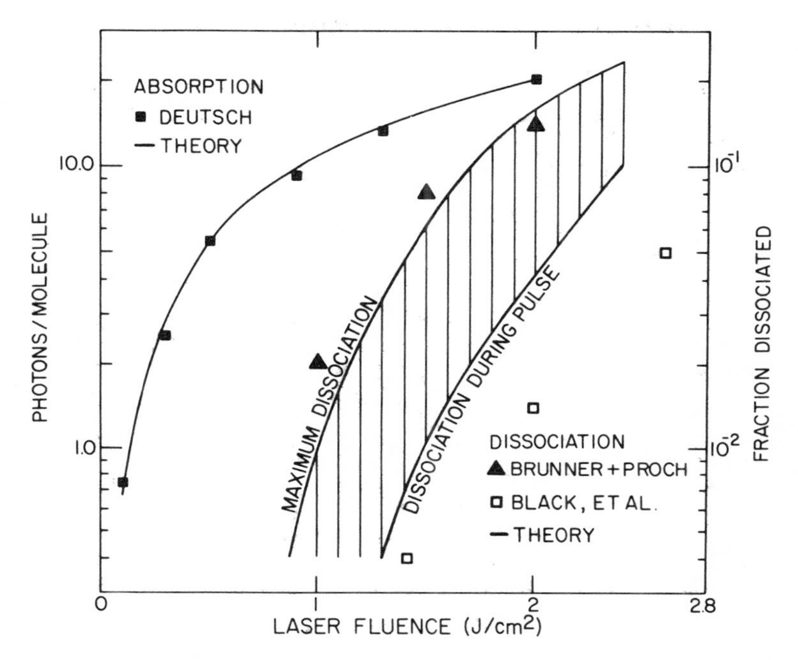

Figure 1.5. A comparison of our theory with experiment for mpe and mpd is shown. Left axis refers to mpe and compares our theory with the data of Deutsch (1977) for $\langle n \rangle$ vs. fluence. Right axis refers to mpd and compares our theory with Brunner and Proch (1978) and Black *et al.* (1979) for fraction dissociated vs. fluence. Both theory and experiment are collisionless and are at 140°K [except the data of Black *et al.* (1979) at 300°K]. The CO_2 pulse is 100 nsec and smooth. Upper curve is RRKM during the pulse plus populations in 33–49. Lower curve is simply the RRKM fraction during the pulse.

Our calculations are compared with the 140°K, $\langle n \rangle$ vs. ϕ data of Deutsch (1977), as calibrated by Judd *et al.* (1979). With minimal parameter adjustment [We could have obtained g solely from Steel and Lam's (1979) data!] we obtain exact agreement with the mpe curve. Our dissociation curves (the upper curve is RRKM during the pulse plus all population above $33\nu_3$ at the end of the pulse; lower is just RRKM during the pulse) are compared with molecular beam (collisionless) data of Brunner and Proch (1978) and the cell data of Black *et al.* (1979). Here the agreement is also quite satisfactory. The model gives a sharp fluence threshold in the dissociation curve at 1 J/cm^2 for 1% dissociation. As we shall see in greater detail in Section 1.5, these data are not sufficiently informative to distinguish between different theories of mpe and mpd. The population distributions in the QC for various fluences are shown in Figure 1.6. As was speculated (but not calculated) in the paper of Black *et al.* (1979), there is trapping in the ν_3 ladder so that, e.g., an average $\langle n \rangle$ of 8 means a long tail

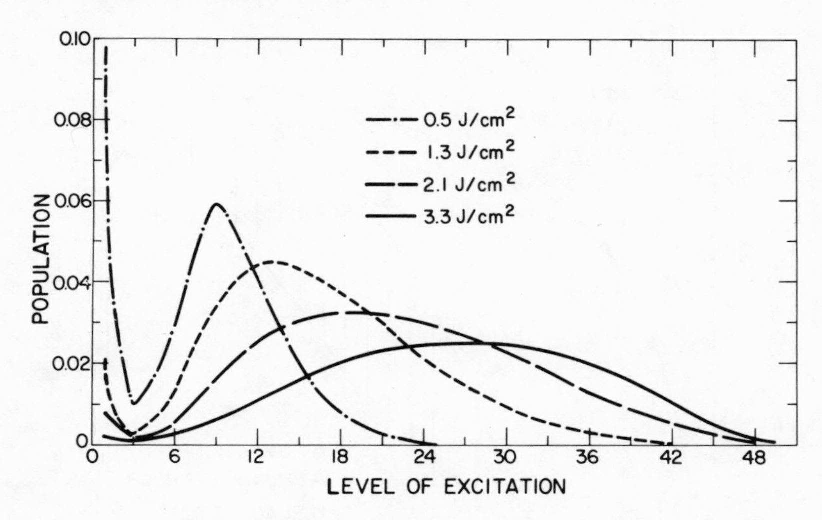

Figure 1.6. Population distributions in QC vs. level of excitation at four fluences.

to very high levels in order to compensate the trapping. The integral of these curves or the fraction reaching the QC as a function of laser fluence is shown in Figure 1.7. The first experimental measurements of this quantity (actually we calculate percent in QC not the fraction pumped) were made by Bagratashvili *et al.* (1979) at $P(16)$ CO_2, and we are in fair agreement

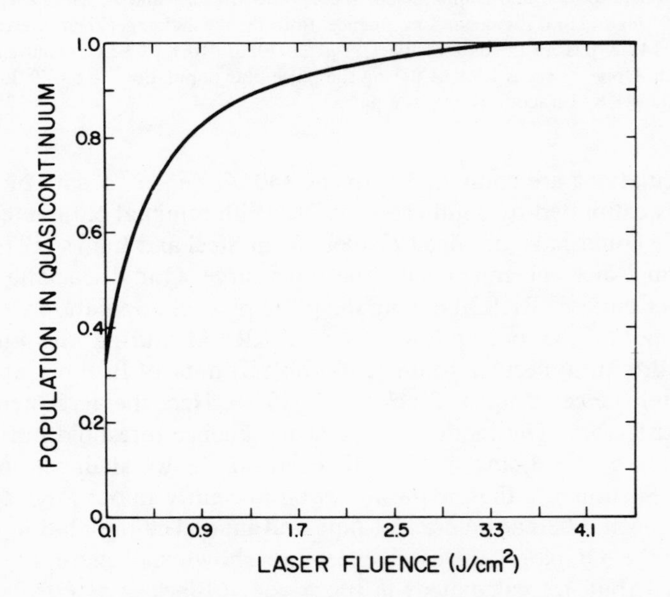

Figure 1.7. Population fraction in QC vs. laser fluence.

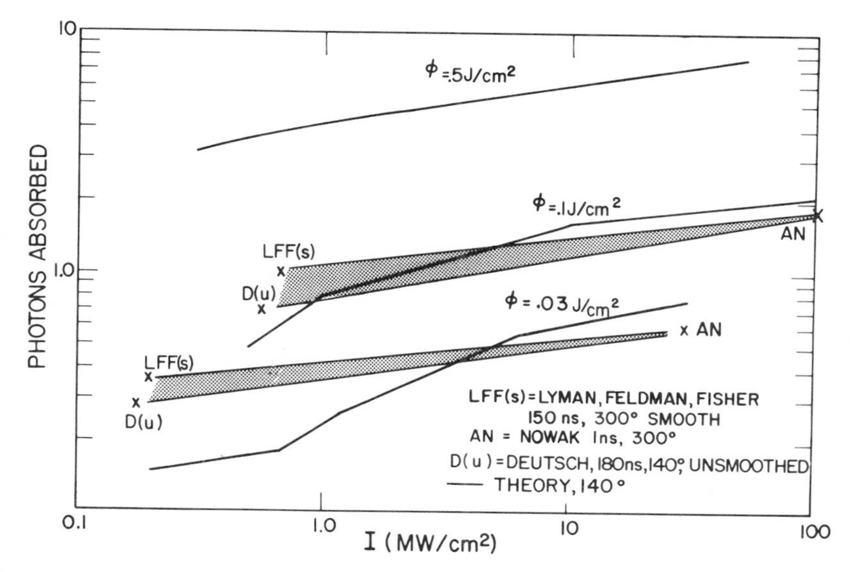

Figure 1.8. Average number of photons absorbed vs. laser intensity at three different laser energies: 0.03 J/cm^2, 0.1 J/cm^2, 0.5 J/cm^2, CO_2-P(20). For comparison, data of Lyman *et al.* (1978), Deutsch (1977), and Nowak (private communication, 1979) are shown.

with the fluence dependence of this more sensitive quantity. Another more difficult test of our theory is the calculation of the intensity dependence of the mpe. Figure 1.8 gives the results of such a calculation compared with three sets of experimental data. The AN (data of A. Nowak in Judd *et al.*, 1979) data are for a 1-nsec pulse for 300°K SF_6 gas at low pressure. The LFF(s) data (Lyman *et al.*, 1978a) are 150 nsec, 300°K smoothed pulse, and D(u) (Deutsch, 1977) is an unsmoothed but cold spectrum, also at low pressure. The theoretical curves show small intensity dependence in the absorption despite the ν_3 ladder with its multiple-photon resonance structure. We see slightly (consider the scale of the log–log plot) greater intensity dependence theoretically than the experiments, however, at 300°K there is considerable hotband pumping [especially at $P(20)$]. Since the hotbands are thermally closer to the QC, they should show less intensity dependence. The curve at 0.5 J/cm^2 represents a prediction of our model. Figure 1.9 shows the population distributions that correspond to the 0.1-J/cm^2 case. As can be seen in Figure 1.9, even though the $\langle n \rangle$ does not change greatly, our model predicts much more population in overtones of the ν_3 ladder for the 1-nsec pulse. Presumably this could be tested in two laser absorption experiments.

To summarize, our model can quite easily accommodate all of the known mpe and mpd data with the adjustment of one effective anharmonic

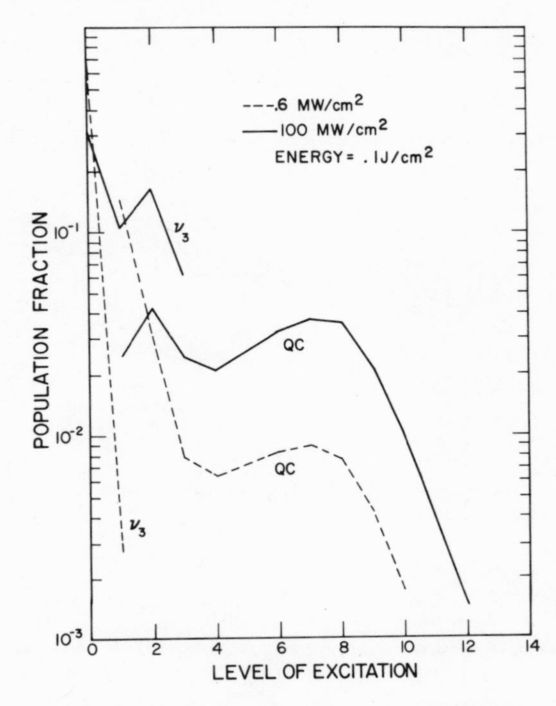

Figure 1.9. Population fraction vs. level of excitation at fixed laser energy, 0.1 J/cm^2, and for two laser intensities, 0.6 MW/cm^2 and 100 MW/cm^2. The ν_3 ladder populations are shown separately from the QC populations.

coupling parameter. Our model is in agreement with mpe and mpd data as well as the more difficult "fraction pumped" and intensity dependent absorption data. We have suggested that quite different population distributions are obtainable with short pulses, even at fairly low fluence levels. Below we list the parameters of our model as it applies to SF_6:

g	anharmonic coupling	0.035 cm^{-1}
$\bar{g}$	k-manifold self-coupling	10^{-4} cm^{-1}
X_{hb}	hotband anharmonicity (averaged)	1 cm^{-1}
Δ_{min}	minimum detuning (standard deviation in X_{hb})	2 cm^{-1}

1.4. Application to S_2F_{10}

Recently Lyman and Leary (1978) have reported laser excitation and dissociation in disulfur decafluoride. S_2F_{10} has a fundamental S–F stretching vibration at 938 cm^{-1} (Jones and Ekberg, 1979) whose vibration–

rotation–absorption band is similar in appearance (intensity and frequency dependence) with the ν_3 band of SF_6. S_2F_{10} is of special interest since it is the largest molecule whose multiple-photon spectrum has been studied. We begin with a discussion of the spectroscopic "facts" such as they are, then proceed to the multiple-photon experiments, and end with our analysis of this system.

The S_2F_{10} molecule consists of two SF_5 subclusters in D_{4d} symmetry, bound through an S–S bond of energy 58 kcal/mol (Benson, 1978). The current FT–IR and matrix spectra of Jones and Ekberg (1979) indicate that the mode being pumped in the mpe experiments is the doubly degenerate E, radial S–F stretch, denoted ν_9. This is a new assignment of the 938-cm^{-1} band based on measurements made at higher resolution than previous data (Wilmshurst and Bernstein, 1957; Dodd *et al.*, 1957). Although this question remains to be settled definitely, we take the recent assignment for use in our analysis. From the electron diffraction studies of Harvey and Bauer (1953) we obtain the rotational B values. The dipole moment is inferred from the paper of Lyman and Leary (1978) so that we have an essentially complete first-order picture of S_2F_{10}. Other quantities that are needed for our analysis are the anharmonicities X_{ij} and the change in B with excitation which we take from the data on SF_6, i.e., all second-order spectroscopic parameters are set equal to the corresponding values of SF_6.

The multiple-photon experiments show some similarities and some important differences from the SF_6 data. As in SF_6 the multiple-photon absorption spectrum shows a red shift from the low-power spectrum. This shift, presumably because of vibrational anharmonicities, X_{ij}, leads to a peak at ~ 933 cm^{-1}, a red shift of 5 cm^{-1}. In our calculations we have included anharmonic (hotband) shifts identical to those for SF_6, and this will lead to such red shifts in the mpe spectrum, although our calculations were carried out primarily at only one frequency, 938 cm^{-1}. Fundamentally, the differences are more striking than the similarities: (a) Onset of dissociation is 22 CO_2 photons; (b) absorption cross section does not fall with $\phi^{-1/3}$ (fluence $= \phi$) but remains approximately constant until dissociation; (c) the fluence for the dissociation threshold (1% dissociated) is 0.013 J/cm^2 (150-nsec pulses) at which about five photons on average have been absorbed; and, finally, (d) experiments were performed only at 300°K (below 200°K the S_2F_{10} condenses) so that most of the molecules ($\sim 80\%$) are thermally in vibrationally excited states.

In view of this last fact, we will treat S_2F_{10} as a purely quasicontinuum molecule, i.e., rates are computed for our general model using the densities of states of S_2F_{10} with g and $\bar{g}$ as parameters, as before. The thermal population distribution is taken to be the prelaser population distribution, and the rate equations are integrated over the pulselength of

150 nsec. The thermal population distribution, plotted as a function of energy (in units of ν_9 quanta) is shown in Figure 1.10. Therefore, at 300°K in S_2F_{10} we have, thermally, an average of $1.17\nu_9$ quanta. The average number of photons absorbed from the laser field is then measured from this reference.

Since the laser-pumped mode is doubly degenerate in this case we have

$$R_k(n) = \frac{\gamma_k(n)/4}{\Delta_k{}^2(n) + \gamma_k{}^2(n)/4}\, \bar{g}\chi_k(n)\frac{k+2}{2}\,\alpha_{01}^2 \tag{1.18}$$

for the incoherent rate of manifold "k" (having $k\nu_9$ quanta) at energy $n\nu_9$, with $\gamma_k(n)$ given as before by Eq. (1.8). Again α_{01} is the $0 \rightarrow 1\nu_9$ Rabi frequency, and g and $\bar{g}$ remain to be determined. The total rates up and down at level n are defined in Eq. (1.12). High in the molecule $\Gamma_k(n) \gg \Delta_k(n)$ so that resonant pumping prevails. There Eq. (1.18) becomes

$$R_k \sim \frac{\bar{g}}{g^2}\,\frac{\chi_k(n)}{2\pi\chi_k(n+1)}\,\frac{k+2}{2}\,\alpha_{01}^2 \tag{1.19}$$

so that the upper quasicontinuum cross sections are proportional to the

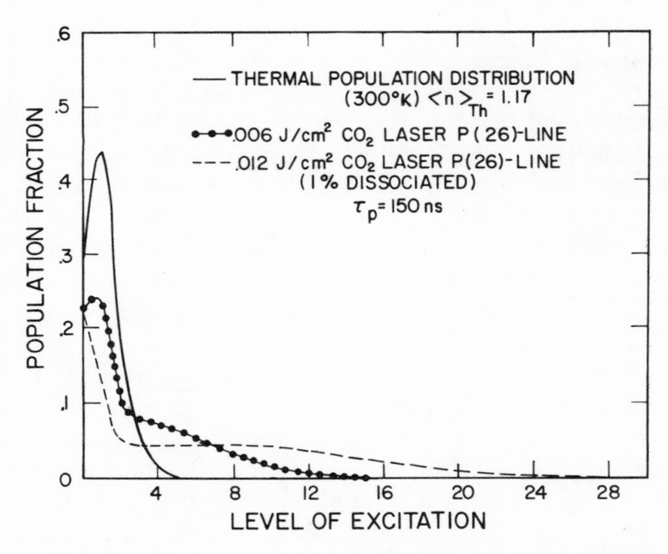

Figure 1.10. Population fraction vs. level of excitation for S_2F_{10}. Graphs of initial Boltzmann at 300°K, $\langle n \rangle = 1.17$, after a 0.006 J/cm^2 pulse and a 0.012 J/cm^2 pulse are shown.

ratio $\bar{g}/g^2$. Recall that $\bar{g}$ is the anharmonic coupling *within* a k manifold and is higher order than g:

$$\bar{g} = \left(\frac{X_{99}}{\omega_9}\right)^{\gamma} g \equiv fg \tag{1.20}$$

Since there are many modes of S_2F_{10} having similar frequencies, we expect stronger self-coupling than in SF_6. Taking g to be one order larger than $\bar{g}$, $\gamma = 1/2$ gives $f \sim 0.07$. Experimental absorption cross sections are $\sim 10^{-17}$ cm^2, which gives $f/g \sim 14$ (for SF_6 $f/g \sim 0.1$) or $g = 0.005$ cm^{-1}, smaller than the SF_6 value by a factor of 7. The theoretical up and down cross sections and the absorption cross sections from our model are shown in Figure 1.11. Observe that the absorption cross sections become constant above $\sim 8\nu_9$. Hence if we use the approximation of Black *et al.* (1979) for the density of states we obtain a Boltzmann-like tail to the population

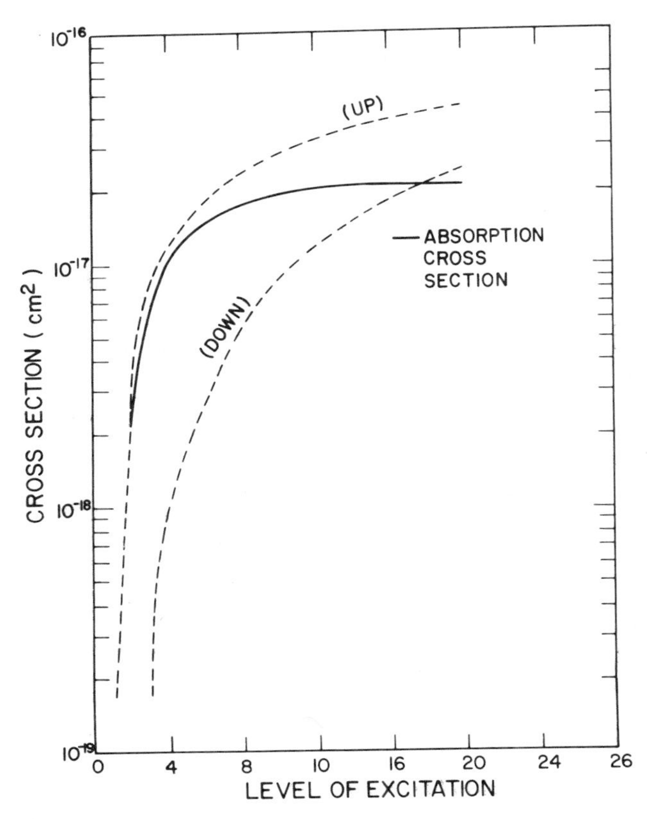

Figure 1.11. QC cross section vs. level of excitation for S_2F_{10}. Absorption cross section at level n is the [(up)–(down)] from level n.

distributions for this molecule. Also in S_2F_{10} there is no peak observed in the rates as in SF_6. This occurs because of the large densities of states involved, i.e., when resonant pumping occurs, effective states also play a role and there is no region of $(1/\Gamma)$ falloff. This result is also because of our assumption that self-mixing is only one vibrational order down from the general anharmonic mixing described by g. Further experiments are needed to test this assumption.

The constant cross sections observed by Lyman and Leary (1978) translate into the absorption versus fluence data of Figure 1.12. With the parameters g and f as discussed, we find excellent agreement with these measurements. We picked only the region from 1 to 13 mJ/cm^2 for our calculations since below 1 mJ/cm^2 the spectrum requires a more precise treatment of the V–R spectroscopy, and above 13 mJ/cm^2 there is a rapid rise in the dissociation rate.

It is quite straightforward to understand how the S_2F_{10} absorbs photons at such low fluences. Consider Figure 1.13 showing the up and down rates from the effective level at $m\nu_9$. The large (relative to SF_6) difference between up vs. down rates follows from $\chi_0(m+1)$ [states of no ν_9 character at $(m+1)\nu_9$], which get populated via FGR leakage out of $\chi_1(m+1)$. These χ_0 states, however, cannot re-radiate back to the $m\nu_9$ manifold and lead to a substantial increase in up vs. down rates. This result follows only because χ_0 is a large number relative to the total χ (viz., at $10\nu_9$, $\chi_0 \sim 5 \times 10^{10}$ and

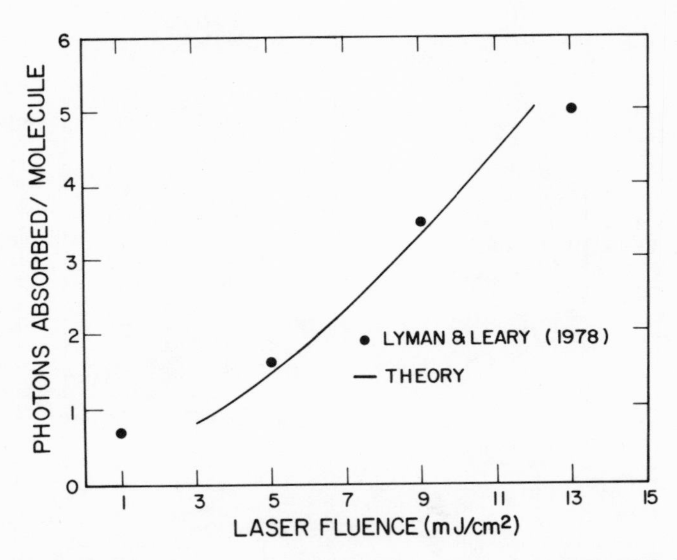

Figure 1.12. Average number of photons absorbed vs. laser fluence for S_2F_{10}, $CO_2P(26)$. Data of Lyman and Leary (1978) are shown along with theoretical curve.

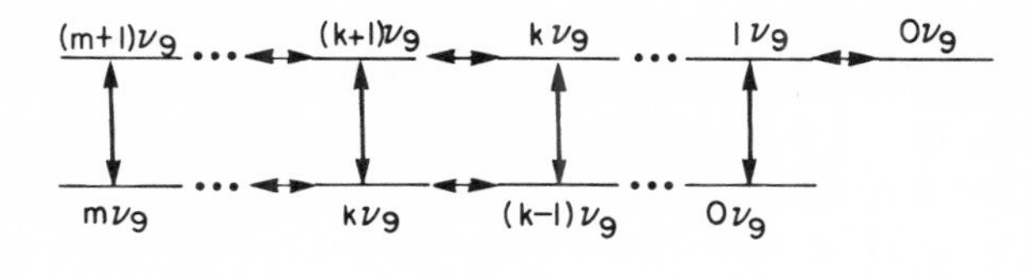

		S_2F_{10}		SF_6	
		χ	χ_0	χ	χ_0
m	2	680	554	2.6	2
	5	3.6×10^6	3×10^6	752	400
	10	8×10^{10}	5×10^{10}	3×10^5	7×10^4
	20	3.5×10^{16}	1.5×10^{16}	6×10^8	4×10^7

Figure 1.13. Upper portion of figure shows that population fraction in $0\nu_9$ manifold cannot be pumped downward in molecule. Statistical equilibrium at isoenergies means χ_0 with respect to χ determines the difference between up and down QC rates. Lower portion of figure shown χ, χ_0 for both SF_6 and S_2F_{10} at different levels of excitation in the molecules. S_2F_{10} is easier to pump than SF_6 in the QC since χ_0 is a major portion of the total density of states in S_2F_{10}.

$\chi \sim 8 \times 10^{10}$), which is not the case for smaller molecules like SF_6. We should emphasize here that this follows in our model *only because* of the *densities of states involved* and does not depend on our values for g and f, which we take to be approximate at this point.

As a final point of discussion it is interesting to consider again the laser-induced population distributions shown in Figure 1.10. Even at 6 mJ/cm^2 of laser fluence we see a long tail forming in the distribution. The average number of photons absorbed there is 1.8, which when added to the 1.7 thermal gives roughly 3 for the average $\langle n \rangle$. This value, however, does not reflect the long tail that is seen in the figure (we have population reaching $16\nu_9$ at 6 mJ/cm^2). At 12 mJ/cm^2, we find 1% dissociated (adding all populations above $22\nu_9$ to the RRKM dissociation) in agreement with experiment (Lyman and Leary, 1978).

1.5. Comparison with Another Complete Model

Over the past several years multiple-photon excitation and dissociation modeling has been done at the University of Southern California by Stone, Goodman, Dows, and more recently with Horsley (Exxon Research Laboratories), and Thiele. Their work as well as ours is based on a microscopic description of the molecule. Since our overall perspectives are related, we feel a description of their work including a discussion of the differences between our models will prove useful.

They have recently improved their original work (Stone *et al.*, 1976) to include V–V relaxation-induced incoherence in the laser excitation. We will discuss only this most recent work (Horsley *et al.*, 1979; Stone and Goodman, 1979; Stephenson *et al.*, 1979). While their model, like ours, can describe different molecules with different characteristics, we will emphasize only their applications to SF_6 (Horsley *et al.*, 1979).

The ν_3 ladder is described by Bloch equations where the QC is determined to begin at $4\nu_3$. The excitation step from $3\nu_3$ to $4\nu_3$ is described by a rate, coupling the population in levels $3\nu_3$–$4\nu_3$. The incoherence in the first few discrete excitation steps is because of T_2 dephasing that is calculated from a density of states broadening.

Rotational broadening in the discrete excitation steps is taken into account by calculating the multiple-photon resonances from the Hamiltonian. A path selection is made such that the most resonant three-photon pathway determines the population dynamics, reducing the discrete excitation to a single ladder.

The QC is described, as in our model, in Figure 1, where the states at energy $m\nu_3$ are subdivided according to their ν_3 character. The coupling between these substates, H_{anh}, at $m\nu_3$ is independent of the states' ν_3 character; i.e., there are no V–V selection rules, and (as in our model) H_{anh} is also independent of the energy in the molecule. The broadening of a state at energy $m\nu_3$ because of its coupling to all of the other states at energy $m\nu_3$ is

$$\Gamma_m = H_{\text{anh}}^2 \chi(m) \tag{1.21}$$

where $\chi(m)$ is the total density as defined by Eq. (1.11). The factor 2π that appears in Eq. (1.2) can be considered to be contained in this model in the definition of H_{anh}.

The laser transition rates from $n\nu_3$ to $(n+1)\nu_3$ at energy $m\nu_3$ in the QC Eq. (1.7) are for this model,

$$R_n(m) = \frac{D_n{}^2(m)\gamma_n(m)}{\Delta_n{}^2(m) + \gamma_n{}^2(m)} \tag{1.22}$$

where

$$\gamma_n(m) = \frac{\Gamma_m + \Gamma_{m+1}}{2} \qquad \text{(independent of } n\text{)} \tag{1.23}$$

The detunings $\Delta_n(m)$ are given by Eq. (1.9). The squared dipole strength, $D_n{}^2(m)$, is given by Eq. (1.3), except that here $N_n(m)$ is unity. The concept of self-mixing effective states, represented in our model by $\bar{g}$, does not play a role here. All anharmonic coupling leads to dephasing, $1/T_2$ rates, and population dynamics, $1/T_1$ rates. As in our model the $1/T_1$ rates are considered very fast with respect to the laser dynamics such that the energy shell populations are distributed as in Eq. (1.11). The total rate of laser excitation at $m\nu_3$ quanta is therefore

$$R^{\uparrow}(m) = \sum_{k=0}^{m} \frac{D_k^{\,2}(m)\gamma_k(m)}{\Delta_k^{\,2}(m) + \gamma_k^{\,2}(m)} \frac{\chi_k(m)}{\chi(m)} \tag{1.24}$$

Dissociation is treated, as in our model, using RRKM rates. [A new approach to bond selective chemistry is presently being developed at USC, but as of this time it has not been incorporated into their mpd modeling (Thiele *et al.*, 1980)].

The ν_3 ladder part of the problem contains the major spectroscopic information. In our model we include a complete first-order vibration–rotation theory such that all allowed rotational levels are taken into account. The USC group has truncated this multilevel excitation problem by a path selection process reducing the ν_3 mode discrete excitation to a single ladder. Both models average over the rotational Boltzmann distribution after solving the model for each rotational ground state. This difference between the models is purely calculational. The QC coupling in our model, however, occurs at every ν_3 level. We use a FGR leakage rate that is computed from the density of states for those states with one less ν_3 quantum and our anharmonic coupling parameter g. (The QC gets populated via V–V relaxation only.) The USC group by contrast computes the place at which the background density dominates the ν_3-ladder density. At that point they introduce a *laser transition* directly from this last discrete level to the lowest QC level. The ratio of the up-to-down rates is based on the ν_3-ladder degeneracy at this last discrete level as compared with the total density of states at the lowest QC level multiplied by H_{anh}.

In the QC the density of states at each energy $m\nu_3$ is subdivided into groups of states according to their ν_3 character. In point of fact the USC group was first to note the importance of this decomposition of the density of states (Stone *et al.*, 1976). This is a very natural description of the QC since the ν_3 mode carries the oscillator strength in the molecule. In our model we go one step further specifying a selection rule for the V–V anharmonic coupling; i.e., the states most strongly coupled are those that differ by $1\nu_3$ quantum. Changes in ν_3 quanta that are greater than one are considered very small and are neglected. This selection rule is consistent with standard notions of the molecular Hamiltonian (Amat *et al.*, 1971; Shaffer *et al.*, 1939; Darling and Dennison, 1940). While we have not explicitly taken into account the octahedral symmetry of the molecule, we expect that this further selection rule will give overall shift throughout the entire molecule making our fitted value of g representative of the additional effect. The USC model is more approximate in this respect since no selection rules are applied, viz., the V–V coupling at energy $m\nu_3$ is considered equally strong between all states independent of their mode character. Therefore, the coupling rates they calculate require the total density of states at energy $m\nu_3$ Eq. (1.24).

Another major difference between the models is concerned with the computation of the transition dipole strengths in the QC. In Eq. (1.4) we introduce the concept of coupling between states of similar ν_3 character with a reduced stength $\bar{g}$. In the USC model, however, this type of coupling is not introduced. (At USC they are currently introducing a selection rule of this type. Their motivation is based on the fact that their $1/T_2$ widths are too large in the high QC. The laser pumping dies away to zero well before the molecule dissociates in contradiction with experiment. They presently cut off the $1/T_2$ widths at 30 cm^{-1} to avoid this difficulty and fit experiment. In addition, they now recognize that a selection rule of this type can automatically cut off the number of coupled states, reducing $1/T_2$ and allowing the molecule to be pumped and dissociated.) At this point we would like to discuss the validity of using effective states for this self V–V interaction since it requires the distribution of couplings to be of a relatively specific form. If a different distribution is chosen for the coupling, it is possible to destroy the effective states character entirely. The difference between having and not having effective states on the final calculated cross sections is dramatic as shown in Figure 1.14. We see that without effective states the cross sections are reduced by orders of magnitude in the high QC destroying the laser excitation. In addition, the physical implication for the microscopic distribution of coupling parameters is significant.

Without V–V anharmonic coupling each group of states $\chi_n(m)$ in the QC act as new ν_3-ladder ground states. Each state within $\chi_n(m)$ is independent of all the other states such that all the new ν_3 ladders are uncoupled. If for simplicity we neglect the degeneracy of the ν_3 mode and consider each QC transition as only involving two states, then the dipole strength for each single transition can be simply defined in terms of the Rabi frequency, Ω. V–V anharmonic self-coupling for each group of states $\chi_n(m)$ strongly mixes states that are within an interaction distance $\bar{g}$ making them effectively resonant. States that are farther than $\bar{g}$ are responsible for a $1/T_2$ type dephasing, but this effect is negligible as compared with the relaxation induced $1/T_2$ because of coupling between $n\nu_3$ character manifolds. The self-mixing of these states preserves the total dipole strength such that

$$\sum_{\mathrm{l,\,u}} \Omega_{\mathrm{lu}}^2 = \Omega^2 \tag{1.25}$$

where the index l, u refers to the transition's lower, upper states, respec-

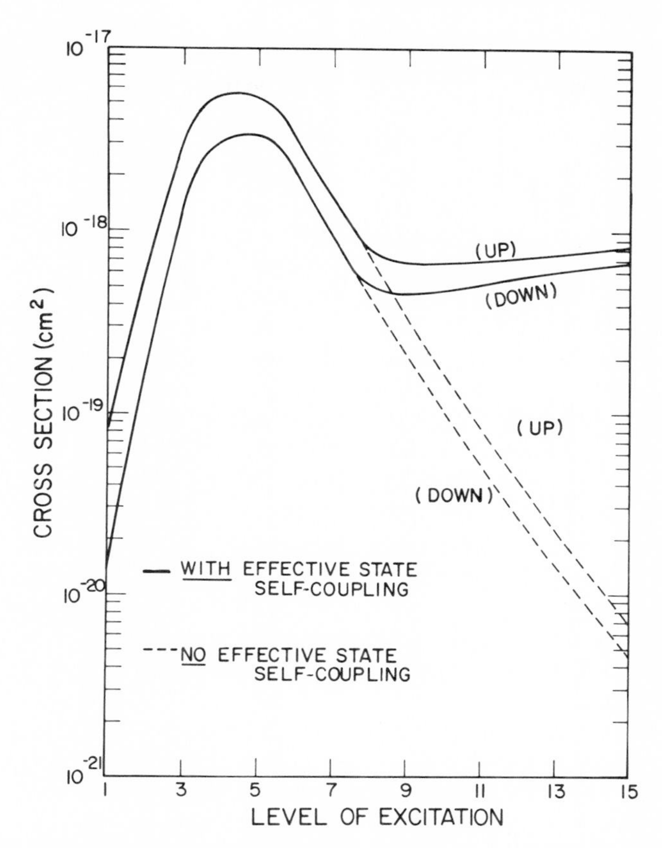

Figure 1.14. Absorption (up) and emission (down) cross sections with and without effective states self-coupling vs. level of excitation. The cross sections not only fall off without self-coupling, but the absorption cross sections [(up)–(down)] are negative above eight quanta of excitation. The difference between this figure and Figure 1.4 is the choice of parameters in the theory. Figure 1.4 is consistent with the parameters given in Section 1.3.

tively. Since the state-to-state self-coupling is not known, we assume it to be random. The choice of randomness with respect to the *phases* of the coupling constants, however, has a major impact on the distribution of dipole moments. We will discuss this randomness only indirectly in the context of the anharmonically mixed transition Rabi frequencies, Ω_{lu}.

The Hamiltonian describing the laser excitation for the transition without loss of generality can be written as

$$H = -\sum_{l} \sum_{u} \Omega_{lu}(\rho_{lu} + \rho_{ul}) \tag{1.26}$$

where we assume the Rabi frequencies Ω_{lu} are real and symmetric. The density matrix operator ρ_{lu} is defined in Dirac notation as

$$\rho_{lu} = |l\rangle\langle u| \tag{1.27}$$

Since $\bar{g}$ is very small with respect to the Rabi frequency, we have neglected Eq. (1.6) the detuning spread in the state's energies. The Schrödinger equation expressed in terms of equations for the state amplitudes is

$$i\dot{C}_l = -\sum_u \Omega_{lu} C_u \tag{1.28a}$$

$$i\dot{C}_u = -\sum_l \Omega_{lu} C_l \tag{1.28b}$$

which after performing a Laplace transformation, $\mathsf{L}(z)$, becomes

$$izS_l(z) = \frac{i}{N^{1/2}} - \sum_u \Omega_{lu} S_u(z) \tag{1.29a}$$

$$izS_u(z) = -\sum_l \Omega_{lu} S_u(z) \tag{1.29b}$$

$S_{l,\,u}(z)$ is the Laplace transforms of $C_{l,\,u}(t)$ with transform variable z. The initial state of the system is such that all ground states are equally populated, $C_l(0) = 1/N^{1/2}$. By iterating Eqs. (1.9) the transforms can be expressed as infinite series,

$$izS_l(z) = \frac{i}{N^{1/2}} \left\{ 1 - \frac{1}{z^2} \sum_u \Omega_{lu} \sum_{l'} \Omega_{l'u} + \frac{1}{z^4} \sum_u \Omega_{lu} \sum_{l'} \Omega_{l'u} \sum_{u'} \Omega_{l'u'} \sum_{l''} \Omega_{l''u'} + \cdots \right\} \tag{1.30a}$$

$$izS_{u}(z) = -\frac{1}{zN^{1/2}} \left\{ \sum_l \Omega_{lu} - \frac{1}{z^2} \sum_l \Omega_{lu} \sum_{u'} \Omega_{lu'} \sum_{l'} \Omega_{l'u'} + \frac{1}{z^4} \sum_l \Omega_{lu} \sum_{u'} \Omega_{lu'} \sum_{l'} \Omega_{l'u'} \sum_{u''} \Omega_{l'u''} \sum_{l''} \Omega_{l''u''} - \cdots \right\} \tag{1.30b}$$

Depending on the distribution of Rabi frequencies, Ω_{lu}, these sums, Eqs. (1.30), give very different solutions for the dynamics. We will consider two cases here: (1) the distribution is random in both sign and magnitude such that $\sum_{u'} \Omega_{lu'} \Omega_{l'u'} = \sum_{u'} \Omega_{lu'}^2 \delta_{ll'} = \Omega^2 \delta_{ll'}$. (2) The distribution is random in magnitude, but all the signs are the same such that $\sum_{u'} \Omega_{lu'} \Omega_{l'u'} = \Omega^2$.

For case (1) Eqs. (1.30) become

$$S_l(z) = \frac{1}{zN^{1/2}} \left\{ 1 - \frac{\Omega^2}{z^2} + \frac{\Omega^4}{z^4} - \cdots \right\}$$

$$= \frac{1}{N^{1/2}} \frac{z}{z^2 + \Omega^2} \tag{1.31a}$$

$$S_u(z) = +\frac{i\sum_l \Omega_{lu}}{z^2 N^{1/2}}\left\{1 - \frac{\Omega^2}{z^2} + \frac{\Omega^4}{z^4} - \cdots\right\}$$

$$= +\frac{i\sum_l \Omega_{lu}}{N^{1/2}}\frac{1}{z^2 + \Omega^2} \tag{1.31b}$$

The inverse Laplace transform of Eqs. (1.31) are well known

$$C_l(t) = \frac{1}{N^{1/2}}\cos \Omega t \tag{1.32a}$$

$$C_u(t) = \frac{i\sum_l \Omega_{lu}}{\Omega N^{1/2}}\sin \Omega t \tag{1.32b}$$

where the population in the collective ground, upper state is obtained by taking the modulus square and summing over l, u, respectively:

$$\sum_l |C_l(t)|^2 = \cos^2 \Omega t \tag{1.33a}$$

$$\sum_u |C_u(t)|^2 = \sin^2 \Omega t \tag{1.33b}$$

The population dynamics are determined by the single transition Rabi frequency. There is no collective behavior giving rise to an effective states dipole strength.

For case (2) Eqs. (1.30) become

$$S_l(z) = \frac{1}{zN^{1/2}}\left\{1 - \frac{N\Omega^2}{z^2} + \frac{N^2\Omega^4}{z^4}\cdots\right\}$$

$$= \frac{1}{N^{1/2}}\frac{z}{z^2 + N\Omega^2} \tag{1.34a}$$

$$S_u(z) = \frac{i}{z^2 N^{1/2}}\sum_l \Omega_{lu}\left\{1 - \frac{N\Omega^2}{z^2} + \frac{N^2\Omega^4}{z^4}\cdots\right\}$$

$$= \frac{i\sum_l \Omega_{lu}}{N^{1/2}}\frac{1}{z^2 + N\Omega^2} \tag{1.34b}$$

The only difference between cases (1) and (2) is that $\Omega^2 \to N\Omega^2$. Therefore, the collective ground, upper-state populations analogous to Eq. (1.33) are

$$\sum_l |C_l(t)|^2 = \cos^2 N^{1/2}\Omega t \tag{1.35a}$$

$$\sum_u |C_u(t)|^2 = \sin^2 N^{1/2}\Omega t \tag{1.35b}$$

The dynamics are determined by the full effective states dipole strength. In our model we calculate this number N using the self-coupling strength $\bar{g}$

multiplied by the appropriate density of states. We recognize that for all the phases to match up such that all the Rabi coupling terms are positive, case (2), it takes a very special V–V anharmonic interaction. We suspect, however, that some amount of self-coupling must exist since without self-coupling the molecule cannot be laser excited past a certain level (see Horsley *et al.*, 1979). If self-coupling is only 50% effective, then $N \rightarrow N/2$, but effective states will still exist to a reduced degree. (Our fit of the coupling parameter $\bar{g}$ can mask the effect such that we cannot distinguish a complete from incomplete self-coupling.) One additional feature that suggests self-coupling is present is that our analytic formulas reduce approximately in the high QC to those calculated by Black *et al.* (1979), in general agreement with typical experimental trends that show an almost constant absorption cross section at high levels of excitation for most polyatomic molecules.

In summary the major conceptual differences between the models are in the QC where we consider a selection rule for the V–V coupling including self-mixing and the group at USC does not. [The selection rule that the USC group is presently considering reduces their value of $1/T_2$ in the QC. It does not introduce self-mixing (private communication, Stone and Goodman)]. Since hard experimental data does not exist at this time that probes the nature of the QC, it is not possible to make any definitive statements about the degree to which self-mixing exists.

1.6. Comparison with a Thermal QC Model

Recently, Black *et al.* (1979) have stated that IR-laser excitation in polyatomic molecules leads to a thermal population distribution. The data supporting this hypothesis are mpe and mpd data of SF_6, plotted in the form of percent dissociated vs. $\langle n \rangle$, i.e., the actual independent variable, the laser fluence, is eliminated from the data. They further state that for most polyatomic molecules the absorption cross section remains approximately constant, independent of laser fluence (Judd, 1980). Using this fact and the density of states function for a degenerate oscillator (one frequency for all modes), we may derive population rate equations that give a thermal distribution. It is speculated in the work of Black *et al.* (1979) that the ν_3-ladder "bottleneck" leads to a slight broadening of this distribution away from thermal but that this effect becomes less and less important at higher laser fluences. In this section we will examine these suppositions in detail. We will derive the thermal model on more fundamental grounds than is presented in the work of Black *et al.* (1979) and show that a fit to the data leads to an unphysically large anharmonic coupling for low levels. In addition, we will join the thermal QC model to our ν_3 ladder to explore in detail the effect of the rotational bottleneck (Ackerhalt and Galbraith, 1979).

To derive the thermal model we need only a strong mixing hypothesis. Suppose we then transform to a basis in which the intramolecular couplings are diagonal (explicitly not normal modes). Dipole couplings exist now between each lower state at energy $n\nu_3$ and all strongly mixed upper states, i.e., $G\chi(n+1)$, with G the anharmonic coupling constant. Clearly if $G\chi \gg 1$, we have transitions from $n \to n+1$ and $n+1 \to n$ via FGR,

$$R_{n\to n+1} = 2\pi s^2 \chi(n+1) \tag{1.36a}$$

$$R_{n+1\to n} = 2\pi s^2 \chi(n) \tag{1.36b}$$

The single-pair dipole strength squared was derived by Stone *et al.* (1976),

$$s^2 = \frac{3\chi_4(n)}{G\chi(n)\chi(n+1)}\alpha_{01}^2 \tag{1.37}$$

where α_{01}^2 is the $0 \to 1$ Rabi frequency and $\chi_4(n)$ is the density of states at $n\nu_3$ with ν_3 treated as a four-dimensional mode. Using Eq. (1.37) in Eqs. (1.36) we obtain

$$R_{n\to n+1} = \frac{6\pi}{G}\frac{\chi_4(n)}{\chi(n)}\alpha_{01}^2 \tag{1.38}$$

and

$$R_{n+1\to n} = \frac{6\pi}{G}\frac{\chi_4(n)}{\chi(n+1)}\alpha_{01}^2 \tag{1.39}$$

which reduce exactly to the thermal rates when the oscillator is assumed fully degenerate. The significance of this derivation follows after a fit to data. Indeed we find that even when coupled to our ν_3 ladder, a fit to experiment gives

$$G \gtrsim 500\ \text{cm}^{-1} \tag{1.40}$$

everywhere in the molecule. Clearly such interaction strengths are not possible at low levels of excitation reflecting the fact that the strong mixing assumption is not valid there.

To compare with our model, consider the $k\nu_3$ submanifold at $n\nu_3$. Asymptotically $\Gamma \gg \Delta$ and Γ is proportional to $\chi_{k+1}(n+1)$,

$$R_k(n) \propto \frac{\chi_k(n)(k+3)/2}{\chi_{k+1}(n+1)}\alpha_{01}^2 \tag{1.41}$$

which gives for the up and down rates

$$R^{\text{up}}(n) \propto \left(\frac{\chi_4(n)}{\chi(n)} - \frac{2}{3}\right)\alpha_{01}^2 \tag{1.42}$$

and

$$R^{\text{down}}(n) \propto \left(\frac{\chi_4(n)}{\chi(n+1)} - \frac{\chi_0(n)}{\chi(n+1)}\right)\alpha_{01}^2 \tag{1.43}$$

giving a "thermal" model high in the QC. Hence, the high QC rates of our model reduce automatically to rates that are very close to those of Black *et al.* (1979). Our model is consistent with standard spectroscopy at the low excitation and with a (reasonable) strong mixing model at the high end and flows smoothly from one limit to the other.

In Figure 1.4 we show the QC transition cross sections (both for absorption and emission) at each level of $n\nu_3$ both for our model and the thermal model fitted to the Deutsch (1977) data and including our ν_3 ladder. (We should note that the thermal model, even with very different rates as shown in the figure, also gives an excellent fit to the data with the ν_3 ladder included.)

In Figure 1.15 we show the effect of the ν_3-ladder bottleneck on the thermal Boltzmann distribution. By adding our ν_3-ladder excitation we obtain a very nonthermal distribution that in fact is *very close* to that predicted in our model (100-nsec pulse, $\langle n \rangle \sim 14$, collisionless).

Black *et al.* (1979) have made the claim that at high intensities the ν_3-ladder bottleneck is not important, and the population distribution is "close" to thermal. Taking their advice we calibrated their thermal QC model with their 500-psec data finding a QC cross section. Using this calibrated QC model with our ν_3-ladder bottleneck we found that this model no longer fit the $\langle n \rangle$ vs. fluence data of Deutsch (1977) for 100-nsec pulselengths. [This data is consistent with the data of Black *et al.* (1979); see Judd (1980).] The QC cross section was too large by a factor of 3. We concluded that the ν_3-ladder bottleneck was fundamental for mpe and mpd of polyatomic molecules at all laser intensities. Therefore, the population distribution (collisionless) is *never* thermal (Arvedson and Kohn, 1979, Thiele *et al.*, 1979).

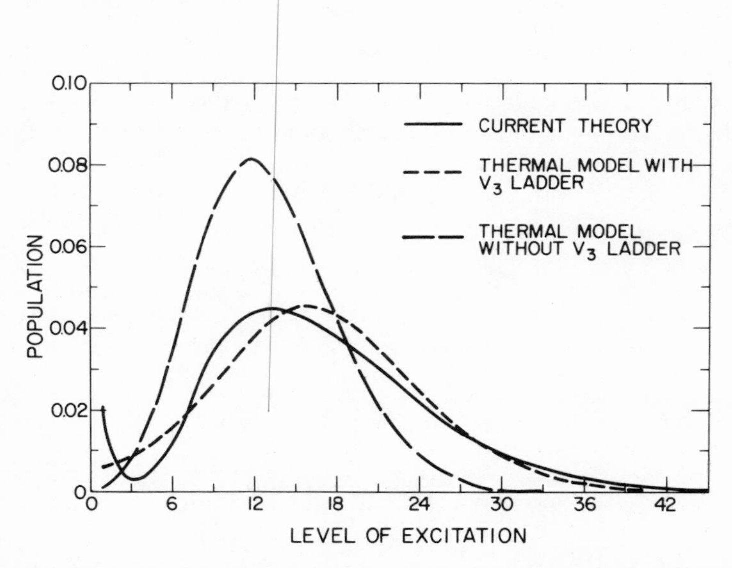

Figure 1.15. Population fraction vs. level of excitation for fixed $\langle n \rangle \sim 14$. Curves for the current theory, thermal model with the ν_3 ladder and without the ν_3 ladder are shown.

ACKNOWLEDGMENTS

We would like to thank Myron Goodman, Jim Stone, and Everett Thiele for communicating results of their work prior to its publication. Our close interaction has enabled us to present Section 1.5.

We would like to acknowledge a communication by D. Proch with respect to his data as presented in Figure 1.5. An overall systematic scaling factor of 1.4 has been used to shift his published data. We were first informed of this error by David Dows and Myron Goodman.

We would also like to thank Margaret Lillberg for her excellent compiling and typing of the manuscript.

This work was performed under the auspices of the U.S.D.O.E.

References

Ackerhalt, J., and Eberly, J., 1976, Coherence versus incoherence in stepwise laser excitation of atoms and molecules, *Phys. Rev. A.* **14**:1705.

Ackerhalt, J., and Galbraith, H., 1978, Collisionless multiple photon excitation of SF_6: A comparison of anharmonic oscillators with and without octahedral splitting in the presence of rotational effects, *J. Chem. Phys.* **69**:1200.

Ackerhalt, J., and Galbraith, H., 1979, Collisionless multiple photon excitation in SF_6: Thermal or not? in *Laser Spectroscopy IV*, Proc. 4th International Conference, Rottach-Egern, Springer Series in Optical Sciences (eds. H. Walther and K. Rothe), Springer-Verlag, New York, Heidelberg, Berlin.

Ackerhalt, J., and Shore, B., 1977, Rate equations versus Bloch equations in multiphoton ionization, *Phys. Rev. A.* **16**:277.

Ackerhalt, J., Flicker, H., Galbraith, H., King, J., and Person, W., 1978, Analysis of $3\nu_3$ in SF_6, *J. Chem. Phys.* **69**:1461.

Akhmanov, A., Baranov, V., Pismenny, V., Bagratashvili, V., Kolomiisky, Yu., Letokhov, V., and Ryabov, E., 1977, Multiple photon excitation of polyatomic molecules from the many rotational states by an intense pulse of ir radiation, *Optics Commun.* **23**:357.

Akhmanov, S., Gordienkv, V., Mikhunko, A., and Panchenko, V., 1978, Dependence of the rate of vibrational–translational relaxation in SF_6 on the intensity of selective laser excitation, *JETP Lett.* **26**:453.

Akulin, V., Alimpiev, S., and Karlov, N., 1978, Mechanism of collisionless dissociation of polyatomic molecules, *Sov. Phys. JETP* **47**:257.

Aldridge, J., Filip, H., Flicker, H., Holland, R., McDowell, R., Nereson, N., and Fox, K., 1975, Octahedral fine-structure splittings in ν_3 of SF_6, *J. Mol. Spectrosc.* **58**:165.

Alimpiev, S., Karlov, N., Sartakov, B., and Khokhlov, E., 1978a, The spectral structure of the low level excitation in the process of the collisionless dissociation of polyatomic molecules, *Optics Commun.* **26**:45.

Alimpiev, S., Bagratashvili, V., Karlov, N., Letokhov, V., Makarov, A., Sartakov, B., and Khokhlov, E., 1978b, Effect of depletion of many rotational states in vibrational excitation of molecules in a strong ir field, *JETP Lett.* **25**:547.

Allen, L., and Eberly, J., 1975, *Optical Resonance and Two-level Atoms*, Wiley, New York.

Amat, G., Nielsen, H., and Tarrago, G., 1971, *Rotation-Vibration of Polyatomic Molecules*, Marcel Dekker, New York.

Ambartzumian, R., Gorokhov, Yu., Letokhov, V., Makarov, G., and Puretskii, A., 1975, Selective dissociation induced by absorption of strong infrared radiation in weak combination bands of polyatomic molecules, *JETP Lett.* **22**:374.

Ambartzumian, R., Gorokhov, Yu., Letokhov, V., and Makarov, G., 1976a, Interaction of SF_6 molecules with a powerful infrared laser pulse and the separation of sulfur isotopes, *Sov. Phys. JETP* **42**:993.

Ambartzumian, R., Furzikov, N., Gorokhov, Yu., Letokhov, V., Makarov, G., and Puretzky, A., 1976b, Selective dissociation of SF_6 molecules in a two-frequency infrared laser field, *Optics Commun.* **18**:517.

Ambartzumian, R., Makarov, G., and Puretzky, A., 1977, Selective dissociation of polyatomic molecules by two infrared pulses, in *Laser Spectroscopy III*, Springer Series in Optical Sciences (eds. J. Hall and J. Carlsten), Springer-Verlag, Berlin, Heidelberg, New York.

Ambartzumian, R., Makarov, G., and Puretzky, A., 1978a, Investigation of multiple photon excitation of OsO_4 by dissociation yield saturation, *Optics Commun.* **71**:79.

Ambartzumian, R., Vasil'ev, B., Grasyuk, A., Dyad'kin, A., Letokhov, V., and Furzikov, N., 1978b, Isotopically selective dissociation of CCl_4 molecules by high power NH_3 laser radiation, *Sov. J. Quantum Electron.* **8**:1015.

Ambartzumian, R., Zubarev, I., Logansen, A., and Kotov, A., 1978c, Investigation of the kinetics of the vibrational excitation of the UF_6 molecule by the IR–UV resonance method, *Sov. J. Quantum Electron.* **8**:910.

Ambartzumian, R., Knyazev, I., Lobko, V., Makarov, G., and Puretzky, A., 1979, Multiple infrared photon absorption in OsO_4, *Appl. Phys.* **19**:75.

Arvedson, M., and Kohl, D., 1979, An electron diffraction study of laser pumped SF_6, *Chem. Phys. Lett.* **64**:119.

Bagratashvili, V., Knyazev, I., Letokhov, V., and Lobko, V., 1976, Optoacoustic detection of multiple photon molecular absorption in a strong ir field, *Optics Commun.* **18**:525.

Bagratashvili, V., Dalzhikov, V., and Letokhov, V., 1979, Kinetics of the ir absorption spectra of SF_6 molecules vibrationally excited by an intense CO_2 laser pulse, *JETP* **1**:49.

Benson, S. W., 1978, Thermochemistry and kinetics of sulfur-containing molecules and radicals, *Chem. Revs.* **78**:23.

Bertsev, V., Kolomiitseva, T., and Tsyganenko, N., 1974, Infrared spectra of cryosystems. 2: Sulfur hexafluoride, *Opt. Spektrosk.* **37**:463.

Black, J., Yablonovitch, E., Bloembergen, N., and Mukamel, S., 1977, Collisionless multiphoton dissociation of SF_6: A statistical thermodynamic process, *Phys. Rev. Lett.* **38**:1131.

Black, J., Kolodner, P., Shultz, M., Yablonovitch, E., and Bloembergen, N., 1979, Collisionless multiphoton energy deposition and dissociation of SF_6, *Phys. Rev. A.* **19**:704.

Bloembergen, N., and Yablonovitch, E., 1977, Collisionless multiphoton dissociation of SF_6: A statistical thermodynamic process, in *Laser Spectroscopy III*, Springer Series in Optical Sciences (eds. J. Hall and J. Carlsten), Springer-Verlag, Berlin, Heidelberg, New York.

Bordé, C., Ouhayoun, M., and Bordé, J., 1978, Observation of magnetic hyperfine structure in the infrared saturation spectrum of $^{32}SF_6$, *J. Mol. Spectrosc.* **73**:344.

Bordé, C., Ouhayoun, M., Van Lerberghe, A., Salomon, C., Avrillier, S., Cantrell, C., and Bordé, J., 1979, Fine and hyperfine structure studies in the infrared spectra of spherical top molecules, in *Laser Spectroscopy IV*, Proc. 4th International Conference, Rottach-Egern (eds. H. Walther and K. Rothe), Springer Series in Optical Sciences, Springer-Verlag, New York, Heidelberg, Berlin.

Bott, J., and Jacobs, T., 1969, Shock tube studies of sulfur hexafluoride, *J. Chem. Phys.* **50**:3850.

Brock, E., Krohn, B., McDowell, R., Patterson, C., and Smith, D., 1979, The structure of Q-branches in infrared-active fundamental-type bands of spherical top molecules, *J. Mol. Spectrosc.* **76**:301.

Brunner, F., and Proch, D., 1978, The selective dissociation of SF_6 in an intense ir field: a molecular beam study on the influence of laser wavelength and energy, *J. Chem. Phys.* **68**:4936.

Cantrell, C., and Galbraith, H., 1976, Towards an explanation of collisionless multiple photon laser dissociation of SF_6, *Optics Commun.* **18**:513.

Cantrell, C., and Galbraith, H., 1977, Effects of anharmonic splitting upon collisionless multiple photon laser excitation of SF_6, *Optics Commun.* **21**:374.

Cantrell, C., Galbraith, H., and Ackerhalt, J., 1977, On the influence of molecular structure upon the collisionless laser photodissociation of SF_6, in *Multiphoton Processes* (eds. J. H. Eberly and P. Lambropoulos), John Wiley, New York, Chichester, Brisbane, Toronto.

Cantrell, C., Freund, S., and Lyman, J., 1979, Laser induced chemical reactions and isotope separation, in *Laser Handbook III*, North Holland, Amsterdam.

Chedin, A., and Cihla, Z., 1974, Mechanization of operations within a noncommutative algebraic structure: Application to the unitary transformation of the Hamiltonian of a polyatomic molecule, *J. Mol. Spectrosc.* **49**:289.

Childs, W., and Jahn, H., 1939, A new Coriolis perturbation in the methane spectrum III. Intensities and optical spectrum, *Proc. Roy. Soc.* **A169**:451.

Coggiola, M., Schulz, P., Lee, Y., and Shen, Y., 1977, A molecular beam study of multiphoton dissociation of SF_6, *Phys. Rev. Lett.* **38**:17.

Darling, B., and Dennison, D., 1940, The water vapor molecule, *Phys. Rev.* **57**:128.

Deutsch, T., 1977, Optoacoustic measurements of energy absorption in CO_2 TEA-laser-excited SF_6 at 293 and 145K, *Optics Lett.* **1**:25.

Deutsch, T., and Brueck, S., 1978, Collisionless intramolecular energy transfer in vibrationally excited SF_6, *Chem. Phys. Lett.* **54**:258.

Dodd, R. E., Woodward, L. A., and Roberts, H. L., 1957, Molecular vibrations of group 6 decafluorides, Part 1.—Infrared and Raman spectra of disulphur decafluoride and ditellurium decafluoride, *Trans. Faraday Soc.* **53**:1545.

Duperrex, R., and Van den Bergh, H., 1979a, Competition between collisions and optical pumping in unimolecular reactions induced by monochromatic infrared radiation, *Chem. Phys.* **40**:275.

Duperrex, R., and Van den Bergh, H., 1979b, Temperature dependence in the multiphoton dissociation of $^{32}SF_6$, *J. Chem. Phys.* **70**:5672.

Emanuel, G., 1979, A simple model for multiphoton molecular absorption, *J. Quant. Spectrosc. and Radiat. Transfer* **21**:147.

Fox, K., 1977, Rotational–vibrational transition moments in excited states of spherical-top molecules, *Optics Lett.* **1**:214.

Fox, K., 1978, $3\nu_3$ vibrational spectra of SF_6 and UF_6, *J. Chem. Phys.* **68**:25.

Fox, K., and Person, W., 1976, Transition moments in infrared-active fundamentals of spherical top molecules, *J. Chem. Phys.* **64**:5218.

Fox, K., Krohn, B., and Shaffer, W., 1979, Cubic and quartic anharmonic potential energy functions for octahedral XY_6 molecules, *J. Chem. Phys.* **71**:2222.

Frankel, D., 1976, Rapid intramolecular v–v energy transfer in SF_6, *J. Chem. Phys.* **65**:1696.

Frankel, D., and Manuccia, T., 1978, Collisionless 16 μm fluorescence in SF_6 following 10 μm CO_2 laser pumping: Comments on the vibrational quasicontinuum, *Chem. Phys. Lett.* **54**:451.

Freund, S., and Lyman, J., 1978, Multiple photon isotope separation in MoF_6, *Chem. Phys. Lett.* **55**:435.

Fuss, W., 1979, Rate equations approach to the infrared collisionless multiphoton excitation, *Chem. Phys.* **36**:135.

Fuss, W., and Hartmann, J., 1979, IR absorption of SF_6 excited up to the dissociation threshold, *J. Chem. Phys.* **70**:5468.

Galbraith, H., 1978, Single-photon transition moments in excited states of spherical-top molecules, *Optics Lett.* **3**:154.

Galbraith, H., and Ackerhalt, J., 1978a, Calculation of the temperature dependence of the multiphoton absorption spectrum of SF_6, *Optics Lett.* **3**:109.

Galbraith, H., and Ackerhalt, J., 1978b, Comparison of multiple-photon excitation models, *Optics Lett.* **3**:152.

Galbraith, H., and Cantrell, C., 1977, Structure of the vibrational states in the ν_3-fundamental and its overtones in SF_6 and multiphoton absorption effects, in *The Significance of Nonlinearity in the Natural Sciences*, Plenum Press, New York.

Goldberg, M., and Yusek, R., 1970, High-resolution inverted lamb-dip spectroscopy on SF_6, *Appl. Phys. Lett.* **17**:349.

Goodman, M., Stone, J., and Dows, D., 1976, Laser induced rate processes in gases: Dynamics of polyatomic systems, *J. Chem. Phys.* **65**:5052.

Gordinko, V., Mikheenko, A., and Panchenko, V., 1978, Vibrational–translational relaxation in SF_6 at high excitation rates, *Sov. J. Quantum Electron* **8**:1013.

Gower, M., and Billman, K., 1977, Collisionless dissociation and isotopic enrichment of SF_6 using high powered CO_2 laser radiation, *Optics Commun.* **20**:123.

Gower, M., and Gustafson, T., 1977, Collisionless dissociation of SF_6 using two resonant frequency CO_2 laser fields, *Optics Commun.* **23**:69.

Grant, E., Schulz, P., Sudbo, Aa., Coggiola, M., Lee, Y., and Shen, Y., 1977a, Multiphoton dissociation of polyatomic molecules studied with a molecular beam, in *Laser Spectroscopy III*, Springer Series in Optical Sciences (eds. J. Hall and J. Carlsten), Springer-Verlag, New York, Heidelberg, Berlin.

Grant, E., Coggiola, M., Lee, Y., Schulz, P., Sudbo, Aa., and Shen, Y., 1977b, The extent of energy randomization in the infrared multiphoton dissociation of SF_6, *Chem. Phys. Lett.* **52**:595.

Grant, E., Schulz, P., Sudbo, Aa., Coggiola, M., Shen, Y., and Lee, Y., 1977c, Molecular beam studies on multiphoton dissociation of polyatomic molecules, in *Multiphoton Processes* (eds. J. Eberly and P. Lambropoulos), John Wiley, New York, Chichester, Brisbane, Toronto.

Ham, D., and Rothchild, M., 1977, Transmission measurements of multiple photon absorption in SF_6, *Optics Lett.* **1**:28.

Harvey, R., and Bauer, S., 1953, An electron diffraction study of disulfur decafluoride, *J. Am. Chem. Soc.* **75**:2840.

Hecht, K., 1960, The vibration–rotation energies of tetrahedral XY_4 molecules, *J. Mol. Spectrosc.* **5**:355.

Heller, D., and Mukamel, S., 1979, Theory of vibrational overtone line shapes of polyatomic molecules, *J. Chem. Phys.* **70**:463.

Hodgkinson, D., and Briggs, J., 1976, Dissociation of polyatomic molecules by intense infrared laser radiation, *Chem. Phys. Lett.* **43**:451.

Horsley, J., Stone, J., Goodman, M., and Dows, D., 1979, A unified model for multiple photon dissociation of SF_6, *Chem. Phys. Lett.* **66**:461.

Horsley, J., Rabinowitz, P., Stein, A., Cox, D., Brickman, R., and Kaldor, A., 1980, Laser chemistry experiments with UF_6, *IEEE J. Quant. Elec.* **QE-16**:412.

Houston, P., and Steinfeld, J., 1975, Low temperature absorption contour of the ν_3 band of SF_6, *J. Mol. Spectrosc.* **54**:335.

Isenor, N., Merchant, V., Hallsworth, R., and Richardson, M., 1973, CO_2 laser-induced dissociation of SiF_4 molecules into electronically excited fragments, *Can. J. Phys.* **51**:1281.

Jahn, H., 1938a, A new Coriolis perturbation in the methane spectrum I. Vibrational–rotational Hamiltonian and wavefunctions, *Proc. Roy. Soc.* **A168**:469.

Jahn, H., 1938b, A new Coriolis perturbation in the methane spectrum II. Energy levels, *Proc. Roy. Soc.* **A168**:495.

Jahn, H., 1939, Coriolis perturbations in the methane spectrum IV. Four general types of Coriolis perturbation, *Proc. Roy. Soc.* **A171**:450.

Jensen, C., Person, W., Krohn, B., and Overend, J., 1977, Anharmonic splittings and vibrational energy levels of octahedral molecules: Application to the $n\nu_3$ manifold of $^{32}SF_6$, *Optics Commun.* **20**:275.

Jensen, C., Steinfeld, J., and Levine, R., 1978, Information theoretic analysis of multiphoton excitation and collisional deactivation in polyatomic molecules, *J. Chem. Phys.* **69**:1432.

Jensen, C., Anderson, T., Reiser, C., and Steinfeld, J., 1979, Infrared double resonance of SF_6 with a tunable diode laser, *J. Chem. Phys.* **71**:3648.

Jones, L., and Ekberg, S., 1979, Vibrational spectrum and potential constants for S_2F_{10}, preprint submitted to *J. Chem. Phys.*

Judd, O., 1979, A quantitative comparison of multiple photon absorption in polyatomic molecules, *J. Chem. Phys.* **71**:4515.

Karl, R., and Lyman, J., 1978, Investigation of the multiple-photon dissociation of SF_5Cl with a real time chlorine atom diagnostic, *J. Chem. Phys.* **69**:1196.

Kay, K., 1974, Theory of vibrational relaxation in isolated molecules, *J. Chem. Phys.* **61**:5205.

Kiang, T., Estler, R. C., and Zare, R. N., 1979, Upper and lower bounds on the F_5S–F bond energy, *J. Chem. Phys.* **70**:5925.

Kildal, H., 1977, Second vibrational overtone absorption spectrum of the ν_3 mode of SF_6, *J. Chem. Phys.* **67**:1287.

Kim, K., Person, W., Seitz, D., and Krohn, B., 1979, Analysis of the ν_4 (615 cm^{-1}) region of the Fourier transform and diode laser spectra of SF_6, *J. Mol. Spectrosc.* **76**:322.

Knyazev, I., Letokhov, V., and Lobko, V., 1978, Role of weak transitions in multiphoton excitation of molecules by ir laser radiation, *Optics Commun.* **25**:337.

Kolodner, P., Winterfeld, C., and Yablonovitch, E., 1977, Molecular dissociation of SF_6 by ultra-short CO_2 laser pulses, *Optics Commun.* **20**:119.

Kolomiiskii, Yu., and Ryabov, E., 1978, Frequency characteristics of isotopically selective dissociation of BCl_3 in a strong infrared CO_2 laser field, *Sov. J. Quantum Electron.* **8**:375.

Kuz'min, M., 1978, Dynamics of many photon excitation of molecular vibrations, *Sov. J. Quantum Electron.* **8**:438.

Kwok, H., and Yablonovitch, E., 1978, Collisionless intramolecular vibrational relaxation in SF_6, *Phys. Rev. Lett.* **41**:745.

Lamb, W., 1977, Multiphoton dissociation of polyatomic molecules: Quantum or classical?, in *Laser Spectroscopy III*, Springer Series in Optical Sciences (eds. J. Hall and J. Carlsten), Springer-Verlag, New York, Heidelberg, Berlin.

Lamb, W., 1979, Classical model of SF_6 multiphoton dissociation, in *Laser Spectroscopy IV*, Proc. 4th International Conference, Rottach-Egern, Springer Series in Optical Sciences (eds. H. Walther and K. Rothe), Springer-Verlag, New York, Heidelberg, Berlin.

Larsen, D., 1976, Frequency dependence of the dissociation of polyatomic molecules by radiation, *Optics Commun.* **19**:404.

Larsen, D., and Bloembergen, N., 1976, Excitation of polyatomic molecules by radiation, *Optics Commun.* **17**:254.

Letokhov, V., 1977, Multiphoton excitation and dissociation of molecules and isotope separation by intense infrared laser radiation, in *Multiphoton Processes* (eds. J. Eberly and P. Lambropulos), John Wiley, New York, Chichester, Brisbane, Toronto.

Letokhov, V., 1978, Laser spectroscopy VII. Multiphoton and multistep methods, *Opt. Laser Technol.* **10**:247.

Letokhov, V., and Makarov, A., 1976, "Leakage" effect as an exciting mechanism of high vibrational levels of polyatomic molecules by a strong quasi-resonant laser ir field, *Optics Commun.* **17**:250.

Letokhov, V., and Makarov, A., 1978, Excitation of multilevel molecular systems by laser ir field, *Appl. Phys.* **16**:47.

Loëte, M., Clairon, A., Frichet, A., McDowell, R., Galbraith, H., Hilico, J., Moret-Bailley, J., and Henry, L., 1977, Constantes spectrales de la bande ν_3 de la molécule $^{32}SF_6$ calculées à partir du spectre d'absorption saturée, *Comptes Rendus* **B285**:175.

Lyman, J. L., 1977, A model for unimolecular reaction of sulfur hexafluoride, *J. Chem. Phys.* **67**:1868.

Lyman, J., and Leary, K., 1978, Absorption of infrared radiation by a large polyatomic molecule: CO_2 laser irradiation of disulfur decaluoride, *J. Chem. Phys.* **69**:1858.

Lyman, J., Feldman, B., and Fischer, R., 1978a, Effect of CO_2 laser-pulse mode quality on multiple photon absorption in SF_6, *Optics Commun.* **25**:391.

Lyman, J., Anderson, R., Fischer, R., and Feldman, B., 1978b, Absorption of pulsed CO_2 laser radiation by SF_6 at 140K, *Optics Lett.* **3**:238.

Lyman, J., Danen, W., Nilsson, A., and Nowak, A., 1979, Multiple photon excitation of difluoroamino sulfur pentafluoride: A study of absorption and dissociation, *J. Chem. Phys.* **71**:1206.

Lyman, J., Quigley, G., and Judd, O., 1980, Single-infrared-frequency studies of multiple-photon excitation and dissociation of polyatomic molecules, in *Multiple-Photon Excitation and Dissociation of Polyatomic Molecules* (ed. C. Cantrell), Springer-Verlag, New York, Heidelberg, Berlin.

Marcott, C., Golden, W., and Overend, J., 1978, 3-quantum transition probabilities in the ν_3 manifold of SF_6 and the assignment of the observed $3\nu_3$ transition in the infrared spectrum, *Spectrochim. Acta* Part A **34**:661.

McDowell, R., Galbraith, H., Krohn, B., Cantrell, C., and Hinkley, E., 1976a, Identification of the SF_6 transitions pumped by a CO_2 laser, *Optics Commun.* **17**:178.

McDowell, R., Aldridge, J., and Holland, R., 1976b, Vibrational constants and force field of sulfur hexafluoride, *J. Phys. Chem.* **80**:1203.

McDowell, R., Galbraith, H., Cantrell, C., Nereson, N., and Hinkley, E., 1977, The ν_3 Q branch of SF_6 at high resolution: Assignment of the levels pumped by $P(16)$ of the CO_2 laser, *J. Mol. Spectrosc.* **68**:288.

McDowell, R. S., Galbraith, H., Cantrell, C., Nereson, N., Moulton, P., and Hinkley, E., 1978, High-J assignments in the 10.5 μm SF_6 spectrum: Identification of the levels pumped by CO_2 $P(12)$ and $P(22)$, *Optics Lett.* **2**:97.

Moulton, P., Larsen, D., Walpole, J., and Mooradian, A., 1977, High resolution transient-double-resonance spectroscopy in SF_6, *Optics Lett.* **1**:51.

Mukamel, S., 1978, On the nature of intramolecular dephasing processes in polyatomic molecules, *Chem. Phys.* **31**:327.

Mukamel, S., 1979, Reduced equations of motion for molecular multiphoton processes, *Phys. Rev. Lett.* **42**:168.

Mukamel, S., and Jortner, J., 1976, A model for isotope separation via molecular multiphoton photodissociation, *Chem. Phys. Lett.* **40**:150.

Narducci, L., Mitra, S., Shatas, R., and Coulter, C., 1977, Selective multiple-photon absorption by an anharmonic molecule, *Phys. Rev. A.* **16**:247.

Ouhayoun, M., and Bordé, C., 1977, Frequency stabilization of CO_2 lasers through saturated absorption in SF_6, *Metrologia*, **13**:149.

Pine, A., and Robiette, A., 1979, Doppler limited spectroscopy of the $3\nu_3$ band of SF_6, *J. Mol. Spectrosc.* **80**:388.

Platonenko, V., 1978, Mechanism of the excitation of high vibrational–rotational states and dissociation of polyatomic molecules by an infrared field, *Sov. J. Quantum Electron.* **8**:1010.

Proch, D., and Schröder, H., 1979, IR laser photochemistry of O_3 and OCS. The first multiphoton dissociation of triatomic molecules, *Chem. Phys. Lett.* **61**:426.

Quack, M., 1978, Theory of unimolecular reactions induced by monochromatic infrared radiation, *J. Chem. Phys.* **69**:1282.

Quigley, G., 1978, Collisional effects in multiple photon ir absorption, in *Advances in Laser Chemistry*, Springer Series in Chemical Physics (ed. A. H. Zewail), Springer-Verlag, New York, Heidelberg, Berlin.

Quigley, G., 1979, Optoacoustic studies of two frequency infrared-laser pumping of SF_6, *Optics Lett.* **4**:84.

Rabinowitz, P., Keller, R., and La Tourrette, J., 1969, Lamb-dip spectroscopy applied to SF_6, *Appl. Phys. Lett.* **14**:376.

Rabinowitz, P., Stein, A., and Kaldor, A., 1978, Infrared multiphoton dissociation of UF_6, *Optics Commun.* **27**:381.

Robinson, P. J., and Holbrook, K. A., 1972, *Unimolecular Reactions*, Wiley-Interscience, New York.
Rothchild, M., 1979, Multiple photon processes in SF_6, Ph.D. Thesis, Institute of Optics, University of Rochester.
Rothchild, M., Tsay, W., and Ham, D., 1978, Threshold behavior of multiple photon dissociation of SF_6, *Optics Commun.* **24**:327.
Sazonov, V., and Finkel'shtein, V., 1977, Analysis of models of radiative dissociation of polyatomic molecules in the field of laser radiation, *Sov. Phys. JETP* **46**:687.
Shaffer, W., Nielsen, H., and Thomas, L., 1939, The rotation–vibration energies of tetrahedrally symmetric pentatomic molecules. I, *Phys. Rev.* **56**:895.
Shimizu, F., 1969, Absorption of CO_2 laser lines by SF_6, *Appl. Phys. Lett.* **14**:378.
Shuryak, E., 1978, Nonlinear mechanics of molecular vibrations and mechanism of collisionless dissociation in a high intensity laser field, *Sov. J. Quantum. Electron.* **8**:1018.
Smith, S., Schmid, W., Tablas, F., and Kompa, L., 1979, Time and intensity dependence of the infrared absorption of SF_6: Measurements with an injection-locked single mode TEA CO_2 laser, in *Laser-Induced Processes in Molecules*, Springer Series in Chemical Physics (eds. K. L. Kompa and S. D. Smith), Springer-Verlag, New York, Heidelberg, Berlin.
Stafast, H., Schmid, W., and Kompa, K., 1977, Absorption of CO_2 laser pulses at different wavelengths by ground-state and vibrationally heated SF_6, *Optics Commun.* **21**:121.
Steel, D., and Lam, J., 1979, Two photon coherent transient measurement of the nonradiative collisionless dephasing rate in SF_6 via Doppler-free degenerate four wave mixing, *Phys. Rev. Lett.* **43**:1588.
Steinfeld, J., 1974, *Molecules and Radiation*, M.I.T. Press, Cambridge, Mass.
Steinfeld, J., and Jensen, C., 1976, Double resonance and energy transfer in sulfur hexafluoride, in *Proceedings of the Conference on Tunable Lasers and Applications*, Loen, Norway, Springer-Verlag, New York, Heidelberg, Berlin.
Steinfeld, J., Jensen, C., Anderson, T., and Reiser, Ch., 1979, Double Resonance Spectroscopy of Multiple-Photon Excited Molecules, in *Laser Spectroscopy IV*., Proc. 4th International Conference, Rottach-Egern, Springer Series in Optical Sciences (eds. H. Walther and K. Rothe), Springer-Verlag, New York, Heidelberg, Berlin.
Stephenson, J. C., King, D. S., Goodman, M. F., and Stone, J., 1979, Experiment and theory for CO_2 laser-induced CF_2HCl decomposition rate dependence on pressure and intensity, *J. Chem. Phys.* **70**:4496.
Steverding, B., Dudel, H., and Gibson, F., 1978, The distribution of vibrationally excited states in resonant radiation and the rate of photodissociation, *J. Appl. Phys.* **49**:852.
Stone, J., and Goodman, M. F., 1979, A re-examination of the use of rate equations to account for fluence dependence, intramolecular relaxation, and unimolecular decay in laser driven polyatomic molecules, *J. Chem. Phys.* **71**:408.
Stone, J., Goodman, M., and Dows, D., 1976, A model for laser isotope separation in SF_6, *J. Chem. Phys.* **65**:5062.
Tamir, M., and Levine, R., 1977, The multiphoton collisionless dissociation of polyatomic molecules: An intramolecular amplification mechanism, *Chem. Phys. Lett.* **46**: 208.
Taylor, R., Znotins, T., Ballik, E., and Garside, B., 1977, A vibrational-bath model for the dynamics of SF_6 absorption near 10.4 μm as a function of wavelength and absorbed energy, *J. Appl. Phys.* **48**:4435.
Thiele, E., 1963, Smooth curve approximation to the energy-level distribution for quantum harmonic oscillators, *J. Chem. Phys.* **39**:3258.
Thiele, E., Stone, J., Goodman, M., 1979, On the nature of laser-induced energy distributions in polyatomic molecules, *Chem. Phys. Lett.* **66**:457.
Thiele, E., Goodman, M., and Stone, J., 1980, Can lasers be used to break chemical bonds selectively?, *Laser Appl. to Chem.* issue of *Opt. Eng.* **19**:10.
Tiee, J., and Wittig, C., 1978a, The photodissociation of UF_6 using infrared lasers, *Optics Commun.* **27**:377.

Tiee, J., and Wittig, C., 1978b, Isotopically selective ir photodissociation of SeF_6, *Appl. Phys. Lett.* **32**:236.

Tsay, W., Riley, C., and Ham, D., 1979, Thermal enhancement of multiple photon absorption by SF_6, *J. Chem. Phys.* **70**:3558.

Van Vleck, H., 1929, On σ-type doubling and electron spin in the spectra of diatomic molecules, *Phys. Rev.* **33**:467.

Walker, R., 1979, The classical mechanics of ir multiple photon absorption dynamics of XY_6 molecules. I., in preparation.

Walker, R., and Preston, R., 1977, Quantum versus classical dynamics in the treatment of multiple photon excitation of the anharmonic oscillator, *J. Chem. Phys.* **67**:2017.

Whitten, G. Z., and Rabinovitch, B. S., 1963, Accurate and facile approximation for vibrational energy-level sums, *J. Chem. Phys.* **38**:2466.

Wilmshurst, J., and Bernstein, H., 1957, The infrared and raman spectra of disulfur decafluoride, *Can. J. Chem.* **35**:191.

Woodin, R. L., Bomse, D. S., and Beauchamp, J. L., 1978, Multiphoton dissociation of molecules with low power continuous wave infrared laser radiation, *J. Am. Chem. Soc.* **100**:3248.

Yablonovitch, E., 1977, Laser pulse requirements for coherent and mode-selective excitation in the quasicontinuum of polyatomic molecules, *Optics Lett.* **1**:87.

Zi-Zhao, G., Guo-Zheng, Y., Ke-An, F., and Hi-Yi, H., 1978, Theory of multi-photon photodissociation of polyatomic molecules in an intense infrared laser field, *Acta Phys. Sin.* **27**:664.

Note Added in Proof. Since the original writing of the manuscript new data have appeared which influence our modeling of SF_6. We should point out, however, that our general picture of multiphoton absorption in large molecules remains as presented in the manuscript.

We are now aware that the measurement of Steel and Lam (1979) fits into a more general category of V–V relaxation experiments, all of which appear to be measuring a dephasing time related to their laser bandwidth (Dave Ham and O'Dean Judd, private communication).

The $3\nu_3$ data of Pine and Robiette (1980) have been analyzed by Patterson, Krohn, and Pine (1980), giving definitive values for the molecular parameters X_{33}, G_{33} and T_{33}. We now know that tensor splitting plays an important role in mpe and mpd of SF_6 such that levels as high as $7\nu_3$ in the ν_3 ladder are excited, and QC leakage is not required to occur (as previously believed) as low as 1–$3\nu_3$. In addition the $\Delta R = 0$ selection rule is violated, eliminating our ability to model each rotational ground-state ladder independently. This implies that the ν_3 ladder portion of our model should occupy a larger number of levels and that coherence will play a somewhat greater role than was anticipated in our calculations (Ackerhalt and Galbraith, 1980).

More seriously, recent data by Nowak and Ham (1980) on coherent self-focusing effects in SF_6 have cast doubt on the validity of *all* mpe and mpd experiments carried out to date.

One of us, JRA, would like to thank Carl Moser and the CNRS for supporting his tenure at the CECAM workshop on "Laser Excitation and Dynamics of Highly Excited Polyatomics" held in Orsay, France, July 1980 which was an aid in the formulation of Section 1.5.

Ackerhalt, J., and Galbraith, H., 1980, Multiple photon excitation modelling in SF_6, paper D. 8 at the Eleventh International Quantum Electronics Conference, June 23–26, Boston, Massachusetts.

Patterson, C., Krohn, B., and Pine, A., 1980, The ν_3 vibrational ladder of SF_6, *Optics Lett.*, submitted for publication.

Nowak, A., and Ham, D., 1980, Detailed measurements of multiple-photon absorption by SF_6, paper D. 4 at the Eleventh International Quantum Electronics Conference, June 23–26, Boston, Massachusetts.

2

Multiphoton Infrared Excitation and Reaction of Organic Compounds

Wayne C. Danen and J. C. Jang

2.1. Introduction

The field of infrared laser photochemistry has burgeoned in the past several years. Excitation with intense, pulsed, infrared laser radiation has been shown to promote molecules to high vibrational levels of the ground electronic state as the result of the absorption of many infrared photons, frequently in the absence of any collisions. As a consequence, the phenomenon is sometimes referred to as megawatt infrared photochemistry or photochemistry in the electronic ground state. Much interest has centered on laser isotope separation, probing the multiphoton absorption process, and observation of reactions that occur when a molecule finds itself suddenly immersed in a sea of infrared photons.

Of the total number of publications that have appeared in this new area, only a comparatively small number have been concerned with the chemistry of relatively large molecules. Much work has been conducted on small, inorganic molecules and polyhalogenated methanes and ethanes; SF_6 has received the most attention and may be considered the prototype of infrared laser photochemistry–photophysics. A number of recent review articles covering various aspects of this new field are available (Ambartzumian and Letokhov, 1977; Letokhov and Moore, 1977; Fuss *et al.*, 1977;

Wayne C. Danen and J. C. Jang · Department of Chemistry, Kansas State University, Manhattan, Kansas 66506

Cantrell *et al.*, 1979; Grunwald *et al.*, 1978; Bloembergen and Yablonovitch, 1978; Kompa *et al.*, 1979; Ronn, 1979; Schulz *et al.*, 1979; Danen, 1980.)

In this chapter, we will restrict coverage to large, usually organic molecules. The term "large" is very relative, of course. To a theorist, anything with more than a few electrons is horrendously complex, while an organic chemist may conjure visions of a polynuclear organic compound or some natural or synthetic polymer when confronted with the term "large molecule." For our purposes, any compound having five to six or more atoms other than hydrogen or halogen and possessing a number of low-frequency distortions will be termed large. More specifically, a large molecule is considered one that possesses a density of vibrational and rotational states exceeding $\sim 10^3$–10^4/cm^{-1} at room temperature. The significance of this criterion will become apparent later.

The objective of this chapter is to give an overall qualitative–semiquantitative view of the pulsed infrared photochemistry of large organic molecules as defined in the previous paragraph from the vantage point of a physical–organic chemist. Following a brief outline of the main differences between large and small molecules, a selected literature survey is given to show some of the large systems investigated to date and to illustrate the types of results that have been observed. A section is devoted to the use of chemical thermometry that addresses the distinction between a thermal process and an unique nonequilibrium laser process. This is followed by the presentation of some experimental results for ethyl acetate, a prototype large molecule. Finally, one section describes the results of simple computational modeling of the multiphoton absorption/reaction process with emphasis on the effects of experimental conditions related to the irradiation process and collisional effects. Particularly in this last section, the differences are stressed between small, rigid molecules and the relatively large, flexible molecules considered in this chapter.

It is hoped that this chapter will prove palatable to most modern organic chemists as well as to researchers more familiar with this exciting new field. The details of the final section dealing with the modeling studies may pose some difficulty to the former, although the main features and conclusions should emerge.

2.2. Features Distinguishing Large from Small Molecules

It would appear advantageous to outline at the outset some of the features that distinguish large from small molecules in multiphoton infrared laser photochemistry. These differences will be elaborated more fully in the last section of this chapter, which is devoted to the modeling studies.

Since more effort has been devoted to the study of the multiphoton excitation/reaction of small molecules, these processes are better characterized than for large molecules. Although there are still major gaps in our understanding and considerable controversy surrounds the topic, the main qualitative features of multiphoton excitation/reaction appear to be established (Quack, 1978; Mukamel, 1979; Stone and Goodman, 1979; Black *et al.*, 1977). Most discussions of this topic refer to various regions or subsections of the rotational–vibrational energy quantum levels for small polyatomic molecules. At low energies such molecules have discrete vibrational and rotational states, and it usually requires fairly intense laser radiation to excite a molecule over these discrete states via a coherent, resonant absorption process. Various mechanisms to overcome the inherent anharmonicity‡ of the vibrational levels have been proposed and excitation through this discrete regime appears to depend upon the laser power rather than the laser energy.§ The density of states increases very rapidly with an increase in energy, and once the requisite number of photons (typically thought to be ~ 3–4) are absorbed, the molecule reaches a second region termed the vibrational quasicontinuum. In this regime, there is such a high density of states that the anharmonicity problem vanishes since there is always a rotational–vibrational level in resonance with the monochromatic laser light. Once in the quasicontinuum, the laser energy rather than the laser power appears to be responsible for further exciting the molecules up to or even beyond the reaction threshold level. Once excited to the threshold, reaction can occur in competition with further excitation beyond the threshold. Since such excitation is time dependent, the laser power rather than energy again becomes important in determining the ultimate level of excitation achieved before reaction becomes so rapid that additional uppumping is impossible. Since most current lasers are capable of producing a pulselength of ~ 10^{-7} sec (100 nsec), once a molecule acquires sufficient

‡ Anharmonicity refers to the unequal spacings between the vibrational energy levels in a polyatomic molecule; the spacings become closer together as one goes to higher vibrational levels. This poses a problem in the multiphoton excitation of a molecule since a highly monochromatic laser tuned to the lowest vibrational energy transition ($v_0 \rightarrow v_1$) presumably would be out of resonance for higher transitions.

§ The distinction between laser energy and power is important but sometimes confused. The *laser energy* is the total energy output per pulse independent of the pulselength; it is usually expressed in J/cm^2. The laser energy is frequently referred to as the fluence or, sometimes, dose. The *laser power* is the rate at which energy is radiated and is dependent on the pulselength; it is usually expressed in W/cm^2. By definition, 1 J/sec = 1 W; therefore, 1 J/100 nsec = 10^7 W or 10 MW, a typical output for a pulsed TEA CO_2 laser.

energy to react with a rate constant of $\sim 10^7$–10^8 sec^{-1}, reaction will, indeed, compete with any further up-pumping.

Although there has been persistent speculation (see Section 2.3.1.4) that the vibrational excitation imparted by the laser might remain localized in the vibrational mode pumped during and even after the excitation process, most studies are interpreted as showing that the energy is statistically distributed throughout the molecule at least at higher excitation levels. Accordingly, the statistical RRKM method of describing unimolecular reaction kinetics appears capable of accounting for the main features of the unimolecular reaction of infrared laser excited molecules.

On the basis of the above summary and elaboration to follow in this chapter, we can list some of the characteristic features of the multiphoton excitation–reaction of large molecules as follows.

1. The high density of vibrational–rotational states possessed by large molecules essentially places them in or very near the quasicontinuum at room temperature.

2. As a consequence of (1), large molecules will frequently exhibit low laser threshold energies and high reaction probabilities. There are data indicating essentially 100% reaction per laser pulse at relatively moderate fluences for some organic molecules.

3. "Hole burning" is not observed for large molecules. This is an observed phenomenon for many small molecules in which intense laser power can effectively deplete the particular rotational–vibrational state pumped.

4. Because of the relatively large number of degrees of freedom and concomitant high density of states of large molecules, reaction rates of molecules excited to just beyond the threshold energy can be quite low.

5. As a consequence of (4), collisional quenching of vibrationally excited large molecules by unexcited reactant molecules or added bath gases can frequently compete efficiently with reaction, particularly at low or moderate laser fluences.

6. Also as a result of (4), reaction of the excited molecules in the irradiated volume produced initially by the laser pulse is nearly always in competition with relaxation via a complex interplay of cooling processes in both the irradiated and surrounding volumes. Only with low pressures and sufficiently intense laser radiation to ensure essentially complete reaction during the laser pulse will such cooling not be significant.

7. The heat capacity of a large molecule is comprised largely of the vibrational component. Thus, intramolecular collisional relaxation of vibrational excitation into translational and rotational energy results in only a relatively small decrease of the effective vibrational temperature of the molecule.

8. The vibrational energy population resulting from absorption of the laser radiation is probably relatively broad, and we will assume that it can be approximated by a Boltzmann distribution as depicted in Figure 2.1. It must be stressed, however, that the true distribution is not known with certainty, and this characteristic of the multiphoton excitation/reaction phenomenon is one of continuing research.

Some elaboration of Figure 2.1 may be beneficial for the reader unfamiliar with such descriptions. On the left is shown the vibrational energy distribution for an ensemble of ethyl acetate ($CH_3CO_2CH_2CH_3$) molecules at room temperature. The mean energy is ~ 3–4 kcal/mol, which is typical for an organic molecule of four to six carbons. The extremely broad curve in Figure 2.1 depicts a Boltzmann distribution corresponding to a temperature of ~1400°K, which is equivalent to absorption of ~15 CO_2 laser photons (1050 cm^{-1}) by the room temperature distribution. It is seen that, although the mean level of excitation is ~45 kcal/mol, some molecules have absorbed many more photons and can possess up to 90 kcal/mol or more vibrational excitation. Since the threshold energy, E_0, of reaction of ethyl acetate is 48 kcal/mol, that fraction of the molecules above the threshold will possess enough internal energy to react providing they are not collisionally relaxed before reaction occurs. This fraction is indicated by

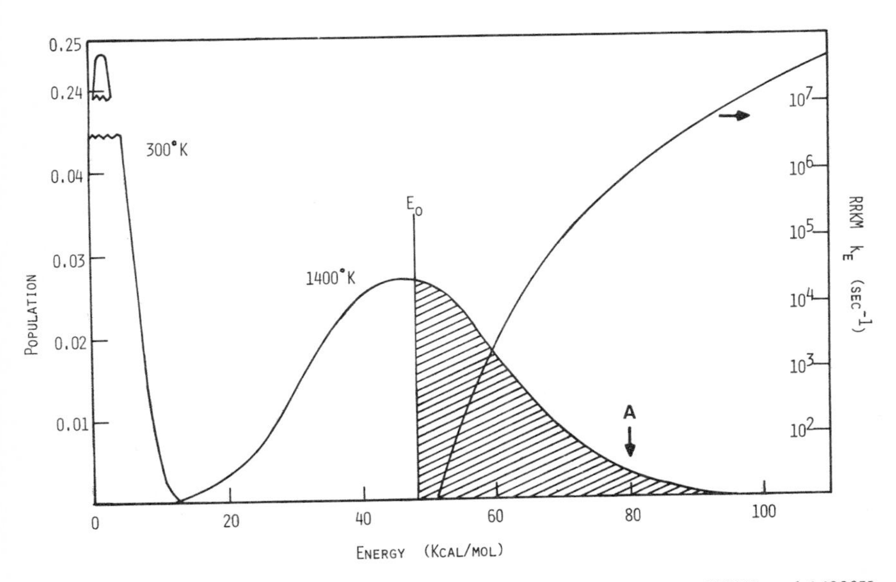

Figure 2.1. Boltzmann vibrational energy distributions of ethyl acetate at 300°K and 1400°K. The right-hand scale and corresponding curve show the RRKM rate constant, k_E, calculated versus the level of excitation (kcal/mol). E_0 is the threshold energy (48 kcal/mol) for the unimolecular reaction of ethyl acetate.

hatched lines in Figure 2.1 and comprises approximately 50% of the total number of molecules. Those molecules possessing $\sim$ 80 kcal/mol (point A) can be calculated by RRKM theory to react at a rate $\sim 10^6$ sec^{-1}, and all would undergo transformation before becoming collisionally deactivated providing the pressure is $\sim$ 0.1 Torr or less. Those molecules with even $\sim$ 60 kcal/mol excitation energy, however, have k_E only $\sim 5 \times 10^3$ sec^{-1} while the collision rate is $\sim$200 times larger than k at $\sim$0.1 Torr pressure; a few collisions would cause deactivation. Thus, the great majority of the molecules excited above the threshold have small rate constants, and collisional processes will be important even if the pressure is sufficiently low that no collisions occur during the laser pulse.

The characteristic features of the multiphoton excitation/reaction of large molecules listed above cannot be considered experimentally verified in all systems. We will, however, take them as the framework of a working hypothesis to be applied to experimental studies of large molecules, but flexibility and openmindedness must be maintained. This is an area of research that is developing rapidly with few firmly established concepts, and tomorrow's experiment frequently requires a modification of today's theory.

2.3. *Selected Literature Survey*

In this section we will present and discuss briefly some of the literature reports describing organic chemical reactions induced by pulsed, infrared laser photolysis. The coverage will be selective rather than exhaustive. For a more thorough compilation of organic systems subjected to intense infrared radiation, the interested reader is directed to the comprehensive 1965–1980 literature survey of laser-induced chemical reactions found in Chapter 4 of this volume. The literature examples of the present chapter were chosen since, in the opinion of the authors, they represent the main organic reaction types investigated to date or otherwise illustrate some unique characteristic.

The great majority of these infrared laser-induced reactions have been conducted on relatively low-pressure gas-phase samples. The apparent necessity of working in the gas phase is dictated by rapid vibrational relaxation in condensed phases that would presumably negate the possibility of any unique nonequilibrium, nonthermal chemistry. Some reports, however, have appeared on infrared laser-induced reactions in solids or at gas–solid interfaces.

Most of the reported laser-induced organic reactions have involved unimolecular processes such as rearrangements, eliminations, and homolytic bond fissions. The vibrational excitation produced by the pulsed infrared laser is particularly effective in driving unimolecular reactions, which accounts for the preponderance of such studies. In a bimolecular or higher-order process, there is always a competition between reaction and vibrational energy relaxation to the reaction partner. For a polyatomic molecule interacting with a nonexcited partner, the latter is particularly rapid and usually competes effectively and detrimentally with chemical reaction. Studies involving attempted laser-induced bimolecular processes with relatively large molecules will be discussed in Section 2.3.2.

2.3.1. *Unimolecular Reactions*

2.3.1.1. *Controlling Chemical Equilibria of Isomers*

We will begin our survey by considering laser-induced isomerization processes since these represent some of the simplest organic reactions. A problem not infrequently encountered in organic chemistry is the synthesis and isolation in pure form of a particular organic isomer. Isomeric compounds abound in organic chemistry in the form of structural isomers, geometric isomers, stereoisomers, etc., and a frequent challenge to a synthetic chemist is not necessarily the preparation of the desired isomer but the separation of the wanted compound from other isomers formed in the reaction. By their very nature, isomers usually have very similar chemical and physical properties and are frequently notoriously difficult to separate. Furthermore, the desired isomer may be thermodynamically less stable than an unwanted isomer. Unless the desired compound is produced in a kinetically controlled reaction, there is no conventional method by which this compound can be made even the major component of the mixture. Figure 2.2 illustrates an example in which compounds A and B are interconverted by a unimolecular process; A is thermodynamically more stable than B by 5.0 kcal/mol. At 25°C, an equilibrium mixture of A and B would contain greater than 99.9% A and less than 0.1% B. A temperature in excess of 800°C would be required to produce a mixture containing only 10% of B; few organic compounds can be heated to such temperature without extensive or total decomposition. A catalyst can lower the energy of activation as indicated by the dashed line in Figure 2.2 and enhance the rate of interchange of isomers, but the position of equilibrium would remain unchanged.

With the advent of tunable, pulsed infrared lasers, it has been demonstrated, at least in principle, that a synthetic chemist no longer need always be limited by thermodynamic control of the reaction. It is possible

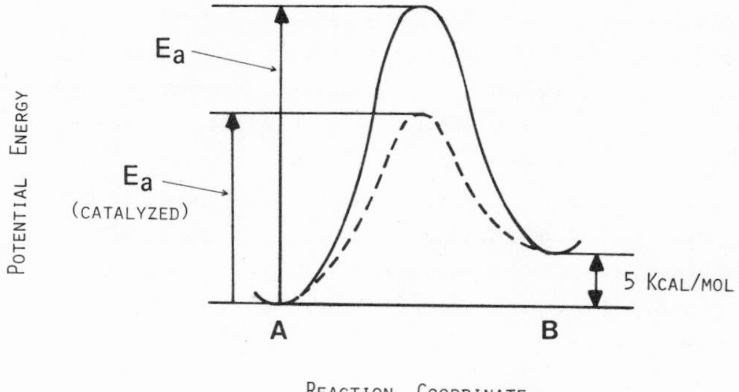

Figure 2.2. Potential energy vs. reaction coordinate diagram for the catalyzed and uncatalyzed endoergic transformation of A to B.

to irradiate selectively one isomer in a mixture and convert the photolyzed compound into the other isomer or to transform it into a nonisomeric structure that may be more easily separated from the desired component. For example, providing isomer A in Figure 2.2 has an infrared absorption band unobscured by B, it may be possible to irradiate A selectively and transform it into the less-stable isomer B in high or even quantitative yield.

Relatively early, an example of such a process was reported by Glatt and Yogev (1976), who investigated the reversible Cope rearrangement of 1,5-hexadiene labeled with deuterium at appropriate positions to remove the degeneracy of the reaction [Reaction (2.1)]. Yogev's group had

$$\underset{\mathbf{1}}{CH_2{=}C(D){-}CD_2{-}CD_2{-}C(D){=}CH_2} \;\underset{\not\leftarrow}{\xrightarrow{10.8\ \mu m}}\; \underset{\mathbf{2}}{CD_2{=}C(D){-}CH_2{-}CH_2{-}C(D){=}CD_2} \qquad (2.1)$$

demonstrated in previous work (Yogev and Loewenstein-Benmair, 1973) that the chemical equilibrium of the *cis–trans* isomerization of 2-butene could be interrupted by selective decomposition of the absorbing component of the reaction mixture. In the present case, a relatively clean conversion of **1** to **2** was achieved in high yield with the formation of only a small amount of volatile products, presumably reflecting decomposition of **1** into two allylic radicals. The terminal CH_2 groups in **1** provided a strong absorption of 10.8-μm radiation while **2** was practically transparent at this wavelength. This nearly degenerate [3S, 3S] sigmatropic shift has $E_a = 34.3$

kcal/mol for the unlabeled isomer and a secondary deuterium isotope effect of 10% per atom at 25°C in favor of **1**. The authors demonstrated that an equilibrium mixture of **1** : **2** of 1.0 : 0.9 was obtained by heating **2** at 300°C for 3 hr. It is apparent that the laser was perturbing the equilibrium thermodynamics of this system by allowing **1** to become highly vibrationally excited and reacting in a unimolecular, single-step manner to form **2**. The reverse reaction presumably occurred during the experiment but had to compete with collisional deactivation of **2** at the relatively high pressures of 5–16 Torr utilized.

A more dramatic example of laser control of chemical equilibrium closely resembling Figure 2.2 has been reported by Yogev and Benmair (1977). In this case, the unimolecular ring opening of hexafluorocyclobutene (**3**) to the less stable isomer hexafluoro-1,3-butadiene (**4**) was effected by a pulsed CO_2 laser [Reaction (2.2)]. The cyclic structure is more

$$\mathbf{3} \underset{E_a = 35.4\ \text{kcal/mol},\ A = 10^{12}\ \text{sec}^{-1}}{\overset{E_a = 47.1\ \text{kcal/mol},\ A = 10^{14}\ \text{sec}^{-1}}{\rightleftarrows}} \mathbf{4} \qquad (2.2)$$

stable than the open form by 12 kcal/mol. When 0.5 Torr of neat **3** was pulsed under a variety of conditions, a maximum yield of 60% of **4** was produced. Presumably a steady state, far removed from equilibrium, was achieved controlled by a complex interplay of multiphoton absorption and collisional energy deactivation. A rather striking effect was observed on the addition of helium as a bath gas. Not unexpectedly, the overall efficiency of the conversion **3** → **4** dropped from, e.g., 24% per given number of pulses with no helium to 15% when 14 Torr of helium was added. This effect can be attributed to V–T,R vibrational relaxation of excited **3** and energy transfer to helium competing with conversion to **4**. On the other hand, the reverse nonlaser reaction **4** → **3** was quenched completely upon addition of 16 Torr of helium allowing the quantitative production of the thermodynamically less-stable isomer. In this isomerization **4** must be produced with at least 35-kcal/mol excess energy; it is likely to be considerably greater than this amount. Although some **4** undoubtedly reverted to **3**, collisional relaxation of **4** rapidly depopulated the highly vibrationally excited states preventing return to the more stable isomer.

These workers also studied the wavelength and fluence dependency of the reaction. A strong enhancement of yield with increasing laser fluence was noted and a rather pronounced redshift of 20–25 cm^{-1} was observed. The latter refers to the maximization of yield at constant laser fluence at a longer wavelength than the single-photon absorption peak. Both phenomena are rather common in the infrared multiphoton photochemistry of organic compounds.

A more recent report (Buechele *et al.*, 1979) further illustrates the potential of nondestructive, isomer-selective laser chemistry. The pulsed CO_2 laser excitation and reaction of the three geometric isomers of 2,4-hexadiene, **5**–**7**, and the two structurally related 1,3-hexadienes **8** and **9** demonstrated that simple equilibrium thermal chemistry could not be occurring. For example, excitation of the *cis–trans* isomer **6** produced more of the *cis–cis* isomer **7** than the thermodynamically more stable *trans, trans* isomer **5** [Reaction (2.3)]. A mass balance of $> 85\%$ for the 300-pulse

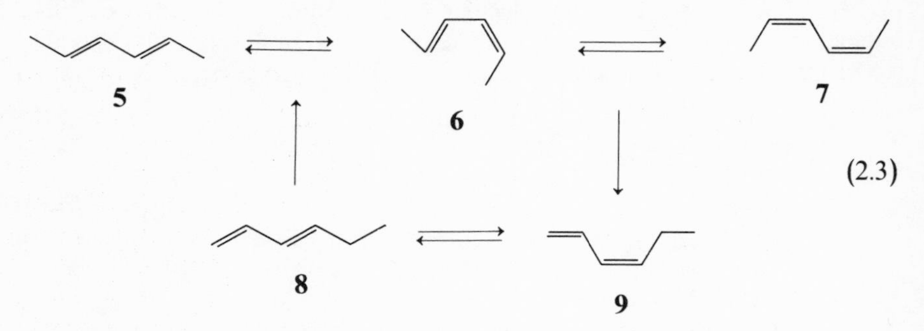

(2.3)

irradiation of **5** demonstrated that extensive fragmentation, as frequently observed in simple monoalkene isomerizations, was not occurring. Over 98% of the thermodynamically most stable isomer, **5**, was converted into less strongly absorbing and/or less efficiently isomerizing species; **7**, **8**, and/or **9** (gplc analytical method could not resolve **8** from **9**) were the predominant products, although the product ratio varied as a function of fluence level and unspecified quencher pressure. Furthermore, the product ratios appeared indicative of two or more successive isomerization processes occurring within or shortly after a single laser pulse (60-nsec FWHM initial spike followed by ~400-nsec tail). For example, single-pulse irradiation of **5** could produce **6** and **7** in a single step, but the formation of **9** would presumably require a second step involving a 1,5 hydrogen shift from **6** or **7**. There may be sufficient vibrational excitation residing in **6** and **7** immediately upon their formation to result in the relatively facile ($E_a = 32.5$ kcal/mol; log A = 10.8 sec^{-1}; Frey and Pope, 1966) isomerization to **9**. Alternatively, vibrationally excited **6** and/or **7** might absorb sufficient additional infrared photons from the tail of the single laser pulse to induce their transformation to **9**. This phenomenon of secondary excitation of primary reaction products during the laser pulse that generated the products is difficult to distinguish from the primary processes and may pervade other studies of large molecules.

From the results presented in this section, it is obvious that the utilization of pulsed infrared laser radiation to control chemical equilibria has been demonstrated in principle. Most of the studies have involved relatively small samples of gas-phase reactants at ~ 1 Torr or less to avoid V–V and V–T,R energy scrambling to the nonirradiated component(s) of the sample. To scale up such a process, it is always possible to employ longer cells, flow systems, multireflection arrangements, etc., to increase the number of molecules processed during a given period of time, and such techniques might find applications in laser chemical systems dedicated to a single process as might be utilized in industry. The present requisite low pressures, however, would seem to preclude the employment of the pulsed infrared laser as a routine benchtop instrument for the processing of gram-scale quantities of chemicals for the research chemist in a realistic length of time. The technique of "vibrationally insulating" molecules by means of relatively high pressures of inert gas (discussed in Section 2.3.2) might be advantageous in some instances, although this technique may frequently merely offset the gain in higher operating pressures with a loss in reaction efficiency. Probably the single most promising means of scaling-up infrared laser-induced chemical processes that are limited by collisional energy relaxation is the development of shorter pulse lasers. For example, if a process is shown to occur with satisfactory selectivity at 1 Torr using a 100-nsec laser pulse as is presently routine, it should be possible to scale-up the process to 100 Torr utilizing a 1-nsec laser pulse of the same fluence or to a 1000 Torr with a 100-psec (0.1-nsec) pulse. Such pressures would be much more attractive to industrial and synthetic organic chemists. This reasoning, however, assumes only that the collisional time remains relatively constant as compared to the laser pulsewidth. For a given laser energy, shortening the pulse results in a higher laser power, and it has already been demonstrated that at least some laser-induced chemistry varies as a function of laser power. Thus, even though collisional relaxation may become less of a problem at shorter pulses and higher power, there is no assurance that the chemistry will remain the same as under lower pressure with lower laser power conditions. Infrared laser photochemistry is not sufficiently well understood at this time to forecast accurately how sample pressure will scale vs. laser pulsewidth. On the whole, however, shorter pulse lasers will almost certainly be advantageous to both scale-up and possible mode-selectivity.

2.3.1.2. *Elimination Reactions of Monofunctional Compounds*

Elimination reactions have been, to date, the most extensively investigated organic transformations for infrared photochemists, with the possible exception of homolysis of polyhalogenated methanes and ethanes.

There are numerous well-characterized, gas-phase, unimolecular elimination reactions reported in the literature so the popularity of this type of reaction for probing the reaction dynamics and photophysics of multiphoton excitation and reaction of large molecules is understandable. Another reason for the popularity of elimination reactions is that molecular products are formed in the primary process that eliminates the complications inherent when reactive free radicals or carbenes are generated as intermediates.

The molecular fragment most usually eliminated is a hydrogen halide or carboxylic acid resulting in the concomitant formation of an alkene or alkyne. Unimolecular, gas-phase, β-elimination of hydrogen halides usually occur in a concerted, four-centered *cis* manner [Reaction (2.4)].

$$\underset{\substack{|\quad\ |\\ R_2C-CR_2}}{H\quad X} \longrightarrow \left[\underset{\substack{\vdots\quad\ \vdots\\ R_2C{=\!=}CR_2}}{H\cdots X}\right]^{\ddagger} \longrightarrow R_2C{=}CR_2 + HX \qquad (2.4)$$

X = F, Cl, Br, I

Activation energies range from ~ 60 kcal/mol for HF elimination to ~ 45 kcal/mol for HI, although the exact value depends upon the nature of the R groups in Reaction (2.4). There is evidence (Hassler and Setser, 1966; Johnson and Setser, 1967), that C–H bond lengthening in the transition state may precede C–X breaking, but the process is still quite satisfactorily classified as concerted.

Some organic halides react under laser stimulation via a three-center, α-elimination process [Reaction (2.5)]. The intermediate carbene thus generated usually rearranges to an alkene or alkyne. This pathway is less common than β-elimination [Reaction (2.4)].

$$R_2CH-\overset{\overset{X}{|}}{C}HR \longrightarrow \left[R_2CH-CR\!\!\begin{array}{c}\cdot\!\cdot X\\ \vdots\\ \cdot\!\cdot H\end{array}\right]^{\ddagger} \qquad (2.5)$$

$$\longrightarrow R_2CH-\ddot{C}R \xrightarrow{\quad} R_2C{=}CHR$$

Organic esters undergo clean, unimolecular, *cis*-eliminations to generate the corresponding carboxylic acid and alkene via a six-membered ring transition state [Reaction (2.6)]. Activation energies parallel the stability of the alkene produced and range from ~ 40 kcal/mol for a highly substituted alkene such as *t*-butyl acetate to 48 kcal/mol for formation of ethylene

$$\mathrm{R{-}\overset{O}{\overset{\|}{C}}{-}OCH_2CH_2R'} \longrightarrow \left[\mathrm{R{-}C}\begin{matrix} {}^{2}\mathrm{O}\cdots\mathrm{H}{}_{3} \\ \qquad\quad \mathrm{CHR'} \\ \mathrm{O}\cdots_{1}\mathrm{CH_2} \end{matrix}\right]^{\ddagger} \tag{2.6}$$

$$\longrightarrow \mathrm{R{-}\overset{O}{\overset{\|}{C}}OH} + \mathrm{CH_2{=}CHR'}$$

from ethyl acetate. There is evidence that the extent of bond lengthening in the transition state of Reaction (2.6) is in the order $1 > 2 > 3$ (Taylor, 1975).

2.3.1.2a. Organic Esters. We will discuss certain aspects of the elimination reactions of organic esters first since we have devoted considerable study to this class of compounds at Kansas State University. Reactions (2.7)–(2.10) illustrate several of the esters that undergo elimination to produce only a single olefin and for which we have quantitative data; other esters will be discussed elsewhere in this chapter.

$$\underset{\mathbf{10}}{\mathrm{CH_3\overset{O}{\overset{\|}{C}}OCH_2CH_3}} \xrightarrow[\log A\,=\,12.6\ \mathrm{sec^{-1}}]{E_a\,=\,48.0\ \mathrm{kcal/mol}} \mathrm{CH_3CO_2H + CH_2{=}CH_2} \tag{2.7}$$

$$\mathrm{CH_3\overset{O}{\overset{\|}{C}}OCH_2CH_2CH_2CH_3} \xrightarrow[\log A\,=\,12.2\ \mathrm{sec^{-1}}]{E_a\,=\,46.0\ \mathrm{kcal/mol}} \mathrm{CH_3CO_2H} + \mathrm{CH_2{=}CHCH_2CH_3} \tag{2.8}$$

$$\mathrm{FCH_2\overset{O}{\overset{\|}{C}}OCH_2CH_3} \xrightarrow[\log A\,\sim\,12.6\ \mathrm{sec^{-1}}]{E_a\,\sim\,46.7\ \mathrm{kcal/mol}} \mathrm{FCH_2CO_2H + CH_2{=}CH_2} \tag{2.9}$$

$$\mathrm{CH_2{=}CH\overset{O}{\overset{\|}{C}}OCH_2CH_3} \longrightarrow \mathrm{CH_2{=}CHCO_2H + CH_2{=}CH_2} \tag{2.10}$$

Most of the esters were irradiated utilizing a laser frequency of ~ 1050 cm^{-1}, which corresponds to the O–CH_2 stretching mode, although in molecules of this complexity there frequently are other modes underlying and contributing to the absorption band. All the esters underwent clean eliminations to the products shown in Reactions (2.7)–(2.10) when irradiated utilizing moderate fluences. Yields per pulse of $\sim$ 10–50% could easily be achieved with fluences $\sim$ 1–3 J/cm, and yields approaching 100% were observed in many instances at higher fluences. We have conducted quantitative studies of laser fluence and other variables on the reaction probabilities, absorption cross sections, and absorbed energies for a number of the esters. These largely unpublished data at the time of this writing will be discussed in some detail in Section 2.5 with particular emphasis on ethyl acetate [Reaction (2.7)].

In 1977, we reported a study utilizing mixtures of ethyl acetate [**10**, Reaction (2.7)] and isopropyl bromide [**11**, Reaction (2.11)] designed to

$$\underset{\mathbf{11}}{CH_3\overset{\overset{\displaystyle Br}{|}}{C}HCH_3} \xrightarrow[\log A\,=\,13.6\ \text{sec}^{-1}]{E_a\,=\,47.8\ \text{kcal/mol}} CH_3CH{=}CH_2 + HBr \qquad (2.11)$$

distinguish between a selective, nonequilibrium laser-induced reaction and an energy-randomized thermal process (Danen *et al.*, 1977). At the time of that work, several reports had been published claiming nonequilibrium behavior in systems at suspiciously high pressures. Our approach was to irradiate selectively the ester **10** in the presence of **11**, the latter serving as a type of chemical thermometer to monitor thermal effects. If intermolecular V–V transfer from vibrationally excited **10** to cold **11** was rapid and extensive compared to unimolecular elimination [Reaction (2.7)], then HBr elimination from **11** would be observed [Reaction (2.11)]. If excitation and reaction of **10** were faster than intermolecular relaxation, then only ethylene, but no propylene, should be detected. The choice of a suitable detector molecule, such as **11**, is critical for monitoring intermolecular relaxation; a more detailed discussion of such chemical thermometry in laser-induced reaction systems will be presented in Section 2.4.

For reaction of **10** monitored by **11**, it was demonstrated that low reactant pressure and high laser fluences were important to enhance the nonequilibrium pathway. Figure 2.3 depicts a plot of the ratio of propylene to ethylene as a function of laser fluence for a 5-Torr mixture of **10** and **11**. Under the conditions of the experiment, a ratio of ~ 5 would indicate a simple, randomized, thermal process; this was observed at fluences < 1 J/cm^2. As the laser fluence was increased, however, selective excitation and reaction of **10** became competitive with intermolecular relaxation of the deposited laser energy and the propylene-to-ethylene ratio dropped below 1.0. More recent studies of this type have demonstrated that a high degree of nonequilibrium behavior can be maintained at lower fluences than illustrated in Figure 2.3 providing that the reactant pressure is reduced (see Section 2.4). These results demonstrate that, as the collision rate is reduced, the rate at which thermal equilibrium is reached is decreased, allowing the nonequilibrium laser-induced pathway to become increasingly important. Increasing the laser fluence at constant reactant pressure has a beneficial effect on the nonequilibrium pathway by driving the excited molecules further above the reaction threshold and increasing the rate constants which allows reaction to be more competitive with collisional energy relaxation.

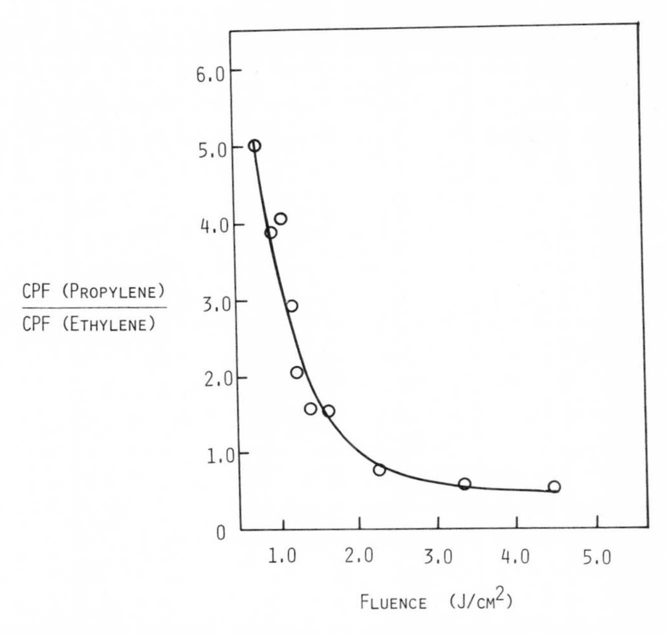

Figure 2.3. CPF(propylene)/CPF(ethylene) vs. laser fluence for 5 Torr of a 3 : 1 mixture of ethyl acetate and isopropyl bromide. CPF = conversion per flash. (After Danen *et al.*, 1977.)

From the study of laser excitation of **10** in the presence of **11**, it was deduced from the effect of pressure on the reaction selectivity that reaction of **10** must occur at a rate of $\sim 10^7$ sec^{-1}. Assuming that the vibrational energy deposited in **10** was intramolecularly randomized, RRKM calculations of the rate constant for reaction of **10** indicated that an energy of $\sim$ 95 kcal/mol was required to produce a decomposition rate of $\sim 10^7$ sec^{-1}. This indicated the need for the absorption of a significant number of infrared photons above the threshold energy of 48 kcal/mol. Similar conclusions suggesting excitation above the threshold level have been demonstrated for other systems (Colussi *et al.*, 1977; Richardson and Setser, 1977). These conclusions are clouded somewhat by more recent observations that quenching of the reaction may not be controlled by collisional effects but instead by less well-characterized "cooling" phenomena (see Section 2.4). Such competitive experiments, however, do set time limits for overall intermolecular relaxation.

Since the nonequilibrium laser-induced reaction of **10** followed the unimolecular pathway of lowest energy and since the laser reaction was competitive with intermolecular relaxation in the 1-Torr range, intramolecular relaxation of the absorbed laser energy was implied. It would be difficult to envision how a complex six-centered elimination involving the breaking of three different bonds as depicted in Reaction (2.6) could be driven if the absorbed laser energy remained localized in only the pumped alkyl C–O band.

Tsang and co-workers (Tsang *et al.*, 1978) utilized **10** and **11** in a similar manner as above to determine the effect of relatively high pressures of helium on the nonequilibrium component of the reaction. A typical result is as shown in Figure 2.4 in which an ethylene–propylene ratio of ~ 0.075 would be expected if the system were at thermal equilibrium. The maintenance of a reasonably high degree of specificity at pressures of helium approaching 1 atm can be explained in terms of the inert bath gas providing a sink for the excitation energy deposited in the ethyl acetate preventing the transmission of energy to the monitor molecule **11**. As anticipated, helium also is efficient in deactivating vibrationally excited **10** as evidenced by the greater than tenfold decrease in absolute yields in going from 0 to 200 Torr of helium.

An unexpected observation was the decrease in specificity at pressures of helium below 200 Torr, which was opposite to the effect at higher pressures. A simple two-level kinetic model was suggested that qualitatively accounted for the major observations. A key feature of this model is that the isopropyl bromide does not decompose via simple heating but through a photodecomposition of isopropyl bromide that becomes vibrationally excited by collisions with helium or ethyl acetate. This model assumes that all dissociation takes place during the laser pulse; the excess helium buffer gas relegates any postpulse phenomena to minimal importance. Since

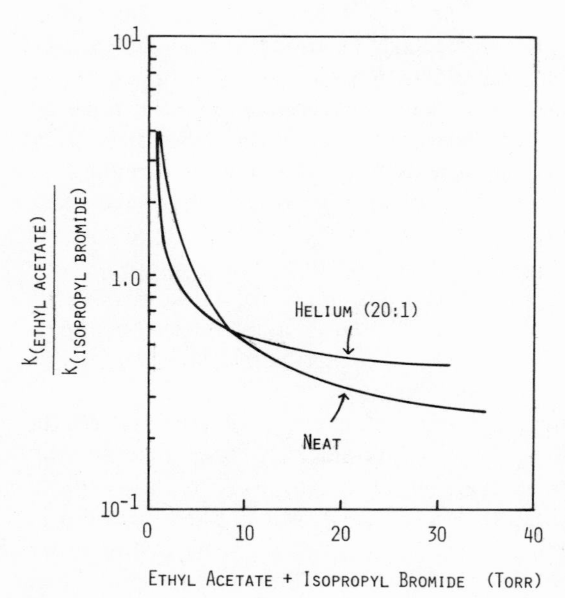

Figure 2.4. Relative rates of decomposition of a 3 : 2 mixture of ethyl acetate and isopropyl bromide both neat and with added helium. The helium/reactant ratio was maintained at 20 : 1. (After Tsang *et al.*, 1978.)

energy transfer processes are more efficient at higher levels of excitation, it is assumed that the molecules in the upper manifold are equilibrated yet the upper manifold can be out of equilibrium with the lower manifold. Direct calorimetry measurements by the authors appear to support the concept of a bottleneck in the absorption process (Braun *et al.*, 1978). These results will be discussed at greater length in Section 2.4.3.2.

2.3.1.2b. Organic Halides. The elimination of HX (X = F, Cl, Br, I) from a number of organic halides has been induced by pulsed infrared laser radiation. Reactions (2.12)–(2.15) list some of the simpler systems investigated. It is apparent from the lack of molecular complexity that probing

$$(CH_3)_2CHCH_2X \longrightarrow HX + (CH_3)_2CH{=}CH_2 \qquad (2.12)$$
$$X = Cl, Br, I \qquad + \text{ other products}$$

$$CH_2{=}CHX \longrightarrow HX + HC{\equiv}CH \qquad (2.13)$$
$$X = F, Cl$$

$$CF_3CH_3 \longrightarrow HF + CF_2{=}CH_2 \qquad (2.14)$$

$$CH_3CH_2F \longrightarrow HF + CH_2{=}CH_2 \qquad (2.15)$$

the multiphoton absorption process and related photophysics were the primary objectives in these studies rather than the development of synthetically useful procedures. Much valuable information on the infrared multiphon excitation/reaction phenomenon has been obtained from these studies since the unimolecular reaction parameters have been well established in many cases and real-time monitoring of reaction products is often feasible. We will attempt only to outline some of the highlights of these studies. Other organic halides that can undergo elimination via more than one channel are considered in Section 2.3.1.3b.

The time-resolved infrared emission from vibrationally excited HF can be relatively easily monitored, and this technique has been employed in several of the systems listed above. Quick and Wittig (1978a) demonstrated that HF elimination from vinyl fluoride [Reaction (2.13), X = F] is a collisionless process and probably occurs via an α,β-elimination. As will be discussed below, however, HCl elimination from vinyl chloride [Reaction (2.13), X = Cl] has been recently demonstrated (Reiser *et al.*, 1979) to occur primarily via an α,α-elimination. Time-resolved emission from both HF and acetylene demonstrated that a nonstatistical partitioning of energy into the products occurred in the laser-induced reaction of vinyl fluoride with more of the available energy preferentially channeled into the HF fragment than into acetylene.

More recent results from this group have been reported using a fast infrared detector (< 20-nsec risetime) and a short (30-nsec FWHM) laser pulse with a very fast cutoff and no tail to measure in real time the unimolecular decomposition rate of 1,1-difluoroethane (Quick *et al.*, 1979). The results showed a marked difference in HF production during and after the laser pulse. With the laser on, dissociation occurred primarily from excited states characterized by lifetimes ≤ 3 nsec; with the laser pulse terminated, molecules at lower levels of excitation were responsible for the bulk of the dissociation. A "center of gravity" distribution of energies of the dissociating molecules produced a unimolecular lifetime of ~ 150 nsec. These lifetimes are much shorter than those experienced by large molecules subjected to laser irradiation at moderate fluences (see Section 2.4.4). It is anticipated that this fast-fall laser pulse technique will yield much additional useful quantitative data in future applications.

In a related study Quick and Wittig (1978b) utilized a time-resolved HF fluorescence method to probe collisional effects in the multiple-photon reaction of vinyl fluoride, 1,1-difluoroethylene, and 1,1-difluorethane. They found that the addition of an inert gas actually enhanced the dissociation yield. These results complement nicely similar work at Kansas State University on 1,1,1-trifluoroethane [Reaction (2.14)] and fluoroethane [Reaction (2.15)] (Jang and Setser, 1979). These workers noted a correlation between the laser pulselength and the extent of collisional quenching and/or collisional enhancement. Figure 2.5 illustrates that at high fluences the yield of Reaction (2.14) is enhanced by the addition of helium only if a short (100-nsec FWHM, no tail) laser pulse is employed. A long (100-nsec FWHM spike followed by ~ 1.3-μsec tail) pulse showed only collisional deactivation resulting in a decrease in yield at all pressures of added helium. At low fluences, the addition of helium or nitrogen always reduced the reaction yield, but the degree of quenching increased strongly with increasing pulselength (Figure 2.6).

The explanation advanced for the rate enhancements at high power levels (short pulse) and low pressures of inert gas was rotational "hole filling" in which collisions during the laser pulse continually repopulate the absorbing levels of CH_3CF_3 allowing the absorption of more energy than in the absence of collisions. If the laser energy is deposited over a long period of time (long pulse), or if the fluence is too low to cause saturation in the first place, or if the pressure of bath gas is too high, only collisional deactivation causing a decrease in reaction probability can be observed.

Such observations, however, earmark these molecules as being "small" rather than "large"; only the latter are of relevance to this chapter. Similar effects involving collisional enhancement of reaction have not been observed with ethyl acetate, a "large" molecule, in the authors' laboratory. A plausible explanation for the lack of collisional enhancement in large molecules is that the high density of states in such molecules does not allow "hole burning" as is possible in small molecules. This will probably

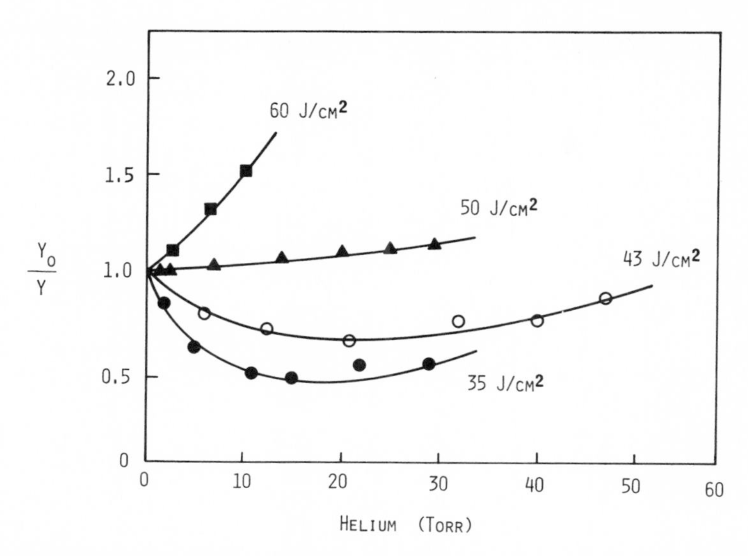

Figure 2.5. Plot of reaction yield ratio from 0.4 Torr of CH_3CF_3 without helium (Y_0) to that with helium (Y) vs. added helium pressure under highly focused conditions for various laser pulselengths: ●, short (100-nsec FWHM) pulse and 43 J/cm^2; ○, short pulse and 35 J/cm^2; ▲, medium (100-nsec FWHM spike followed by ~1.0-μsec tail containing ~35% of the energy) pulse and 50 J/cm^2; ■, long (100-nsec FWHM spike followed by ~1.3-μsec tail containing ~45% of the energy) pulse and 60 J/cm^2. It was shown that the Y_0/Y curves were influenced primarily by the laser pulselength rather than fluence. (After Jang and Setser, 1979; the laser fluences noted herein are increased by a factor of 1.7 over those originally reported.)

be a general result for large molecules, but more experimental data addressing the interaction of added bath gas and laser pulse duration for such molecules are needed.

Many other hydrogen halide elimination reactions from organic halides have been induced by pulsed infrared irradiation and are tabulated in the literature survey found elsewhere in this book to where the interested reader is referred. Several organic halides and esters that can react by more than one reaction channel are considered in the next section.

2.3.1.3. *Polyfunctional Reactants*

2.3.1.3a. Polyfunctional Halides and Esters. The organic halides and esters in Reactions (2.16)–(2.21) can undergo elimination by more than one reaction channel, and these systems have been utilized to probe the various parameters influencing the choice of reaction channel. A rather consistent picture emerges. Pulsed infrared laser induction usually results in reaction

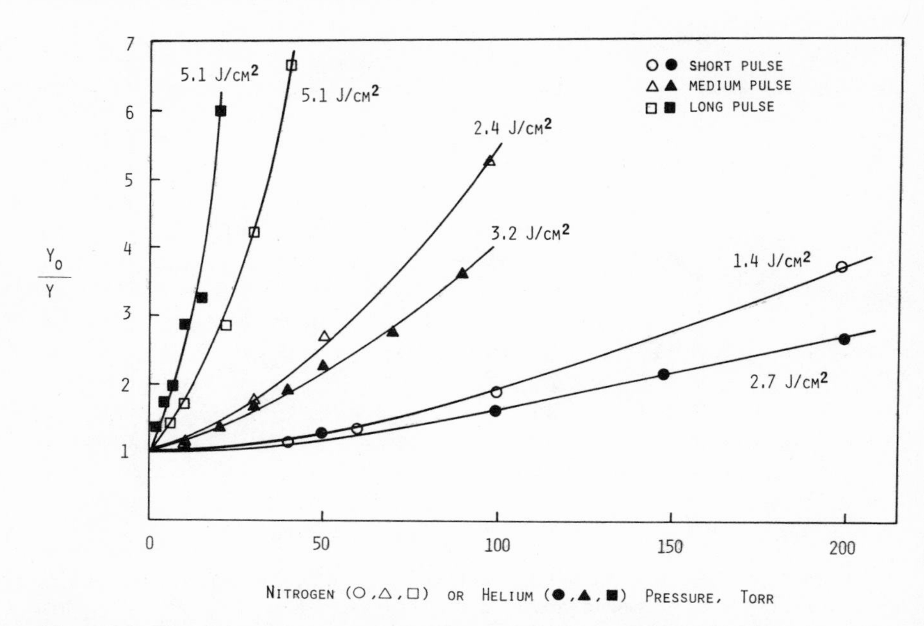

Figure 2.6. Plot of the ratio of reaction yield from 0.4 Torr of CH_3CF_3 without inert bath gas (N_2 or He) to that with inert gas for relatively low-laser fluences and various laser pulselengths. Pulselengths as described in caption to Figure 2.5. (After Jang and Setser, 1979; the laser fluences noted herein are increased by a factor of 1.7 over those originally reported.)

$$CH_2FCH_2Br \longrightarrow HBr + CHF{=}CH_2 \quad (2.16a)$$
$$CH_2FCH_2Br \longrightarrow HF + CH_2{=}CHBr \quad (2.16b)$$
$$CH_2DCH_2Cl \longrightarrow HCl + CH_2{=}CHD \quad (2.17a)$$
$$CH_2DCH_2Cl \longrightarrow DCl + CH_2{=}CH_2 \quad (2.17b)$$
$$CF_2ClCH_3 \longrightarrow HCl + CF_2{=}CH_2 \quad (2.18a)$$
$$CF_2ClCH_3 \longrightarrow HF + CFCl{=}CH_2 \quad (2.18b)$$

$$(D)(H)C{=}C(H)(Cl) \xrightarrow{\text{isomerization}} (H)(D)C{=}C(H)(Cl) \quad (2.19a)$$
$$(D)(H)C{=}C(H)(Cl) \xrightarrow{\alpha,\beta\text{-elimination}} HCl + DC{\equiv}CH \quad (2.19b)$$
$$(D)(H)C{=}C(H)(Cl) \xrightarrow{\alpha,\alpha\text{-elimination}} HCl + (D)(H)C{=}C{:} \longrightarrow DC{\equiv}CH \quad (2.19c)$$

$$CH_3CHICH_2CH_3 \longrightarrow HI + CH_2{=}CHCH_2CH_3 \quad (2.20a)$$

$$CH_3CHICH_2CH_3 \longrightarrow HI + CH_3CH{=}CHCH_3\ (cis\text{-}trans) \quad (2.20b)$$

$$CH_3CHICH_2CH_3 \longrightarrow I\cdot + CH_3\dot{C}HCH_2CH_3 \xrightarrow[-CH_3\cdot]{} CH_3CH{=}CH_2 \quad (2.20c)$$

$$CH_3CHBrCO_2CH_2CH_3 \longrightarrow CH_3CHBrCO_2H + CH_2{=}CH_2 \quad (2.21a)$$

$$CH_3CHBrCO_2CH_2CH_3 \longrightarrow HBr + CH_2{=}CHCO_2CH_2CH_3 \quad (2.21b)$$

via all channels with similar activation energies although frequently producing product ratios considerably different from those obtained by conventional pyrolytic activation. In some cases the product ratio is strongly influenced by variation of experimental conditions, whereas for others, variation of conditions has relatively little effect.

In spite of such observations, all data apparently can be explained without assuming any type of mode-selectivity or inhibited intramolecular randomization of vibrational energy. The observed results can be accommodated within the framework of the RRKM statistical theory for unimolecular reactions according to the specific rate constants for each reaction channel as a function of randomized energy level. It is convenient to discuss the two possible extreme cases qualitatively in terms of different Arrhenius parameters for the different reaction channels even though this is not rigorously correct since such parameters imply a Boltzmann equilibrium distribution that may not be fully compatible with laser-induced vibrational excitation. One extreme, but fully plausible, case is that in which the two accessible reaction channels for a molecule have different Arrhenius activation energies but similar A-factors. A plot of the unimolecular rate constants for both channels as a function of excitation energy would produce curves as depicted in Figure 2.7a. The other extreme case would be as shown in Figure 2.7b in which the two reaction channels had considerably different E_a and A factors. Unlike the first case, in Figure 2.7b there is a crossover in specific rate constants for a sufficiently high excitation level. Since laser excitation can produce molecules with vibrational excitation well in excess of the threshold energy, it is understandable that reaction via a high E_a and high A factor channel can sometimes be very competitive with the lower E_a channel, which may be the normal reaction course under conventional thermal conditions. Examples of both extremes can be found in Reactions (2.16)–(2.21).

For example, Colussi *et al.* (1977) studied the intramolecular isotope effect in the infrared laser-induced elimination from CH_2DCH_2Cl [Reaction (2.17)]. An HCl/DCl ratio of 2.6 ± 0.1 was observed that was independent of fluence, beam geometry, laser frequency, number of pulses, and

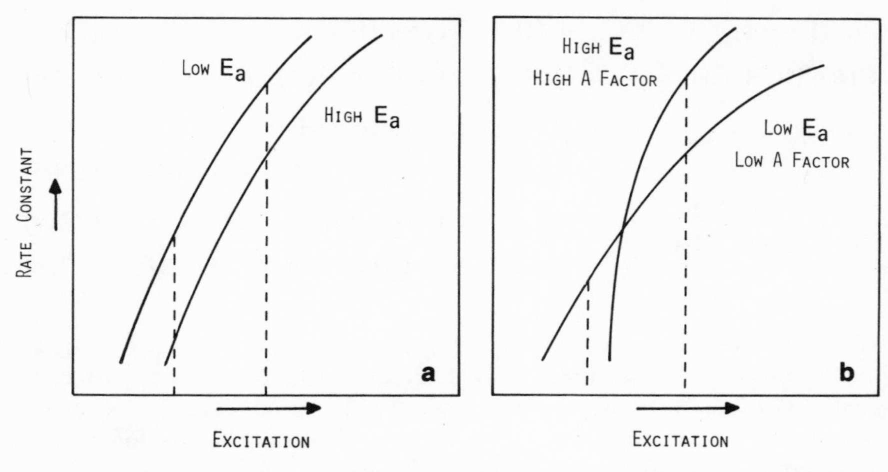

Figure 2.7. (a) Plot of rate constant vs. level of excitation for two different reactants possessing similar A factors but different activation energies. (b) Plot of rate constant vs. level of excitation for two different reactants, one possessing a high E_a and a high A factor while the other exhibits a low E_a and a low A factor.

pressure. It was argued that an average excess energy of 15 ± 2 photons or 35–46 kcal/mol above the threshold energy of 55 kcal/mol for HCl elimination was in accord with the observed isotope effect. This case resembles Figure 2.7a. The reported lack of any effect on changing the laser fluence is surprising since it will be shown below that the branching ratio is normally quite strongly dependent upon the level of excitation. This appears, however, to be an experimental artifact resulting from attenuation of the laser beam prior to focusing to a point within the reaction cell. This does *not* change the average excitation level but merely reduces the number of molecules subjected to sufficiently high fluence levels to react without changing the range of intensities seen by the molecules (Schulz *et al.*, 1979). From the RRKM results necessitating an internal energy centered at 88 ± 5 kcal/mol to account for the observed isotope ratio, it can be concluded that the mean lifetime of molecules in this energy range lie between $10^{-8.5}$ and $10^{-9.3}$ sec for the two pathways.

Similar effects were observed by Richardson and Setser (1977) for Reaction (2.16). Elimination of HBr from CH_2FCH_2Br is more facile by $\sim$6 kcal/mol than HF elimination. If intramolecular energy randomization did not occur or was minimal, laser excitation of the C–F stretching mode would be expected to favor HF elimination. In fact, experiments demonstrated a pronounced favoring of HBr over HF elimination consistent with intramolecular randomization of the absorbed energy prior to reaction. The experimental value of HBr/HF ratio was 10 ± 1 corresponding to a mean internal energy of $\sim$ 75 kcal/mol if energy relaxation within

CH_2FCH_2Br is assumed for $E_a \sim 54$ kcal/mol for Reaction (2.16a) and ~ 60 kcal/mol for Reaction (2.16b). As concluded above, it is apparent that the laser must provide significantly more than just the threshold energy if unimolecular reaction occurring via a RRKM mechanism is expected to compete with relaxation processes at these relatively low pressures of ~ 1 Torr or less.

Steinfeld's group (Reiser *et al.*, 1979; Lussier and Steinfeld, 1977) has investigated the infrared photochemistry of several halogenated ethylenes including vinyl chloride [Reaction (2.13), X = Cl] and *trans*-β-d_1-vinyl chloride [Reaction (2.19)]; results from the latter and related compounds proved particularly valuable because the multichannel reaction pathways offered an example of both types of behavior demonstrated in Figure 2.7. Calculated RRKM rate constants for *cis–trans* isomerization [Reaction (2.19a), $E_a = 61$ kcal/mol], α,β-elimination [Reaction (2.19b), $E_a = 69$ kcal/mol] and α,α-elimination [Reaction (2.19c), $E_a = 73$ kcal/mol] for *trans*-β-d_1-vinyl chloride are as shown in Figure 2.8. As anticipated from the lower E_a, infrared induced isomerization [Reaction (2.19a)] of *trans*-β-d_1-vinyl chloride always competed effectively with elimination. At low incident energy, isomerization was the only observable reaction, while at higher fluences elimination became increasingly important. The experimental results were reproduced quite well with average rates calculated from RRKM theory (Figure 2.8) using either a relatively narrow Poisson or broader Boltzmann-like (maximal entropy) distribution function for $\langle E \rangle$. The results did not permit a choice to be made between these two rather extreme distribution types.

Using both CHD=CHCl and CH_2=CDCl, it was determined that $\sim 80\%$ of reaction occurred via an α,α-elimination [Reactions (2.5) and (2.19c)] and only $\sim 20\%$ by means of a four-centered, *cis*-α,β-elimination [Reactions (2.4) and (2.19b)]. Although this was at slight variance with the RRKM results in Figure 2.8, which indicate that α,β-elimination should be somewhat favored at lower levels of excitation, it was observed that the extent of α,α-elimination decreased and α,β-elimination increased as the laser fluence was increased. This trend toward a ratio approaching 1.0 was in agreement with Figure 2.8. Interestingly, the $\alpha,\alpha/\alpha,\beta$-elimination ratio for CHD=CHCl dropped from ~ 80 to $\sim 65\%$ over the pressure range of 1–15 Torr with irradiation at 927 cm^{-1} but remained unchanged when irradiated at 933 cm^{-1}. This is a rather rare example of driving different reaction channels by selective excitation with one or another laser line. The authors stressed, however, that this was not the result of any mode-selective excitation but rather simply reflected different amounts of energy deposition in the system. Irradiation at 927 cm^{-1} apparently was more effective in exciting the vinyl chloride to higher energy, which resulted in α,β-elimination becoming more competitive.

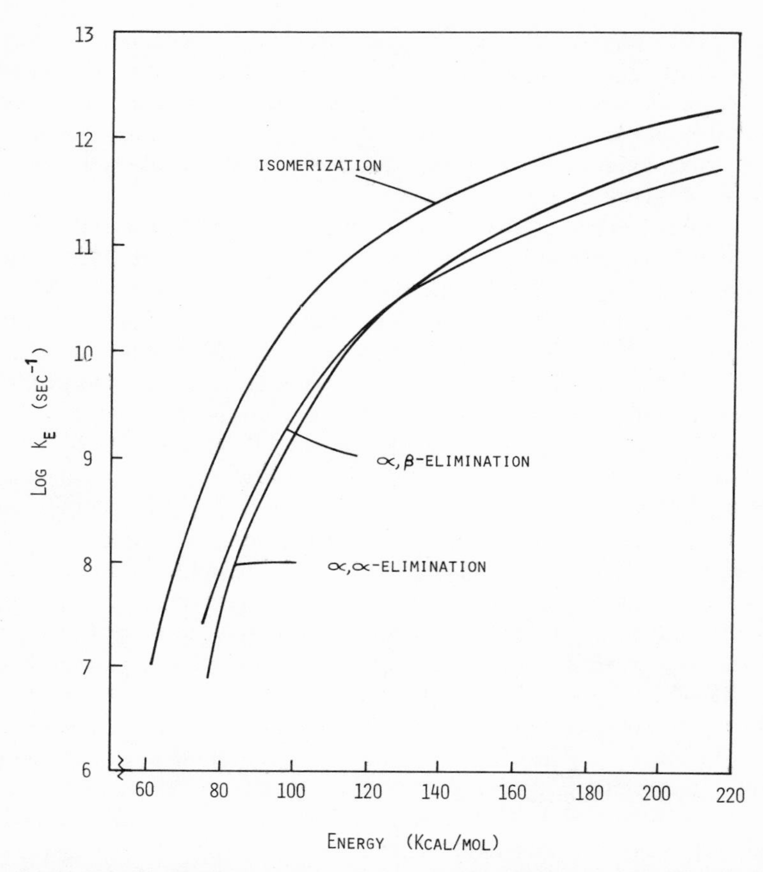

Figure 2.8. Calculated RRKM rate constants for isomerization [Reaction (2.19a), $E_a = 61$ kcal/mol], α,β-elimination [Reaction (2.19b), 69 kcal/mol], and α,α-elimination [Reaction (2.19c), 73 kcal/mol] for *trans*-β-d_1-vinyl chloride vs. excitation energy. (After Reiser *et al.*, 1979.)

We have observed (Danen and Hanh, 1980) somewhat similar results for the pulsed laser reaction of 2-iodobutane [Reaction (2.20)]. Figure 2.9 depicts the RRKM calculated specific rate constants as a function of internal excitation for Reactions (2.20a)–(2.20c). As predicted, formation of the 2-butenes was highly favored at low levels of excitation; heating 2-iodobutane at 312°C produced 83% of *cis*- and *trans*-2-butene [Reaction (2.20b)] and 17% 1-butene [Reaction (2.20a)]. Inducement of reaction with a pulsed CO_2 laser produced a much higher level of vibrational excitation resulting in more equal amounts of 1-butene (~ 35%) and the 2-butenes (~ 43%) being formed as well as propene (~ 22%). Production of the latter presumably resulted from C–I bond homolysis as shown in Reaction

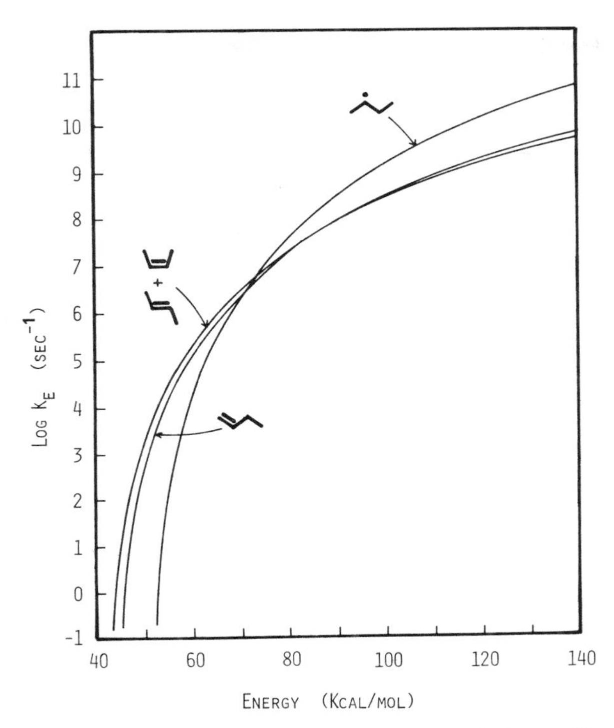

Figure 2.9. Calculated **RRKM** rate constants for the formation of 1-butene [Reaction (2.20a)], *cis*- and *trans*-2-butene [Reaction (2.20b)], and 2-butyl free radicals [Reaction (2.20c)] vs. excitation energy. (After Danen and Hanh, 1980.)

(2.20c), which has a considerably higher threshold energy than the elimination processes. From the crossover of the curves in Figure 2.9, it is apparent that Reaction (2.20c) could become the dominant reaction channel at sufficiently high levels of vibrational excitation.

Under normal thermolysis conditions, an equivalent amount of butane was produced for every equivalent of HI and butene formed via an elimination event. The butane arises from a fast, free radical reaction of HI with 2-iodobutane [Reaction (2.22)].

$$CH_3\overset{\displaystyle I}{\overset{|}{C}}HCH_2CH_3 + HI \longrightarrow CH_3CH_2CH_2CH_3 + I_2 \qquad (2.22)$$

Interestingly, the production of butane is almost totally quenched in the laser reaction since Reaction (2.22) is not a simple unimolecular process and is not effectively induced by means of vibrational excitation.

In contrast to these cases, the ratio of products from the dual channel bromoester shown in Reaction (2.21) was relatively independent of fluence up to ~ 3 J/cm^2 with Channel (2.21a) being favored over Channel (2.21b) by threefold (Danen *et al.*, 1980). This insensitivity to fluence was anticipated since the Arrhenius parameters for the two channels are expected to be similar, and the level of excitation should not provide much discrimination. A similar product ratio was observed from a SiF_4 sensitized experiment at low input energy. At fluences > 3 J/cm^2 for both direct and sensitized experiments, the ethylene/ethyl acrylate ratio increased considerably; this probably resulted from secondary laser-induced reaction of the ethyl acrylate produced in Channel (2.21b). It may be noted that the ester moiety is pumped by the CO_2 laser in these experiments but that the "remote" bromide group also undergoes reaction giving no evidence for any mode-selectivity.

A principal conclusion arising from the studies discussed in this section is that hitherto inaccessible high-energy reaction channels can be successfully induced by intense, pulsed infrared laser radiation. This ability to "instantly heat" and "superexcite" organic molecules beyond the normal reaction threshold offers chemists a novel technique for controlling chemical processes. All the data of this sort accumulated to date appear to be accommodated by means of a statistical, non-mode-specific, RRKM model in which the internal energy distributions are characterized by a mean energy $\langle E \rangle$. The data however, do not specify whether this energy distribution is more narrow or more broad than a Boltzmann, or, perhaps, even bimodal.

2.3.1.3b. Other Polyfunctional Reactants. In this section we will discuss somewhat more complex reactions in which an unsaturated alkene or carbonyl product is produced in a type of elimination process but which do not conveniently fit into any of the categories considered above. In the cases to be considered here, reaction via more than one channel is possible, and the arguments developed above will be seen to apply.

$$\text{CH}_2{=}\text{CH–O–CH}_2\text{CH}_3 \xrightarrow[\log A = 11.6 \text{ sec}^{-1}]{E_a = 44 \text{ kcal/mol}} \text{CH}_3\overset{\overset{\text{O}}{\|}}{\text{C}}\text{H} + \text{CH}_2{=}\text{CH}_2 \quad (2.23a)$$

$$\xrightarrow[\log A = 15 \text{ sec}^{-1}]{E_a = 65.8 \text{ kcal/mol}} \cdot\text{O–CH}{=}\text{CH}_2 + \cdot\text{CH}_2\text{CH}_3 \longrightarrow \text{products} \quad (2.23b)$$

Rosenfeld *et al.* (1977) have studied the infrared laser-induced reaction of ethyl vinyl ether [Reaction (2.23)] at relatively high pressures of 5–440 Torr. The ratio $k_{23a}/k_{23b} \sim 2.7$ was essentially independent of pressure.

Since the chemistry appeared to be that of a collisionally thermalized system, the authors calculated an effective temperature of $\sim 1600°K$ as necessary to reproduce the experimental channel ratio.

Brenner (1978) investigated the same system but under collisionless conditions and reported the interesting observation that the branching ratio depended on the laser pulse duration at constant fluence; i.e., the laser power determined the reaction pathway. These results will be discussed in more detail in Section 2.3.1.4 dealing with mode-selective processes.

$$\text{C}_6\text{H}_9\text{COCH}_2\text{CH}_3\ (\text{O}) \xrightarrow[\log A \sim 12.6\ \text{sec}^{-1}]{E_a \sim 47\ \text{kcal/mol}} \text{C}_6\text{H}_9\text{CO}_2\text{H} + \text{CH}_2{=}\text{CH}_2 \qquad (2.24\text{a})$$

$$\xrightarrow[\log A \sim 15\ \text{sec}^{-1}]{E_a \sim 60\ \text{kcal/mol}} \text{CH}_2{=}\text{CH}{-}\text{CH}{=}\text{CH}_2 + \text{CH}_2{=}\text{CHCO}_2\text{CH}_2\text{CH}_3 \qquad (2.24\text{b})$$

We have investigated a compound with dual reaction channels of similar Arrhenius parameters to ethyl vinyl ether (Danen and Hanh, 1980). 3-Carboethoxycyclohexene [Reaction (2.24)] can react via a low-energy, low A-factor ester elimination [Reaction (2.24a)] or a higher E_a, higher A-factor retro-Diels–Alder process [Reaction (2.24b)]. We have found that the ratio k_{24a}/k_{24b} changes from >90 to 0.5 in varying the laser energy from 1.0 to 4.2 J/cm^2. Unlike ethyl vinyl ether (Brenner, 1978), changing the laser pulse duration at constant fluence had only a very minor effect on the ethylene/1,3-butadiene ratio. The short pulse produced only a slightly lower ratio, as anticipated, since Reaction (2.24b) should become more competitive at higher levels of excitation. The explanation for the different experimental observations for these two, somewhat similar, reaction systems is not obvious. Possibly, the smaller ether [Reaction (2.23)] can be excited to sufficiently high energy that reaction competes with further absorption of energy leading to a power dependence.

We have also reported (Danen *et al.*, 1979) that pulsed CO_2-laser irradiation of *cis*-3,4-dichlorocyclobutene produced *cis, trans*-1,4-dichloro-1,3-butadiene as the only detectable product [Reaction (2.25a)]: This is the isomer predicted by the Woodward–Hoffmann rules of conservation of orbital symmetry for a thermally allowed, ground-state, conrotatory ring opening. Arrhenius parameters for reaction via a higher energy nonallowed channel or biradical intermediate were as estimated in Reaction (2.25b). This system is, thus, similar to that depicted in Figure 2.7a. Even though multiphoton excitation at low pressures presumably drives the reactant to vibrational excitation levels considerably in excess of the normal threshold

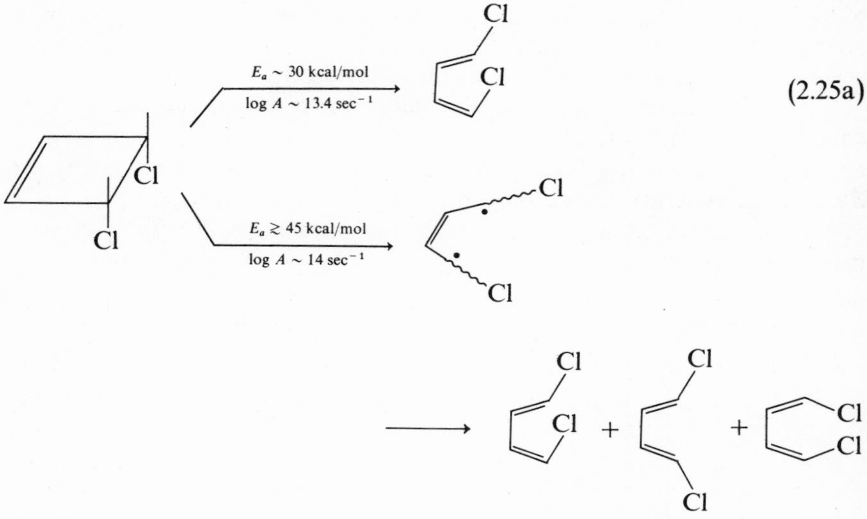

$E_a \sim 30$ kcal/mol, $\log A \sim 13.4\ \text{sec}^{-1}$ (2.25a)

$E_a \gtrsim 45$ kcal/mol, $\log A \sim 14\ \text{sec}^{-1}$ (2.25b)

energy, the only slightly higher A factor for Channel (2.25b) apparently cannot compensate for the relatively large difference in activation energies to allow reaction via Channel (2.25b) to compete efficiently with the lower-energy Channel (2.25a). Utilizing even short (160-nsec FWHM) pulses was ineffective in promoting the nonallowed pathway; intense focusing (> 100 J/cm^2) promoted the degradation into smaller molecules. These results, thus, help define the limits for which laser inducement of a higher-energy channel can be expected to be competitive.

Back and Back (1979) have studied the decomposition of cyclobutanone at pressures from 0.1 to 10 Torr [Reaction (2.26)]; the lower E_a

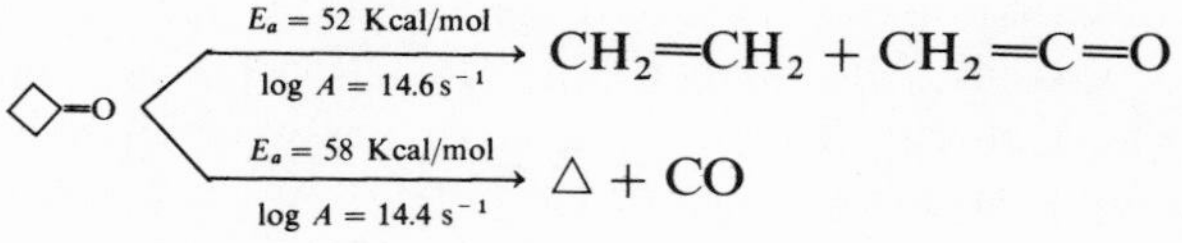

$$\xrightarrow[\log A = 14.6\,\text{s}^{-1}]{E_a = 52\ \text{Kcal/mol}} CH_2{=}CH_2 + CH_2{=}C{=}O \qquad (2.26a)$$

$$\xrightarrow[\log A = 14.4\,\text{s}^{-1}]{E_a = 58\ \text{Kcal/mol}} \triangle + CO \qquad (2.26b)$$

pathway always predominated greatly. Above ~ 3 Torr, k_{26b}/k_{26a} was reasonably constant at ~ 0.017 presumably resulting from a thermally equilibrated system corresponding to an Arrhenius temperature of ~ 830°K. Surprisingly, addition of air or xenon as a bath gas to 0.5 Torr of cyclobutanone resulted in an unexpected increase in the k_{26b}/k_{26a} ratio and an anticipated decrease in the overall yield; no satisfactory explanation was offered for the former effect. At lower pressures the ratio increased rather significantly to ~ 0.1 at 0.5 Torr, which would require an Arrhenius temperature of ~ 1600°K and suggests a nonequilibrium process. Unfortunately, no attempts were made to determine the effect of laser pulselength on the product ratio as in the ethyl vinyl ether system [Reaction (2.23)].

2.3.1.4. *Mode-Selective Chemistry*

As alluded to earlier, perhaps the ultimate application of pulsed infrared laser photochemistry would be to excite selectively a particular bond or functional group within a complex molecule and cause reaction exclusively at that site. Of course, competing with such a process is the rapid intramolecular randomization of the deposited laser energy that would presumably interdict any localization of vibrational energy within a particular mode or restricted set of vibrational modes at least at high levels of excitation. At the time of this writing, apparently few researchers in the area of infrared multiphoton-induced chemistry believe that mode-selective reactions are possible with the presently available lasers that operate with tens of nanosecond pulselengths. There are some experiments, however, that authors have interpreted in terms of restricted intramolecular randomization.

Much of the initial interest in pulsed infrared laser chemistry was spawned by the report of Grunwald and co-workers in 1976 that infrared energy deposited into $CFCl_3$ and CF_3Cl remained localized in a single mode of the molecules for a time sufficiently long for reaction to occur even at pressures as high as 60 Torr (Dever and Grunwald, 1976). Subsequent interpretations (Schulz *et al.*, 1979; Cantrell *et al.*, 1979) have suggested that the early interpretations of the data were almost certainly erroneous and that there was no need to postulate any localization of vibrational energy in these systems.

At about the same time, Zitter and Koster (1976) had reported the surprising result that irradiation of CF_2Cl–CF_2Cl by low-power continuous wave (CW) radiation produced bond-specific chemistry as evidenced by a 160 × faster rate when the laser was tuned to the weaker of two absorption bands. The implication was drawn that significant compartmentalization of energy in vibrational modes was taking place despite V–V transfer. These authors, however, later retracted this interpretation when evidence was acquired demonstrating that the laser energy actually absorbed was not predictable from the low-power absorption coefficients. In fact, the reaction rates were maximum at frequencies $\sim 30\ cm^{-1}$ to the red of the ground-state absorption peaks, and the maximum rates for the two bands actually differed little when irradiated at the red-shifted frequencies (Zitter and Koster, 1977).

A more recent study suggesting that vibrational excitation may not be randomized on the time scale required for reaction has been reported by Brenner (1978) and is shown by Reaction (2.23) in Section 2.3.1.3b. Ethyl vinyl ether can react by at least two different pathways, and Brenner observed a different branching ratio when laser pulses of different *intensities* (power) but constant energy were utilized to drive the reaction. Both Reaction Channels (2.23a) and (2.23b) were observed when a short, 200-nsec pulse with no tail was utilized, but only the lower-energy channel (Reaction (2.23a)] was driven when the same total number of photons were supplied with a ~200-nsec spike followed by ~2000-nsec tail. The ratio k_{23b}/k_{23a} was essentially invariant at 1.5 ± 0.1 at low pressures, and it was not possible to cause the low-energy channel to predominate over the higher energy channel by decreasing the fluence with the short pulse. Even at the threshold fluence of ~0.5 J/cm^2, the higher-energy pathway still predominated. Using statistical RRKM theory it was estimated that absorption of 35 photons (104 kcal/mol) would be required to reproduce the ratio k_{23b}/k_{23a}. Thus, this result appears similar in nature to those discussed earlier in which multiphoton absorption drives the level of vibrational excitation considerably above the threshold level for the lower-energy channel.

However, since it was not possible with the short pulse to induce only the lower-energy reaction at threshold, Brenner argued that the dynamics of multiphoton absorption were not identical for the long and short pulse cases and that a statistical energy distribution may not be realized during pumping with the short pulse. It was suggested that a competition between intramolecular relaxation and laser up-pumping might be made to exist depending upon laser parameters. Stone and Goodman (1979) have recently developed a theoretical model based on rate equations that include the dynamics of intramolecular relaxation as well as intramolecular heat bath feedback to the pumped vibrational mode and laser-induced excitation. If intramolecular relaxation is rapid compared to the reaction rate and if the molecules do not decompose during the pulse, the rate will depend only on laser energy and can be described by statistical RRKM theory. If such relaxation times are long or comparable to the pumping time scale, the reaction will depend on the peak laser intensity (power) and not the total energy. It was suggested that this theory might offer an explanation for Brenner's data on ethyl vinyl ether. Of course, the fundamental question remains: How fast is intramolecular vibrational relaxation in the quasicontinuum and above the threshold level?

Sudbø *et al.* (1979) have noted that, although there is good evidence that the absorption in the quasicontinuum depends only on the laser energy and not power, the level of excitation above the dissociation threshold does depend on the laser power since up-pumping above the threshold is now competitive with dissociation. This was suggested as possibly accounting for the changes observed by Brenner in the branching ratio as a function of laser intensity.

A somewhat related study has been reported by Hall and Kaldor (1979) for the dual channel reaction pathways for cyclopropane [Reaction (2.27)]. The yield of the lower-energy channel, isomerization, to the higher-

$$\triangle \xrightarrow[\log A = 15.2\ \text{sec}^{-1}]{E_a = 65\ \text{kcal/mol}} CH_3CH{=}CH_2 \qquad (2.27a)$$

$$\triangle \xrightarrow[\log A \sim 15\ \text{sec}^{-1}]{E_a \sim 100\ \text{kcal/mol}} CH_2{=}CH_2 + :CH_2 \rightarrow \text{products} \qquad (2.27b)$$

energy channel, fragmentation, was reported to depend on the mode excited. Excitation of the C–H asymmetric stretch with 3.22-μm radiation drove reaction (2.27a) almost exclusively, while CO_2 laser excitation of the CH_2 wag at 9.50 μm gave comparable amounts of both reaction channels. The authors concluded from this and the effects of added argon bath gas that the initial laser excitation leads to a nonstatistical final state distribution that persists for the time scale for unimolecular reaction. The main experimental observation supporting the interpretation of mode selectivity was that the addition of argon to the experiments with 3.22-μm radiation increased the yield of the higher-energy fragmentation channel. The authors argued that collisions with argon cannot add any significant amount of energy to the cyclopropane and, therefore, the effect of argon was to enhance an otherwise relatively slow intermode vibrational energy transfer.

Thiele *et al.* (1980) have applied their theoretical model based on rate equations and including restricted intramolecular vibrational relaxation and conclude that there is no reason to doubt that Hall and Kaldor's experiment [Reaction (2.27)] is a genuine example of selective bond breaking. The theory divides the vibrational modes of a molecule into two groups (e.g., high- and low-frequency modes) and assumes that scrambling of vibrational energy *within* a group occurs rapidly with a time scale of τ_m. Relaxation *between* the two groups was presumed slower, on a time scale of τ_m/α. The parameter α is adjustable, which the authors acknowledge is required to disguise their ignorance of the rates of intramolecular relaxation in the quasicontinuum. For $\alpha = 1$, relaxation between the two groups is rapid and RRKM theory would apply. As $\alpha \rightarrow 0$, relaxation is slower and vibrational energy can accumulate in the pumped mode leading to possible mode-selective reaction. As noted above, however, a knowledge of the intramolecular vibrational relaxation rates in the quasicontinuum and above the threshold levels is required before a model such as advanced by Thiele *et al.* (1980) can be considered realistic.

It appears plausible that the different product ratios in Hall and Kaldor's case can also be accounted for by differences in up-pumping rates and ultimate levels of excitation as discussed earlier. The effect of an added bath gas is very dependent upon factors such as laser fluence and laser power; changing the laser pulselength can alter the bath gas effect quite dramatically as discussed in Section 2.3.1.2b (Jang and Setser, 1979). In this study it was not possible to compare on an absolute basis the laser energy actually absorbed at 3.22 μm vs. 9.50 μm from the opto-acoustical data reported. From Brenner's data (Brenner, 1978) it is apparent that the product ratios in a dual channel reactant are quite sensitive to changes in laser operating parameters, yet in the cyclopropane system two quite different types of lasers were utilized. The 3.22-μm laser radiation had a pulse duration of 40-nsec FWHM and a fluence of ~ 2 J/cm^2 at the focus; the 9.50-μm laser exhibited a 400-nsec pulse FWHM and a focused fluence of ~ 60 J/cm^2. Such differences almost certainly assure considerably different multiphoton excitation–reaction behavior, and the dependence of reaction channel on the laser utilized is not unanticipated.

A very recent study of a large organometallic uranium compound has been reported by Kaldor and co-workers (Kaldor *et al.*, 1979; Cox *et al.*, 1979), and the data were interpreted as necessitating that the vibrational excitation be localized sufficiently long to lead to spatial and site-selective unimolecular dissociation. Irradiation of the hexafluoroacetylacetonate–tetrahydrofuran complex of UO_2^{2+} [**12**, Reaction (2.28)] at 956.2 cm^{-1} excited the asymmetric stretch of the UO_2^{2+} moiety and resulted in high yields of the tetrahydrofuran (THF) dissociation product at quite low fluences (e.g., $\sim 100\%$ dissociation at ~ 0.1 J/cm^2). Reaction (2.28) was estimated to have $E_a \sim 30$ kcal/mol and is reversible. The extent of reaction and temporal concentration of the reactant molecule were monitored by laser-induced fluorescence with a time resolution of ~ 20 nsec.

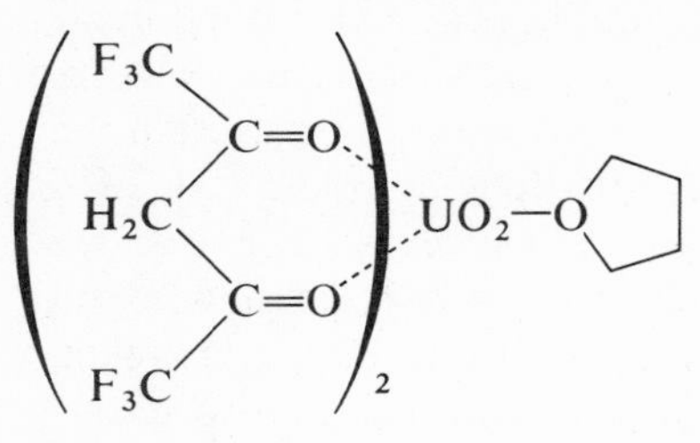

$$\left(\begin{matrix} F_3C \\ \quad C{=}O \\ H_2C \\ \quad C{=}O \\ F_3C \end{matrix}\right)_2 UO_2{-}O(THF)$$

12

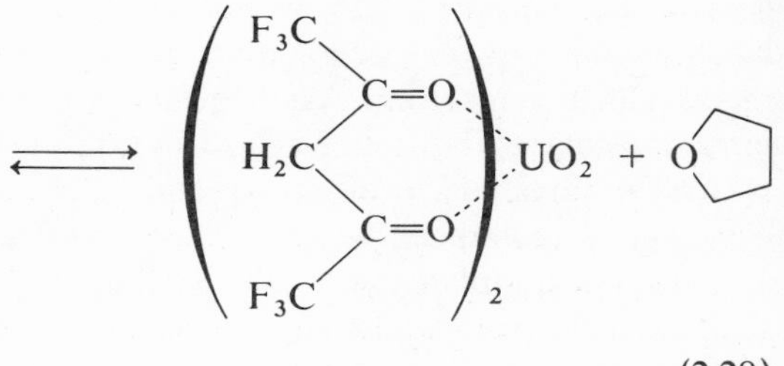

$$\rightleftharpoons \left(\begin{matrix} F_3C \\ \quad C{=}O \\ H_2C \\ \quad C{=}O \\ F_3C \end{matrix}\right)_2 UO_2 + O(THF) \qquad (2.28)$$

The evidence for energy localization in the multiple photon excitation–reaction of **12** was not direct. In essence, the authors argued that the deposited energy must remain localized in a limited number of modes in order to explain the dissociation on the observed time scale at such low fluences. It was estimated that < 10 photons per molecule were absorbed when the dissociation yield was nearly 100%, and it was argued that, if the deposited energy were randomized and statistically dissipated over such a large molecule, there would not be sufficient excitation at the U–THF bond to cause rupture. Instead, the 5–10 photons absorbed by the molecule were postulated to delocalize rapidly (estimated time $\sim 10^{-9}$–10^{-12} sec) over only 5–15 modes of the first bonding sphere that includes the uranyl moiety and all attendant U–O bonds including that bond that complexes the THF ligand. Leakage from this select set of modes into the remainder of the molecule occurred at a much slower, unspecified, time scale but $\gtrsim 10^{-7}$ sec to account for the experimental yields. It was critical, of course, that the U–THF bond that was ruptured in the reaction process be coupled to the select set of modes initially excited. In this regard, the authors also excited the C–H stretch of the THF moiety with a 30 nsec, 10-mJ/cm^2 laser operating at 3 μm but observed no reaction. This was rationalized as demonstrating a lack of strong coupling of the C–H and U–O stretches. The energy deposited into the C–H stretch was thought to couple instead into the THF ring and become so diluted that the entire system vibrationally relaxed before U–O bond fission could occur.

If the authors' interpretation of their data is correct, it certainly is contrary to the majority of the data accumulated to date concerning infrared multiphoton excitation/reaction and to currently "accepted" time scales of intramolecular vibrational relaxation processes. To fail to accept an interpretation simply because it is unconventional or unpopular, however, is totally unscientific unless a more plausible explanation can be advanced. Although a complete critique of this interesting work will not be attempted here, several unsettling points will be noted.

First, the energy absorbed by this unique uranyl compound was not actually experimentally measured. This, of course, is a critical quantity on which rests the entire interpretation of nonstatistical energy distribution. Instead, the absorbed energy was estimated from the energy fluence and single-photon absorption cross section. As discussed more fully in Section 2.6, there is no assurance that the single-photon cross section is applicable to a multiphoton process. In Chapter 1 of this volume, Galbraith and Ackerhalt describe a model that exhibits strongly increasing cross sections at higher levels of excitation (see Figures 1.4 and 1.11) that opens the possibility that the absorbed energy in the uranyl case might be larger than anticipated. An experimentally measured $\langle n \rangle$ would be highly desirable in this unique case.

Second, the high yields per pulse observed for the uranyl compound may not be as atypical of large molecules as suspected by the authors. In our laboratory we have preliminary data on several organic esters which, likewise, indicate that these moderately large molecules undergo a surprisingly efficient reaction. We can induce essentially 100% reaction in some cases with fluences of ~2–3 J/cm^2. Plum and Houston (1979) observed ~10–20% reaction per pulse for cyclic $C_2F_4S_2$ at fluences ~0.3 J/cm^2. Although such fluences are higher than those required for Reaction (2.28), they are still below even the threshold levels observed in many small molecules. The extremely high density of vibrational states even at room temperature for **12**, the largest gas-phase compound yet subject to pulsed infrared laser excitation, would appear to predispose this molecule to a facile laser-induced reaction, and the high conversions per pulse at low fluences may not be so unusual.

The laser-induced reactions of large organic esters are very susceptible to collisional quenching by an inert bath gas. The reaction of **12** was also noted to be easily quenched, but the argon was thought to enhance the intramolecular V–V redistribution process. A more thorough discussion of the interplay of the relatively long lifetimes of vibrationally excited large molecules, collisional effects, and heat capacity influences is given in Section 2.6. Although there are obvious similarities between these data and those of Kaldor *et al.* for the uranyl system and their static cell experiments might be explained by such a phenomenon, the mass spectral flow (Kaldor *et al.*, 1979) and molecular beam (Cox *et al.*, 1979) studies of these workers would appear to preclude any significant collisional effects.

Finally, it should be noted that **12** has only one low-energy reaction channel, and it is precisely this pathway that is induced by the laser even though the U–THF bond was not pumped directly. The ultimate test of infrared, laser-induced, spatial and site-selective chemistry in large molecules would be to investigate a bifunctional reactant that can be selectively irradiated at each functional group. If different reaction channels can be induced by irradiation of different functional groups for the same amount of total absorbed energy in the same amount of time from each laser frequency, then mode-selective chemistry will have been truly verified.

2.3.2. Bimolecular Organic Reactions

The overwhelming majority of infrared laser-induced studies have involved unimolecular reactions. Although there is a small number of documented examples of infrared-enhanced bimolecular processes (Birely and Lyman, 1975), these all involve small, di- or triatomic inorganic molecules and most are simple atom-transfer reactions. We are not aware of any successful infrared laser-induced bimolecular transformations of relatively large organic molecules. On the other hand, there are several negative reports of such attempted transformations.

There are several reasons why bimolecular processes have received less scrutiny and why infrared laser inducement or augmentation of such reactions is less successful than with unimolecular reactants. First of all, irradiation of a molecule with a pulsed infrared laser generates a vibrationally excited species which is all that is required for the molecule to undergo a unimolecular transformation. That is, only vibrational excitation is required to promote a unimolecular reaction. Usually, this is not the case with a bimolecular process in which translational energy and, in some cases, even rotational energy is important, as well as vibrational energy, for the reacting molecules to transcend the activation barrier. Such excitation is not provided, at least in the initial stages of an infrared laser excited system at low pressures. The problem is compounded by the fact that individual bimolecular reactions probably require specific combinations of vibrational and translational energy for efficient reaction, and our present understanding of such reactions does not allow a confident *a priori* prediction of the partitioning factors.

A second important difference between infrared laser inducement of a bimolecular reaction as compared to a unimolecular reaction process is the competing vibrational relaxation of the laser-excited reagent. By definition, a bimolecular reaction requires the collision of two molecules. The collision might result in reaction, but transfer of vibrational excitation from the activated molecule to the "cold" collision partner will usually be very rapid, particularly at high levels of vibrational excitation. It can be estimated that as much as 5–15 kcal/mol can be lost from a large, vibrationally excited, organic molecule on each gas kinetic collision with a large, but cold, collision partner. Laser excitation of both reactant molecules (most often requiring two different laser frequencies) would alleviate much of this second encumbrance to laser-assisted bimolecular reactions, but this does not yet seem to have been attempted with any organic systems.

The objective in a bimolecular system, then, is to demonstrate an enhancement in rate upon infrared laser excitation [k_{laser}, Reaction (2.29a)] over and above the unassisted rate [k, Reaction (2.29b)] plus the quenching rate [$k_{transfer}$, Reaction (2.29c)]; i.e., $k_{laser} > (k + k_{transfer})$. As can be anticipated from the above discussion, the effect of vibrational excitation on a

$$A^* + B \xrightarrow{k_{laser}} C \tag{2.29a}$$

$$A + B \xrightarrow{k} C \tag{2.29b}$$

$$A^* + B \xrightarrow{k_{transfer}} A + B^* \tag{2.29c}$$

bimolecular reaction is not obvious, and examples in simple di- and triatomic systems have been reported in which infrared irradiation enhances, retards, or has no effect on the overall reaction kinetics (Birely and Lyman, 1975). Most of these studies have involved relatively simple atom transfer reactions in which A is a two- or three-atom molecule and B is some atom such as Cl• or H•; the O-atom transfer reaction $NO + O_3 \rightarrow NO_2 + O_2$ in which either reagent has been vibrationally excited has been extensively studied.

Anticipating that V–V transfer [Reaction (2.29c)] would be detrimental to an infrared-induced organic bimolecular reaction, we investigated the HBr-catalyzed bimolecular dehydration of ethanol in which the HBr is both a reagent and a highly efficient catalyst (Danen, 1979). The unimolecular dehydration of ethanol [Reaction (2.30a)] is strongly catalyzed by HBr [Reaction (2.30b)], which presumably behaves in a manner as depicted in Reaction (2.32) (Ross and Stimson, 1960). 2-Propanol [Reactions (2.31a) and (2.31b)] was utilized as a chemical thermometer to monitor for simple

$$CH_3CH_2OH \xrightarrow{E_a \sim 71\ \mathrm{kcal/mol}} CH_2{=}CH_2 + H_2O \quad (2.30a)$$

$$CH_3CH_2OH \xrightarrow{+HBr\ \ E_a = 37.6\ \mathrm{kcal/mol}} CH_2{=}CH_2 + H_2O \quad (2.30b)$$

$$CH_3CH(OH)CH_3 \xrightarrow{E_a \sim 67\ \mathrm{kcal/mol}} CH_3CH{=}CH_2 + H_2O \quad (2.31a)$$

$$CH_3CH(OH)CH_3 \xrightarrow{+HBr\ \ E_a = 33.2\ \mathrm{kcal/mol}} CH_3CH{=}CH_2 + H_2O \quad (2.31b)$$

$$CH_3CH_2OH + HBr \longrightarrow \left[\begin{array}{c} Br\text{—}H \\ H \qquad O\text{—}H \\ CH_2\text{—}CH_2 \end{array}\right]^+ \longrightarrow CH_2{=}CH_2 + H_2O + HBr \quad (2.32)$$

heating of the system. It was demonstrated that the selective, nonthermal inducement of H_2O elimination from vibrationally excited ethanol in the presence of 2-propanol required total reactant pressures of less than ~ 1 Torr; at higher pressures, propylene became the predominant product as a result of intermolecular energy relaxation to 2-propanol. Under conditions in which the unimolecular elimination of ethylene was being induced, however, HBr, instead of catalyzing the reaction in a bimolecular process [Reaction (2.30b)], actually had a detrimental effect as compared to the uncatalyzed unimolecular reaction. From Figure 2.10 it can be seen that the half-quenching pressure of HBr is only ~ 1 Torr, indicating that HBr is much more effective as a vibrational heat sink than it is as a catalyst for the laser-induced dehydration reaction. We concluded that collisional deactivation by the nonexcited reaction partner may be a general result for an

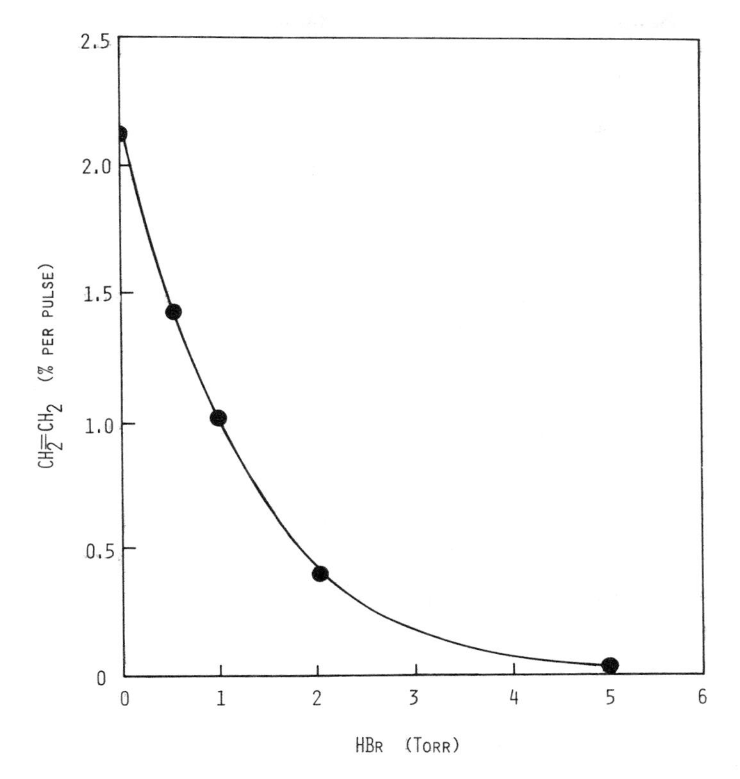

Figure 2.10. Percent yield of ethylene produced in the irradiated volume per laser pulse from 0.5 Torr of ethanol as a function of HBr pressure; laser fluence = 3.0 J/cm^2. (After Danen, 1979.)

infrared laser-induced bimolecular reaction involving even moderate-sized organic molecules, particularly if the reaction has a low Arrhenius *A* factor and if relative translational energy is required to overcome the activation barrier.

Herman and Marling (1979) have, likewise, reported negative results in several attempts to vibrationally stimulate addition reactions between hydrogen halides and unsaturated hydrocarbons [Reactions (2.33)–(2.36)]. No evidence was found for enhancement of any of these addition reactions

$$\underset{(v = 5 \text{ or } 6)}{HCl^*} + Cl_2C{=}CCl_2 \longrightarrow CHCl_2CCl_3 \tag{2.33}$$

$$\underset{(v = 5 \text{ or } 6)}{HCl^*} + CH_2{=}CHCH{=}CH_2 \begin{cases} \longrightarrow CH_2{=}CHCHClCH_3 & (2.34a) \\ \longrightarrow CH_2ClCH{=}CHCH_3 & (2.34b) \end{cases}$$

$$\underset{(v = 5 \text{ or } 6)}{HCl^*} + \text{1,3-cyclohexadiene} \longrightarrow \text{3-chlorocyclohexene} \tag{2.35}$$

$$HX + HC{\equiv}CH^* \longrightarrow CH_2{=}CHX \tag{2.36}$$

in which one reactant was excited to a vibrational state above the activation energy of the corresponding thermal reaction. This study differs from the majority of those discussed in this review since the vibrational excitation was provided, not by an infrared laser, but by one-photon absorption inside a CW dye laser cavity. It was estimated that the rate constants for addition of the vibrationally excited molecules were at least 10–1000 times slower than the pre-exponential factors of the corresponding thermal reactions; i.e., the rates were that much slower than if vibrational excitation totally overcame the activation energy.

As a result of the rather detailed understanding of at least some of the dynamics of HX elimination from organic halides, the concept of microscopic reversibility allows insight into the reverse process, the bimolecular addition of HX to alkenes and alkynes. It has been estimated (Berry, 1974) for some four-centered, α,β-dehydrohalogenation reactions that $\sim$ 15–40% of the available energy is partitioned into HX vibrational energy. The remaining energy is vented into translational, rotational, and vibrational modes of the olefin and translational and rotational modes of HX although the exact partitioning is not known except in a few cases (Holmes and Setser, 1975; Kim and Setser, 1974; Marcoux and Setser, 1978, Subdø *et al.*, 1978). Since vibrationally excited HX is a product of an elimination reaction, microscopic reversibility assures that vibrational excitation of HX to the appropriate level will be required, although not necessarily sufficient, to drive the reverse bimolecular addition of HX to the unsaturated hydrocarbon. In the elimination reaction, the HX distribution declines exponentially with increasing level of excitation and $v = 0$ is the favored level. As a result, it is not usually understood what fraction of the vibrational energy that is deposited in a reactant actually contributes to overcoming the activation energy.

The most plausible explanation for the lack of laser augmentation of the HCl* + olefin reactions in Herman and Marling's study [Reactions (2.33)–(2.35)] is that only vibrational excitation of HCl was provided whereas vibrational excitation of olefin and relative translational energy are also needed to stimulate the reaction. As discussed above, vibrational excitation of *both* reactants might be effective in accelerating the bimolecular addition reaction, although relative translational excitation is also obviously necessary.

The authors rationalized the lack of enhancement of Reaction (2.36) on the fact that the excited C–H stretching of HC≡CH had no component along the reaction coordinate. Unlike reactions induced by multiphoton absorption (which comprise most of the examples in this review), the single-photon vibrational excitation of acetylene does not result in significant populations of other vibrational modes. Use of an infrared laser to pump the C–H bending modes might be more efficient in promoting the bimolecular addition reaction, particularly if simultaneous stimulation of HX was provided.

In a discussion of various parameters for proposed isotope separation schemes based on laser-augmented bimolecular reactions, Bauer (1978) also reported negative results in attempts to add vibrationally excited HF to $CF_2{=}CF_2$, $CH_2{=}CF_2$, $CH_2{=}CHCO_2CH_2CH_3$, $CH_2{=}C{=}CH_2$, $CH_3C{\equiv}CH$, and cyclopentadiene. Experiments were conducted at both 25 and 190°C at ~ 1 Torr using an infrared HF laser (150 mJ per pulse, 2 Hz) to excite HF vibrationally. Bauer concluded, also, that excitation of the olefin might be as essential as excitation of HF to produce a significant rate enhancement for such bimolecular addition reactions.

Douglas and Moore (1979) attempted to add vibrationally excited HF($v = 4$, 43.9 kcal/mol) to isobutene [Reaction (2.37)]. The activation

$$HF^* + (CH_3)_2C{=}CH_2 \longrightarrow (CH_3)_2\overset{\overset{\displaystyle F}{|}}{C}CH_3 \qquad (2.37)$$

energy for this reaction is 45.8 kcal/mol so the vibrational excitation is near the threshold value. It was found that isobutene was very efficient at relaxing vibrationally excited HF, exhibiting a quenching rate constant of at least 1.5×10^{-10} cm^3 mol^{-1} sec^{-1}. Again, quenching processes proved more than competitive with the addition reaction.

Attempts to induce the gas-phase S_N2 displacement Reactions (2.38) and (2.39) have proved futile (Hwang *et al.*, 1979). For Reaction (2.38), no

$$NH_3 + CH_3Br \longrightarrow CH_3\overset{+}{N}H_3Br^- \qquad (2.38)$$

$$NH_3 + CH_3F \longrightarrow CH_3\overset{+}{N}H_3F^- \qquad (2.39)$$

$$CH_3F + BCl_3 \longrightarrow (\overset{+}{C}H_3)(BCl_3F^-) \longrightarrow CH_3Cl + BCl_2F \qquad (2.40)$$

reaction was detected for irradiation of CH_3Br only (940.5 cm^{-1}) or irradiation of both reactants at 973.3 cm^{-1}. Reaction (2.40) was observed to occur, but the authors concluded that a heterogeneous surface reaction rather than a gas-phase process was responsible for the observed products.

It should be noted that Reactions (2.38)–(2.40) are not well-characterized gas-phase processes, and the production of such charged ionic products in the gas phase is expected to require a high activation energy. The authors modeled their studies after the facile $NH_3 + HCl \rightarrow \overset{+}{N}H_4Cl^-$ reaction, but there are distinct differences between a nucleophilic attack of NH_3 on a proton as compared to a tetrahedral carbon atom. In the latter, no significant N–C bond formation can occur without extensive heterolytic C–X (X = Br or F) bond rupture and molecular distortion involving Walden inversion about the tetrahedral carbon. Even in solution where solvation of the developing charges can assist the process, nucleophilic attack of a base upon, e.g., HCl is diffusion-controlled but attack of the same base on CH_3Br is many orders of magnitude slower. In the gas phase, where solvation is absent, the differences in dynamics are expected to be even more pronounced.

In addition to the relatively simple di- and triatomic molecule–atom inorganic reactions mentioned at the beginning of this section, similar bimolecular reactions involving simple organic molecules and halogen atoms have been vibrationally enhanced. For example, Hsu and Manuccia (1978) have utilized a CW CO_2 laser to vibrationally excite CH_2D_2 in the presence of CH_4 and a large excess of argon bath gas. The latter was present essentially to isolate the excited CH_2D_2 from the nonexcited CH_4. Intentional V–T deactivation of $CH_2D_2{}^*$ competed with interisotopic V–V transfer, and the gas sample was enriched in excited CH_2D_2 while the CH_4 remained unexcited on a steady-state basis. A subsequent chain reaction of this mixture with chlorine atoms–molecular chlorine produced deuterated methyl chloride enriched in deuterium by up to 72%. In essence, a gain in isotopic selectivity was realized by sacrificing some fraction of the excited molecules. The pertinent reactions are as shown in Reactions (2.41)–(2.44):

$$CH_2D_2 \xrightarrow{\text{CW infrared laser}} CH_2D_2{}^* \tag{2.41}$$

$$CH_2D_2{}^* + \cdot Cl \xrightarrow{k_{\text{laser}}} HCl + \cdot CHD_2 \tag{2.42a}$$

$$CH_2D_2{}^* + \cdot Cl \xrightarrow{k_{\text{laser}}} DCl + \cdot CH_2D \tag{2.42b}$$

$$CH_4 + \cdot Cl \xrightarrow{k_{\text{thermal}}} HCl + \cdot CH_3 \tag{2.43}$$

$$\cdot CH_3,\ \cdot CH_2D,\ \text{or}\ \cdot CHD_2 + Cl_2 \xrightarrow{\text{fast}} \text{methyl chloride} + \cdot Cl \tag{2.44}$$

For Reactions (2.42) and (2.43), $k_{\text{laser}} > k_{\text{thermal}}$, where k_{laser} is the rate constant for the vibrationally enhanced reaction neglecting any differences because of isotope effects and k_{thermal} is the rate constant for the pure thermal reaction. The reaction involving CH_4 proceeded via the pure thermal rate since CH_4 vibrational excitation was rendered negligible by V–T deactivation to argon. The thermal rate was estimated to require ~ 4000 gas kinetic collisions, which was slower than the estimated 300 collisions for V–V interisotopic energy transfer between $CH_2D_2^*$ and CH_4. The latter was rendered negligible by the ~ 500 : 1 excess of argon to CH_2D_2/CH_4 in the flow reactor.

This example demonstrates that the bimolecular reaction of, at least, simple organic molecules excited by only one quantum of infrared radiation (i.e., from $v = 0$ to $v = 1$ level) can be significantly enhanced provided the reaction partner is an atom. Moreover, such a system need not be restricted to isotope separation and offers the ability to scale the process to high pressures since the only constraint to molecular selectivity is the maintenance of a constant ratio of V–V to V–T rates. This CW laser-enhanced bimolecular system, however, has an exceedingly low E_a of 3.9 kcal/mole (Pritchard *et al.*, 1955) allowing $k_{\text{laser}} > k_{\text{thermal}}$ at only $v = 1$ excitation levels. This is totally unlike the examples of attempted bimolecular reactions discussed above involving multiphoton infrared absorption in high E_a bimolecular processes. The latter are much more common and are usually of more interest to organic chemists.

Thus, at least five different reports have been made of attempts to augment an organic chemical reaction by vibrational excitation of one of the reactants and all have been unsuccessful. There is compelling evidence from the known dynamics of, at least, some of the reverse reactions that vibrational excitation of *both* reactants and translational excitation may be a requirement for successful augmentations of these bimolecular processes, and this would seem to be an obvious experimental objective. This could be accomplished by sensitization (Section 2.3.3) as well as judicious choice of two organic reactants that both absorb the same frequency radiation or, more generally, by use of two different exciting lasers. Although this has been attempted for Reaction (2.38), this was probably not a good system for a definitive test as discussed above.

If relative translational energy is necessary for a bimolecular reaction, as is known for some cases, it is difficult to envision how infrared lasers will significantly impact such processes. The unique ability of an infrared laser to excite a molecular vibration selectively would no longer be sufficient to induce a bimolecular reaction. Once the selectively deposited vibrational excitation is degraded by collisional processes into translational and rotational energy, the equilibrium state would be expected to be similar, if not identical, to that produced by simple heating. Of course, reactions could be conducted in a homogeneous manner by means of the laser, and possibly detrimental wall effects could be avoided and high effective temperatures and short (milli- or microsecond) reaction times could be achieved, but some of the unique advantages of utilizing an infrared laser to excite the reaction system would appear to be lost.

A possible complication arising from attempting to induce a bimolecular reaction by simultaneous irradiation of both reactants is that, if successful, the product molecule might be produced with too much vibrational excitation to be stable. For example, if both HX and an olefin are individually excited with 40–50 kcal/mol of vibrational energy and successfuly undergo an addition reaction that is exothermic by ~ 15 kcal/mol, the organohalide product will be formed with 95–115 kcal/mol excess energy. This would be sufficient to drive a number of unimolecular reactions of the organohalide such as depicted in Reactions (2.45a)–(2.45c).

$$HX^* + R_2C{=}CR_2^* \longrightarrow \left[\begin{array}{c} \text{H} \quad \text{X} \\ | \quad\;\; | \\ R_2C{-}CR_2 \end{array}\right]^{**} \longrightarrow HX + R_2C{=}CR_2 \qquad (2.45a)$$

$$\longrightarrow R_2CH\dot{C}R_2 + X\cdot \qquad (2.45b)$$

$$\longrightarrow R_2\dot{C}H + R_2\dot{C}X \qquad (2.45c)$$

In summary of this section, no complex organic *bimolecular* reactions have as yet been successfully augmented by vibrational excitation of one of the reactants. An obvious experimental approach is to excite both reactants simultaneously, but this might produce additional complications. The application of pulsed infrared lasers to bimolecular processes will certainly not be as straightforward as for unimolecular reactions. This, however, is precisely the area in which the biggest contribution to organic chemistry might be expected since synthetic molecule building is generally much more important than molecular dissociation or rearrangement reactions.

2.3.3. Sensitized Organic Reactions

The first law of photochemistry states that only light that is absorbed by a molecule can be effective in producing photochemical change in the molecule (Calvert and Pitts, 1966). Photosensitization is an important method for carrying out photochemical reactions when a molecule cannot be brought to the desired excited state by direct absorption of light. In this method, an unreactive sensitizer molecule is irradiated and the absorbed energy is transferred, at least in part, to the desired reactant compound; the use of various triplet and singlet sensitizers in visible–ultraviolet photochemistry is commonplace.

For pulsed infrared laser-induced processes, direct excitation of the reactant molecule is usually convenient and efficient when the reactant has a strong absorption band in the tunable range of the laser. Organic molecules frequently possess absorption bands in the 9–11-μm

(1100–900-cm^{-1}) emission region of the CO_2 laser, but oftentimes these are of too low absorption cross section to be pumped efficiently to a reactive state by multiple-photon absorption. In such cases, reaction can still usually be induced by adding a sensitizer. This is a compound that will absorb the infrared radiation strongly and transfer a significant portion of the absorbed energy to the desired reactant molecule without itself decomposing or chemically participating in the reaction process [Reactions (2.46)–(2.48)]. Both SF_6 and SiF_4 have been rather commonly utilized as

$$S \xrightarrow{\text{infrared laser}} S^{**} \tag{2.46}$$

$$S^{**} + R \overset{\text{many collisions}}{\rightleftarrows \;\rightleftarrows\; \rightleftarrows} S^{*} + R^{*} \tag{2.47}$$

$$R^{*} \longrightarrow \text{products} \tag{2.48}$$

sensitizers for organic chemical reactions; SiF_4 is usually the preferred choice. In both the authors' and other laboratories (Olszyna *et al.*, 1977) SF_6 has been found to decompose under conditions of relatively mild irradiation with a pulsed CO_2 laser. On the other hand, SiF_4 is stable and remains unreactive in the presence of a wide variety of substances at considerably higher laser fluence levels. This difference in stability can be understood in terms of the relevant bond dissociation energies: $D_0°(SF_5\text{–}F) = 93$ kcal/mol (Benson, 1978); $D_0°(SiF_3\text{–}F) = 142$ kcal/mol (JANAF, 1971).

Shaub and Bauer (1975) pioneered the use of sensitizers in infrared laser photochemistry. They utilized SF_6 and a CW CO_2 laser to generate high bath temperatures for gas-phase reactions under homogeneous conditions. With total operating pressures in the range of 10–100 Torr, controlled temperatures from 500 to 1500°K could be developed with a modest CW CO_2 laser. The authors termed their method laser-powered homogeneous pyrolysis. Risetimes for heating were estimated to be in the millisecond range, and the reaction times were ~ 10 sec, which makes the laser pyrolysis method complementary to the faster time scale single-pulse shock-tube technique. A key advantage of this technique, as compared to conventional pyrolytic methods, is that complicating catalytic hot wall effects are avoided. This technique will not be considered further in this chapter, however, since it employs a CW laser and our emphasis is on pulsed phenomena (see Chapter 3 for further discussion).

The most extensive application of sensitizers to pulsed CO_2 laser reactions of organic molecules has been reported by the Brandeis University group of Keehn and co-workers (Olszyna *et al.*, 1977; Cheng and Keehn, 1977; Garcia and Keehn, 1978). These workers utilized SiF_4 as an inert sensitizer to isomerize cyclopropane to propene [Reaction (2.49)], allene to methyl acetylene [Reaction (2.50)], convert $CHClF_2$ to $CF_2{=}CF_2$ and HCl [(Reaction (2.51)], and induce a variety of retro-Diels–Alder reactions

[Reactions (2.52)–(2.56)]. The cyclopropane isomerization was studied in most quantitative detail. At typical laser power levels of 0.1–2 MW/cm^2 and

$$\triangle \xrightarrow{SiF_4} CH_3CH{=}CH_2 \quad (2.49)$$

$$CH_2{=}C{=}CH_2 \xrightarrow{SiF_4} CH_3C{\equiv}CH \quad (2.50)$$

$$2\ CHClF_2 \xrightarrow{SiF_4} CF_2{=}CF_2 + 2\ HCl \quad (2.51)$$

$$\xrightarrow{SiF_4} \quad + \ CH_2{=}CH_2 \quad (2.52)$$

$$\xrightarrow{SiF_4} \quad + \ HC{\equiv}CH \quad (2.53)$$

$$\xrightarrow{SiF_4} \ 2 \quad (2.54)$$

$$\xrightarrow{SiF_4} \ 2 \quad (2.55)$$

$$\text{O} \xrightarrow{SiF_4} \text{O} \quad + \ CH_2{=}CH_2 \quad (2.56)$$

fluences of 0.03–0.5 J/cm^2 and SiF_4–reactant pressures of 10–50 Torr, it was found that the cross section of SiF_4 was not constant; i.e., increasing the laser fluence caused a decrease in cross section, and at constant fluence the cross section increased with an increase in either SiF_4 or cyclopropane pressure. The former effect is commonly observed for both small and large molecules, and the phenomenon is discussed in more detail in Section 2.6. The increase in cross section with increasing gas pressure may be attributable to collisional repopulation of the pumped states of SiF_4 (rotational hole filling) or rapid V–V transfer from vibrationally excited SiF_4 to cyclopropane. At 50-Torr total pressure with a 300-nsec laser pulse, a molecule of SiF_4 experiences ~ 100–200 collisions during the pulse. For collisions with relatively large molecules, this number of collisions is almost certainly sufficient to degrade the vibrational excitation initially deposited in the SiF_4 into random thermal energy; i.e., intermolecular V–V and V–T,R relaxation would occur.

Figure 2.11 demonstrates the energy absorbed vs. fluence for various

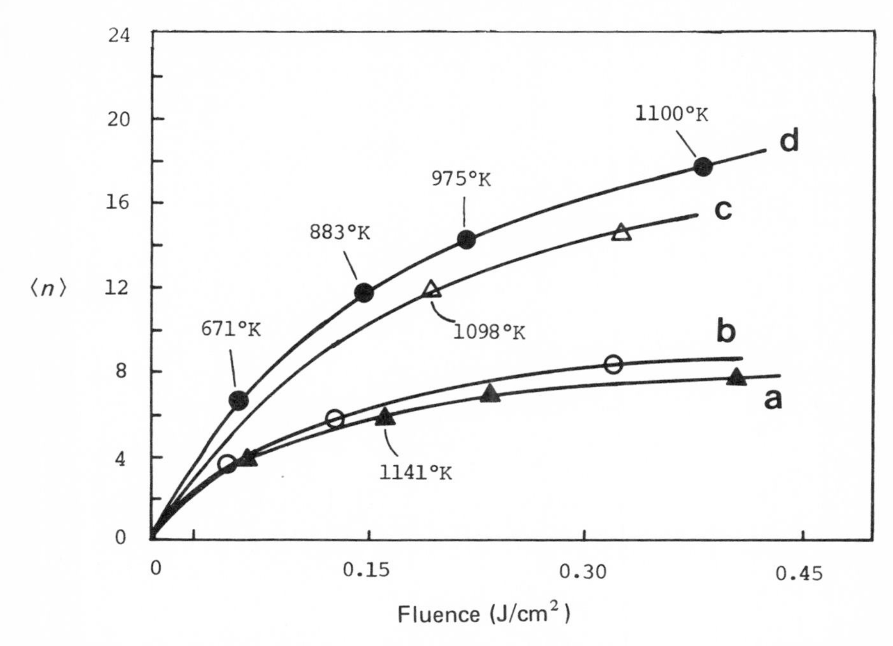

Figure 2.11. Plot of average number of photons absorbed, $\langle n \rangle$, per molecule of SiF_4 (einsteins absorbed per mole of SiF_4) at 1025 cm^{-1} for mixtures of SiF_4 and cyclopropane vs. fluence. (a) 19 Torr SiF_4, no cyclopropane; (b) 20 Torr SiF_4 with 5 Torr cyclopropane; (c) 31 Torr SiF_4 with 30 Torr cyclopropane; (d) 20 Torr SiF_4 with 35 Torr cyclopropane. The temperatures indicated assumed that the absorbed energy was converted adiabatically into thermal energy. The conversion per flash of cyclopropane to propylene increased with temperature for curve (d) as follows: 671°K, 0%; 883°K, 0.08%; 975°K, 3%; 1100°K, 10%. (After Olszyna *et al.*, 1977.)

SiF_4–cyclopropane mixtures (Olszyna *et al.*, 1977). Also shown are the calculated "temperatures" that would be produced if the absorbed energy was converted adiabatically into thermal energy. It can be seen that the temperature can be controlled by varying the SiF_4–cyclopropane pressures or by varying the laser fluence. At a given fluence, increasing the reactant pressure results in a lower effective temperature because of the increased heat capacity of the system. It may be noted that temperatures in excess of 800°C (1100°K) can be achieved at relatively low laser fluences. The percent conversion of cyclopropane scales with temperature as anticipated.

These SiF_4-sensitized reactions appear to be typical high-temperature pyrolyses. No reaction occurred for cyclopropane at low fluences; at intermediate fluences cyclopropane was converted cleanly to propene [Reaction (2.49)]; and at high fluences decomposition of the propylene to methane, ethylene, and acetylene occurred. Likewise, a relatively clean conversion of allene to methylacetylene [Reaction (2.50)] was effected at low fluences, but rather extensive secondary decomposition of methyl acetylene to methane and acetylene was observed at higher fluences. The formation of tetrafluoroethylene from $CHClF_2$ [Reaction (2.51)] presumably occurred via a CF_2 carbene intermediate as in a conventional thermolysis, but pressures > 1 Torr of this transient intermediate were postulated to be generated in the laser process.

The SiF_4-sensitized retro-Diels–Alder Reactions (2.52)–(2.56) also all appeared to resemble typical pyrolytic processes to produce mainly the products indicated. It was observed that many of the pulsed laser-induced transformations were cleaner with respect to minimizing secondary decomposition than conventional thermolyses or CW laser-heated systems provided that relatively low pressures of SiF_4 were employed to avoid excessive vibrational temperatures within the cell. Since the reaction cell never gets warm, the technique was termed "cold pyrolysis."

The most obvious advantage of utilizing a sensitizer as compared to direct laser excitation is that a reactant need not possess a strong absorption band in the wavelength region accessible with currently available lasers. This universitality thus allows any organic molecule with a sufficient vapor pressure to be laser reacted. Garcia and Keehn (1978) have reacted even a solid organic compound, sulfolene, by means of SiF_4 sensitization.

The ability to conduct high-temperature reactions in the absence of hot walls by means of laser radiation is a unique advantage over conventional heating methods shared by both sensitized and direct excitation. Heterogeneous processes occurring at the gas–solid interface in a conventional reactor are almost always complicating and detrimental. Catalytic or quenching effects make kinetic analyses difficult, and the hot walls frequently spawn free radicals or other undesired reactive intermediates that ultimately produce unwanted side products. The homogeneous nature of the infrared laser technique avoids such complications, and this feature might ultimately prove to be one of the most significant advantages, particularly in synthetic applications.

There are several disadvantages of using sensitizers as contrasted to direct excitation. Molecular or isomer selectivity in a mixture of reactants as discussed in Section 2.3.1.1 cannot be expected when a sensitizer is utilized. The sensitization method depends upon collisional energy transfer from the vibrationally excited absorber to the reactant molecule. Although, perhaps in some binary mixtures of small inorganic molecules some sort of selectivity might be achieved because of more nearly resonant V–V energy transfer between the sensitizer and one of the reactants, any such discrimination would be virtually nonexistent with larger organic reactants. In this regard, Grunwald and co-workers (Grunwald *et al.*, 1979; Popok *et al.*, 1979) have demonstrated that direct V–V transfer from laser-excited SiF_4

or $CHClF_2$ to Br_2 is relatively unimportant and that the Br_2 acceptor molecule instead acquires energy from the translationally and rotationally relaxed system. No molecular specificity can be expected under such conditions.

An important question is whether sensitized excitation produces a significantly different *initial* population of excited states as compared to direct pumping. The answer is yes, if one is comparing low-pressure, "collisionless" direct-pumping excitation with a relatively high-pressure collisional sensitized process. The former will produce an organic molecule with a high level of vibrational energy but that will be rotationally and translationally cold. Excitation by multiple energy transfer resulting from collisions between an excited absorber molecule and the reactant molecule will generate a thermalized system in equilibrium at a temperature defined by the amount of energy absorbed and the heat capacity of the system. Although the precise dynamics of such collisional pumping of reactant molecules to highly excited states are not well understood, particularly with respect to the amounts of energy transferred per collision at different levels of excitation of both absorber and acceptor, more than one collision is certainly required. If a maximum of ~ 10–15 kcal/mol of energy is assumed to be transferred per encounter, a very minimum of 4–5 collisions would be required to drive a reactant molecule to a threshold energy of, e.g., 50 kcal/mol. A more realistic estimate might be 10–20 or more collisions since the estimated amount of energy transferred is probably a maximum figure and some of the encounters will deactivate rather than excite the reactant molecules. For a relatively large organic molecule experiencing this many collisions, extensive rotational and translational excitation is inevitable and a thermally equilibrated system will result.

The two different excitation schemes, however, need not lead to a significantly different *final* distribution of vibrational excitation in the reactant molecules. As discussed in some length in Section 2.6, much of the dynamics and chemistry of direct multiphoton excitation of organic molecules reacting by unimolecular mechanisms apparently can be explained quite adequately by a Boltzmann distribution of vibrational excitation. Because of the relatively low rate constants for large organic molecules excited to just above the threshold energy, only a few collisions are necessary to equilibrate the initially deposited vibrational excitation into the rotational and translational degrees of freedom. Of course, the resulting distribution is precisely that anticipated for an excited molecule generated by a multicollisional sensitization process. Moreover, because of the relatively small contribution made by the translational and rotational components to the total heat capacity of a large organic molecule, the distinction between a vibrational-only excited ensemble of molecules and a totally thermalized system may become rather insignificant, at least for unimolecular reactions. Thus, there would be a little difference anticipated for the unimolecular chemical behavior of a molecule excited by either the direct or sensitizer technique. In cases where relative translational energy is important for chemical transformation, sensitized excitation may well prove more efficient in promoting reaction than direct irradiation. This may be the case for many bimolecular reactions as discussed in the previous section. A possible disadvantage of sensitization is that the high, transient, translational temperatures produced may promote secondary reactions, particularly of free radicals.

2.3.4. Low-Intensity CW Infrared Multiphoton Dissociation

Except for the studies discussed in this short section, all of the multiphoton infrared laser-induced results discussed in this chapter have been acquired utilizing high-intensity, pulsed lasers. Since it had been demonstrated for SF_6 (Kolodner *et al.*, 1977) that the dissociation probability in the collisionless regime was independent of the laser pulselength (i.e., the reaction probability was dependent on the laser energy and not the laser power), Beauchamp and co-workers (Woodin *et al.*, 1978, 1979; Bomse *et al.*, 1978, 1979) showed that several organic ions could be induced to react with low-power CW infrared irradiation. Various ions were produced and electromagnetically stored at very low pressures in an ion cyclotron resonance spectrometer and irradiated for up to 2 sec with CW intensities of only 1–100 W/cm^2. Several of the ionic reactions investigated are shown in Reactions (2.57)–(2.59). The diethylether–H$^+$ dimer

$$[(CH_3CH_2)_2O]_2H^+ \xrightarrow{E_a \sim 31\ \text{kcal/mol}} (CH_3CH_2)_2\overset{+}{O}H + (CH_3CH_2)_2O \tag{2.57}$$

$$(CH_3CH_2)_2\overset{+}{O}H \xrightarrow{E_a \sim 27\ \text{kcal/mol}} CH_3CH_2\overset{+}{O}H_2 + CH_2{=}CH_2 \tag{2.58a}$$

$$(CH_3CH_2)_2\overset{+}{O}H \xrightarrow{E_a \sim 53\ \text{kcal/mol}} CH_3\overset{+}{C}H_2 + CH_3CH_2OH \tag{2.58b}$$

$$C_3F_6^+ \xrightarrow{E_a \sim 56\ \text{kcal/mol}} C_2F_4^+ + CF_2 \tag{2.59}$$

shown in Reaction (2.57) could be induced to dissociate with a laser power of only 4 W/cm^2, which may be contrasted to the megawatt power levels usually produced by pulsed TEA CO_2 lasers. The dissociation of $(CH_3CH_2)_2{}^+OH$ could potentially occur by the two channels shown in Reaction (2.58) but, in fact, only the lower E_a channel is driven by 6–8 W/cm^2 CW laser radiation. This behavior is anticipated from the discussion in Section 2.3.1.3 since further laser excitation beyond the lowest energy threshold would be in competition with reaction and would respond to laser power rather than energy; the low-power CW laser would be incapable of exciting the protonated ether to sufficiently high vibrational levels to permit Reaction (2.58b) to become significant. The competitive multiphoton dissociation of $(CH_3CH_2)(CD_3CD_2)^+OH$ exhibited a k_H/k_D isotope effect ≥ 6 indicating that dissociation [Reaction (2.58a)] occurs via a β–H transfer. Such a large isotope effect further supports the conclusion that excitation is very much slower than decomposition above the threshold energy in these low-power CW experiments and that the reaction probably occurs very close to threshold.

The multiphoton dissociation of the dimer $[(CH_3CH_2)_2O]_2H^+$, is considerably more facile than $(CH_3CH_2)_2{}^+OH$, which requires higher laser powers. It appears that the dimer is a large molecule as we have defined in this chapter and would possess $\sim 9 \times 10^4$ states/cm^{-1} after absorbing only two infrared photons. Since the thermal contribution to the internal energy at room temperature is already comparable to excitation by one photon, the absorption of only a single photon would place the dimer in the quasi-continuum. This is not the case for $(CH_3CH_2)_2{}^+OH$ since the absorption of four 1000-cm^{-1} photons would be required to reach a density of states $\sim 2 \times 10^4$/cm^{-1}. This compound, therefore, is more typical of a small molecule. Relatedly, Bomse *et al.* (1979) conducted some model calculations treating the absorption as an incoherent process with a decreasing cross section at higher levels of excitation for the dimer. Although the cross section may decrease at higher fluences for SF_6, we show in Section 2.5.2 that the cross sections for large molecules are much less sensitive to the laser fluence. In Chapter 1 of this volume, Galbraith and Ackerhalt arrive at a rapidly increasing cross section with level of excitation for both SF_6 and S_2F_{10}; the latter fits our definition of a large molecule. However, such behavior of cross sections at high levels of excitation cannot be considered well characterized at this stage, and a decreasing cross section for some molecules may not be unrealistic.

Collisions with added Ar, N_2, or SF_6 were observed to enhance the multiphoton dissociation for $C_3F_6{}^+$ [Reaction (2.59)]. This was interpreted as either collisional enhancement of intramolecular V–V transfer or collisional redistribution of rotational states within the pumped absorption band. Such collisional enhancement is not anticipated for large molecules as discussed above, but $C_3F_6{}^+$ is not a large molecule; a density of states calculation indicated that four quanta of 1000-cm^{-1} infrared radiation were necessary to produce $\sim 2 \times 10^3$ states/cm^{-1}.

The low-intensity CW, ion cyclotron resonance method of investigating the multiphoton absorption/reaction process appears to complement nicely the more common pulsed megawatt infrared technique. Especially attractive are the long confinement and irradiation times that allow for direct monitoring of reactant and product concentrations during laser photolysis. Of course, the inherent low pressure of the ion cyclotron technique does not permit any appreciable scale-up for a possible practical application.

2.4. *Application of Chemical Thermometers in Pulsed Infrared Laser Photochemistry*

2.4.1. *"Thermal" vs. "Nonthermal" Processes*

A problem dating back to some of the earliest infrared multiphoton induced reactions has been whether these reactions should be described as "thermal" or "nonthermal" in nature. A part of the problem is defining precisely what is meant by the terms "thermal" and/or "heat." It has been demonstrated that the pulsed infrared laser need not function as a conventional heat source in the sense that it is capable of "heating" a single component in a mixture, a capability not shared by a flame, heating mantle, shock tube, etc. The multitude of successful infrared laser isotope separation schemes and the molecular selectivity results discussed earlier have led to acceptance of the fact that a pulsed infrared laser is certainly capable of producing results quite different from normal heating. We will describe the use of "chemical thermometers" to monitor for "thermal" effects of this sort in this section.

It is worthwhile to consider the differences between exciting molecules with a short, intense pulse of infrared photons and heating the molecules in a conventional manner. When a chemical is heated, energy is imparted from the hot walls of the reaction vessel first into translational motion of the molecules, i.e., the molecules travel faster. Collisions among molecules result in some of the translational energy becoming rotational excitation; i.e., the molecules begin to spin in space. Finally, the vibrational energy induces the atoms within the molecules to vibrate faster. At this stage one has an ensemble of translationally, rotationally, and vibrationally excited molecules with various energies characterized by a Boltzmann distribution appropriate for the given temperature. Although the total energy of a molecule is the sum of its component types of energy, it is the vibrational energy that leads to unimolecular reactions since interatomic bonds break only when they have vibrational excitation in excess of the threshold energy. It is precisely this type of excitation that is achieved last by conventional heating methods.

On the other hand, pulsed infrared laser excitation produces a vibrationally excited molecule that is rotationally and translationally unexcited. In Figure 2.12 an attempt is made to depict the various stages of excitation

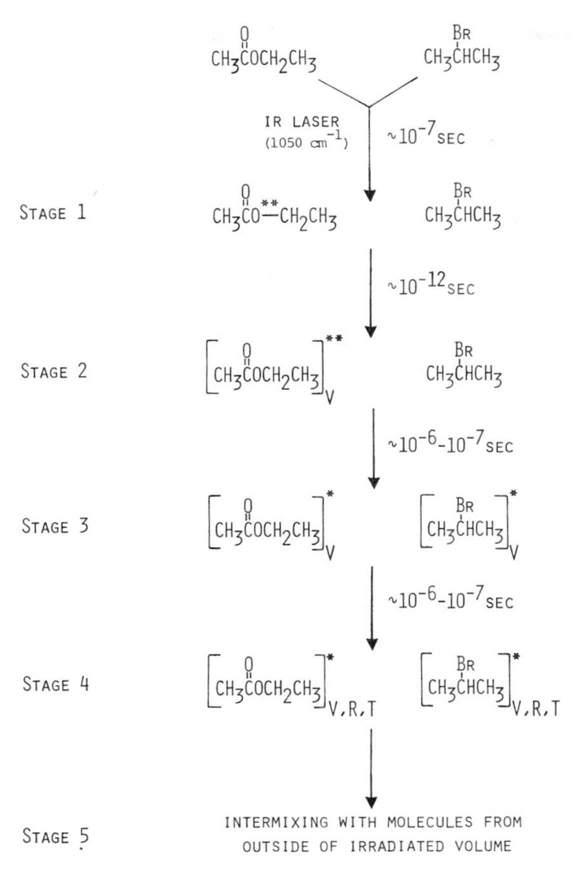

Figure 2.12. Various stages of excitation experienced by multiphoton excitation of ethyl acetate in the presence of isopropyl bromide at 1-Torr total pressure. Stage 1: selectively excited O–CH_2 stretching mode; Stage 2: vibrational excitation statistically distributed intramolecularly throughout ethyl acetate molecule; Stage 3: vibrational excitation distributed intermolecularly between ethyl acetate and isopropyl bromide; Stage 4: both types of molecules within the irradiated volume vibrationally, rotationally, and translationally equilibrated; Stage 5: intermixing of excited molecules from irradiated volume with surrounding cold molecules.

experienced by ethyl acetate excited by infrared laser radiation in the presence of isopropyl bromide at 1-Torr total pressure. At the top of the figure is pictured an unexcited molecule of ethyl acetate that is subjected to a 100-nsec (1×10^{-7} sec) infrared laser pulse in Stage 1 that is tuned to the O–CH_2CH_3 stretching motion at ~ 1050 cm^{-1}. This frequency corresponds to $\sim 3 \times 10^{-14}$ sec per vibration that would also reflect the fastest possible rate of reaction of the molecule. It is seen that this bond experiences $\sim 10^7$ vibrations *during* the laser pulse. Stage 2 depicts the intramolecular randomization of vibrational energy from the initially pumped $O-CH_2CH_3$ stretching mode into other vibrational motions of the molecule. There is evidence from chemical activation studies that this occurs on the order of $\sim 10^{-12}$ sec although, as discussed in Section 2.3.1.4, the rate for a multiphoton laser excitation process is of some controversy. If this rate is correct, it may be noted that intramolecular energy randomization occurs $\sim 10^5$ times faster than energy deposition during the laser pulse to produce a molecule that is vibrationally excited, presumably in a statistical manner; i.e., Stage 2 is produced *during* the laser pulse.

At 1 Torr pressure, an excited molecule will collide with another molecule once every 10^{-7} sec, or so. Collisional transfer of vibrational energy from the initially excited molecule to another molecule will occur in $\sim 10^{-6}$–10^{-7} sec assuming that such transfer occurs at a minimum of once every 10 collisions. The likelihood of transfer and the amount of energy actually exchanged will depend upon the nature of the two colliding molecules. Although at low levels of excitation, intermolecular V–V transfer to produce Stage 3 is usually more rapid than V–T,R relaxation, for large molecules excited to vibrational levels near or exceeding the threshold, V–R and V–T transfer is probably as rapid as V–V transfer. Since the rotational and translational components comprise only a small fraction of the total heat capacity of a large molecule, these degrees of freedom are probably rapidly equilibrated by collisions to produce Stage 4 directly from Stage 2. Thus, after $\sim 10^{-6}$–10^{-7} sec all the molecules in the irradiated volume are expected to be vibrationally, rotationally, and translationally equilibrated with an energy distribution most likely characterized by a Boltzmann function. A last stage involving intermixing and energy transfer from excited molecules within the irradiated volume with other molecules in the cell may also be considered as Stage 5. This is the type of distribution to be anticipated by conventional heating of a sample of ethyl acetate.

It is apparent from Figure 2.12 that there are at least five different stages or degrees of excitation that a molecule might traverse when exposed to a short pulse of infrared radiation, although the intermediacy of Stage 3 is probably unlikely for large molecules. At one extreme would be mode-selective excitation where reaction would occur before *intramolecular* energy relaxation took place. There are some data that have been interpreted in this manner (see Section 2.3.1.4), but most of the work in the field of infrared photochemistry support intramolecular randomization prior to reaction. The other extreme, complete V,T,R equilibration among all the molecules in the sample, (Stage 5) and even the irradiated volume

(Stage 4) can certainly be achieved in a pulsed infrared process at high pressures but can be avoided by working at low pressures as evidenced by successful isotope separation and other molecule-selective processes.

But, exactly what constitutes a high-pressure experiment: 10 Torr? 1.0 Torr? 0.01 Torr? How low in total reactant pressure must one work in order to avoid significant intermolecular V,T,R relaxation? If one assumes that, at low pressures, the laser produces an intramolecularly relaxed, vibrationally excited molecule (Stage 2), the question is whether reaction occurs from that stage or whether collisional processes will give significant *intermolecular* energy transfer prior to reaction (Stage 3, 4, or 5).

To address the problem raised at the beginning of this section, of "thermal" vs. "nonthermal" laser reactions, we choose, for unimolecular reactions, the point at which vibrational energy is equilibrated among all molecules in the irradiated volume (Stage 3) as the differentiation point between nonequilibrium, laser-induced chemistry and thermal chemistry. Thus, if reaction occurs *prior* to Stage 3 in Figure 2.12, the reaction is considered a nonequilibrium laser process, whereas, if it occurs subsequent to intermolecular energy transfer, then it is equivalent to a simple thermal reaction. It is important to note that this definition applies only to unimolecular transformations for which vibrational excitation is both necessary and sufficient for reaction. The definition differs from the usual concept of "heat" which implies a totally equilibrated system of translational and rotational energy in addition to vibrational energy (Stage 5).

The remainder of this section will be devoted to a discussion of the use of chemical monitor molecules to differentiate between an unique, nonequilibrium laser reaction and a simple thermal process as defined herein.

2.4.2. *Choice of Thermal Monitor Molecule*

A relatively simple, but effective, means of distinguishing between a nonequilibrium unimolecular laser process and a thermal reaction as defined above is simply to add to the reaction cell a molecule that does not absorb the laser radiation but that can undergo a unimolecular thermal reaction. Such a technique might be termed "chemical thermometry." If pulsed irradiation of the absorbing reactant molecule results in chemical transformation of the reactant but not of the monitor molecule, then a true laser-specific, nonequilibrium process has been achieved. If, however, irradiation of the reactant also produces chemical change in the monitor molecule, then intermolecular V–V (Stage 3) and, probably, V–T,R energy transfer from the reactant molecule has occurred (Stage 4) and a thermal reaction as we have defined has ensued.

A useful monitor molecule should possess the following characteristics:

1. not react chemically with the reactant molecule;
2. be nonabsorbing at the infrared wavelength employed to excite the reactant molecule;
3. produce an easily detected, nonabsorbing (at the working wavelength) molecular product upon reaction;
4. possess an Arrhenius E_a and A factor very similar to the reactant compound;
5. have a minimum of the problems discussed in Section 2.4.3.

Characteristics 1–3 are obvious and require no elaboration. The necessity of the thermal monitor to possess Arrhenius parameters similar to the target molecule may not be readily apparent. Note that our working definition of a thermal reaction requires that the vibrational energy reaches *equilibrium* among all molecules in the irradiated volume. For relatively large organic molecules, it is unrealistic to expect that absolutely no vibrational energy be transferred to the monitor molecule during the experiment except at extremely high levels of vibrational excitation that would cause all molecules to react during the laser pulse or in a molecular-beam-type experiment. Even at relatively low pressures of 0.10 Torr or less that are frequently cited as being in the "collisionless regime" (no collisions during, e.g., a 100-nsec pulse) some intermolecular processes will always come into play after the laser pulse but before all of the excited molecules have reacted or had the opportunity to traverse to the cell walls that are the ultimate repository of the energy contained within the laser pulse. In the extreme, an exceptionally thermally labile monitor molecule might be employed that would undergo extensive reaction even though it may have experienced only one or a few collisions with vibrationally excited target molecules and may not be in true equilibrium with the remainder of the molecules in the irradiated volume. It is plausible that the target reactant molecule might even behave as an inert sensitizer for such a monitor molecule at a level of excitation below that necessary for unimolecular reaction of the former. Although such a chemical thermometer would demonstrate that collisional energy transfer was occurring sometime during the experiment, it would not address the question of whether or not thermal equilibrium had set in. On the other extreme, the monitor molecule obviously cannot be so thermally stable with respect to the reactant that the latter might react in even an equilibrated system without the monitor becoming sufficiently activated to undergo transformation.

The necessity for having similar thermal reaction rates for the target and monitor molecules should be apparent after the above discussion. The need to have them as similar as possible can be appreciated when one considers that the actual temperature attained in the irradiated volume during a megawatt pulsed infrared laser experiment may not be known with certainty. Therefore, the thermal reaction rates for both the reactant and monitor molecules must respond in the same way to changes in temperature. This can be expected only when the unimolecular E_a and A factors for both chemicals are the same. Table 2.1 illustrates the effect of temperature on rate for two different systems. With ethyl acetate as target molecule and isopropyl bromide as monitor, a good combination is achieved since the rate ratio is virtually independent of temperature over a wide range. The other system shown is a poor combination in which the ratio varies strongly with temperature.

We have demonstrated the use and advantages of using a chemical thermometer molecule in our early study of the CO_2 laser-induced reaction of ethyl acetate (Danen *et al.*, 1977). We have shown that ethyl acetate reacted cleanly when pulsed at 1046.9 cm^{-1} to produce acetic acid and ethylene [Reaction (2.7)]. The elimination of HBr from isopropyl bromide [Reaction (2.11)] was used as the monitor reaction to evaluate thermal effects (see Figure 2.3). For a 1 : 3 mixture of isopropyl bromide and ethyl acetate, the expected product ratio is ~ 4 for an equilibrium thermal process (see Table 2.1). Experimentally, we observed a ratio of 4 ± 1 from a mixture of 5 Torr of isopropyl bromide and 15 Torr of ethyl acetate irradiated at 0.8 J/cm^2. In the absence of ethyl acetate, isopropyl bromide did not undergo any reaction when irradiated at 1046.9 cm^{-1}, an obvious requirement for a monitor molecule.

As a check on the thermal ratio obtained from the Arrhenius parameters, a sensitized experiment utilizing SF_6 as the thermal sensitizer was conducted. Irradiation of 10 Torr of SF_6 and 10 Torr of a 1 : 3 mixture of isopropyl bromide and ethyl acetate at 945.0 cm^{-1} at which only the SF_6 absorbed produced a ratio of 5.5 ± 1. These experiments demonstrated that the majority of ethylene produced from irradiation of 20 Torr samples

Table 2.1. Calculated Ratio of Rate Constants vs. Temperature for Two Reactant–Monitor Combinations

Temperature, °K	$\frac{k_{(EtOAc)}}{k_{(i\text{-}PrBr)}}$	$\frac{k_{(sec\text{-}BuOAc)}}{k_{(t\text{-}BuCl)}}$
300	0.069	0.021
500	0.080	0.063
800	0.086	0.116
1000	0.088	0.141
2000	0.093	0.211

of ethyl acetate under nonfocused conditions arose from thermal processes rather than from any selective, nonequilibrium laser inducement since the sensitized experiment would generate a thermally equilibrated system.

It was necessary to focus the laser beam in order to effect nonequilibrium laser-induced chemistry prior to collisional redistribution of energy at pressures of 5 Torr or greater. The effect produced is displayed by Figure 2.3. At low energies of irradiation, the propylene/ethylene ratio for 5 Torr of a 1 : 3 mixture of isopropyl bromide and ethyl acetate was near the thermal value of ~ 4 as determined by the sensitized experiment noted above. The ratio dropped dramatically as the laser fluence was increased; at 8 J/cm^2, the ratio was only 0.5. These experiments demonstrated that, as the laser fluence was increased, the nonequilibrium laser-induced pathway became increasingly important.

The product ratio was also monitored as a function of reactant pressure at a constant laser fluence of 2.1 J/cm^2. It was observed that, as the number of collisions per second was reduced, the rate at which thermal equilibration was reached was reduced, as anticipated.

As a result of the similarities of Arrhenius parameters, the percent of ethylene produced via the laser-induced reaction and that produced via the thermal process was calculated as follows for low percentage conversion per flash (CPF). It was assumed that the CPF(ethylene)/CPF(propylene) ratio could be partitioned as shown by Equation (2.60); all values of the

$$\begin{aligned}\frac{\text{CPF(ethylene)}}{\text{CPF(propylene)}} &= \frac{[\text{CPF(ethylene)}]_{\text{thermal}} + [\text{CPF(ethylene)}]_{\text{laser}}}{\text{CPF(propylene)}} \\ &= \frac{[\text{EtOAc}][3.89 \times 10^{12} \exp(-48{,}000/RT)]}{[i\text{-PrBr}][3.98 \times 10^{13} \exp(-47{,}800/RT)]} \\ &\quad + \frac{[\text{CPF(ethylene)}]_{\text{laser}}}{\text{CPF(propylene)}}\end{aligned} \tag{2.60}$$

equation were measured experimentally except the $\text{CPF(ethylene)}_{\text{laser}}$. For a typical experiment with a laser intensity of 2.1 J/cm^2 and 5-Torr total reactant pressure, the observed $\text{CPF(ethylene)}_{\text{total}} = 19\%$ and $\text{CPF(propylene)} = 28\%$. For a thermal rate constant ratio of 0.08 (Table 2.1), Equation (2.60) yielded a $\text{CPF(ethylene)}_{\text{laser}} = 12\%$ indicating that 64% of the ethylene was produced via the nonthermal, nonequilibrium laser-induced pathway under these conditions. As discussed in the next section, essentially 100% nonequilibrium chemistry can be achieved at sufficiently low pressures (usually < 0.10 Torr).

The above should suffice to demonstrate the utilization of a chemical thermometer molecule to probe whether intermolecular vibrational equilibration has occurred for a laser-induced unimolecular chemical reaction.

Within the limitations discussed in the next section, this technique is seen to be quite powerful in delineating the extent of nonequilibrium chemistry taking place within the reaction cell. Moreover, it is a general technique that can be applied to virtually any unimolecular reactant providing an appropriate monitor molecule can be obtained. The latter is always more difficult to achieve in practice than in theory. As discussed, a necessary condition is that the monitor must have Arrhenius parameters quite similar to the reactant. One can frequently choose an isomer of the reactant or at least a compound of the same functionality. The extensive, critical compilation of unimolecular rate parameters by Benson and O'Neal (1970) or the text by Robinson and Holbrook (1972) are very valuable in selecting a suitable monitor; the series in "Gas Kinetics and Energy Transfer," volumes 1–3 (Robinson, 1975; Quack and Troe, 1977; Frey and Walsh, 1978) attempt to update continually these compilations. One can experimentally check to determine whether the selected monitor and reactant molecules exhibit sufficiently similar rate behavior with respect to temperature change by conducting a laser-sensitized experiment. The product ratio from the monitor and reactant should not vary significantly from conditions of low laser fluence and low sensitizer pressures to high fluences and relatively high sensitizer pressures. By measuring the absorbed energy it is a simple matter to calculate from the heat capacity of the system a crude effective temperature produced by the laser. By proper choice of laser fluence and sensitizer pressure, it is quite easy to achieve effective temperature differences of 500°K or more. If the product ratio is insensitive to such temperature changes, one can be assured that the Arrhenius parameters are sufficiently well matched to utilize the chosen compound as a chemical thermometer for that particular reactant.

Of course, closely matched Arrhenius parameters are not critical if one wishes only to determine that there is not a significant thermal component to a particular laser-induced reaction. For such purposes, all that is required is that the monitor molecule undergo reaction at a comparable or faster rate than the reactant at all realistic laser-generated temperatures (typically, 800–1500°K). If transformation of the monitor occurs, one can assume that, at least, some intermolecular relaxation of absorbed laser energy has taken place on the time scale of the reaction, but a quantitative assessment of the amount of intermolecular relaxation would normally not be possible since the effective temperature achieved would not be known with certainty.

In this manner, Danen (1979) utilized 2-propanol to monitor for thermal effects in the laser-induced dehydration of ethanol to ethylene. Heating a 1 : 1 mixture of ethanol and 2-propanol or selectively irradiating the 2-propanol resulted in the formation of propylene as the greatly predominant product. The selective excitation/reaction of ethanol using 2-propanol as the monitor restricted the total pressure to < 0.05 Torr to assure nonthermal, nonequilibrium reaction with a minimal thermal component. Approximately equal amounts of ethylene and propylene were produced at 1.0 Torr, and the latter became almost the exclusive product at higher pressures.

2.4.3. Complications in Utilizing Chemical Thermometers

2.4.3.1. Deactivation of Excited Reactant by Monitor Molecule

In spite of the obviously useful information obtained concerning whether a particular laser-induced process is nonequilibrium or thermal in nature, the chemical thermometer technique is not without complications. As defined above, a thermal reaction is assumed to take place when the deposited vibrational energy has reached equilibration among both reactant and monitor molecules within the irradiated volume. As alluded to earlier, the ultimate nonequilibrium process would be that for which the reactant molecule was excited and underwent reaction without experiencing any collisions. Although this is possible in a molecular beam or for high-fluence, low-pressure static cell conditions that would drive the majority of absorbing reactant molecules well beyond the threshold level resulting in significant reaction *during* the laser pulse, it is not realistic for the more common experimental conditions that produce low to moderate ($< 10\%$) conversions of reactants to products in bulb experiments. Large organic molecules are expected to undergo a relatively slow unimolecular transformation, and the bulk of reaction occurs *after* the laser pulse has terminated. Even though a molecule may be excited beyond its threshold energy, its mean reaction rate can be slow and the molecule can have a significant lifetime during which it might be collisionally deactivated. Therefore, a monitor molecule can affect the extent of the primary reaction even though it, itself, might not ever become sufficiently excited to react.

We have extended our earlier studies (Danen *et al.*, 1977) to address this problem in a more quantitative manner (Danen *et al.*, 1980). Our earlier work was at relatively high pressures for which collisional effects are expected to be of prime significance. From Figure 2.3 it may be noted that, for a 5-Torr mixture of a 3 : 1 mixture of ethyl acetate and isopropyl bromide, the CPF(propylene)/CPF(ethylene) ratio asymptotically levels off to ~ 0.5 at high fluences but does not seem to approach zero. We have extended these studies to much lower pressures and observed that a completely selective, nonequilibrium reaction can, indeed, be effected for a 3 : 2 mixture of ethyl acetate–isopropyl bromide irradiated with a fluence of 9

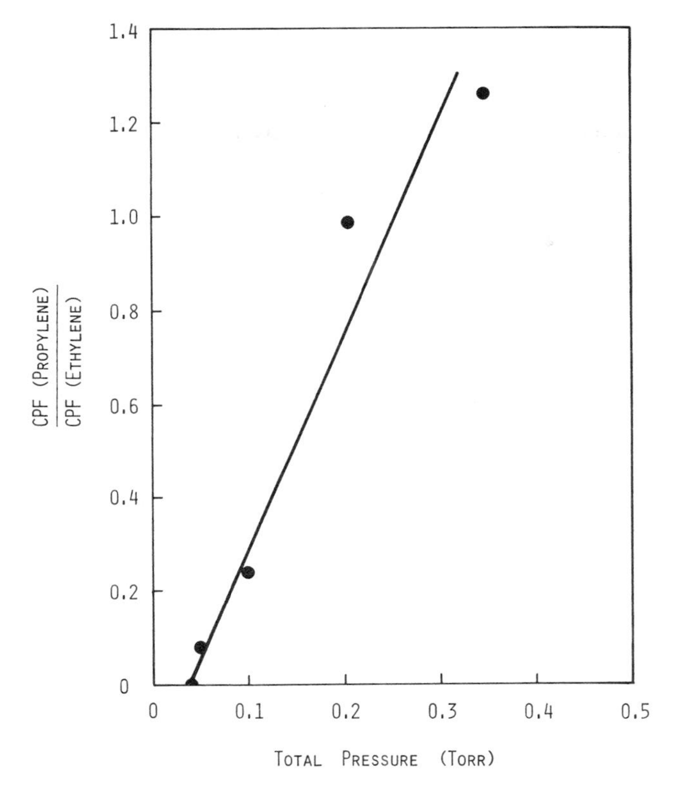

Figure 2.13. CPF(propylene)/CPF(ethylene) vs. total reactant pressure for a 3 : 2 mixture of ethyl acetate and isopropyl bromide. Ethyl acetate was selectively excited at 1046.85 cm^{-1} [$P(20)$ of 9 μ band] with a fluence of 9 J/cm^2.

J/cm^2 at total reactant pressures of 0.04 Torr or less (Figure 2.13). Similar data (Figure 2.14) were obtained for 1 : 1 mixtures of *sec*-butyl acetate [Reaction (2.61)] monitored by *tert*-butyl chloride [Reaction (2.62)]. The latter system was investigated at several fluences and the results demonstrated no strong dependence of thermal effects upon fluence for pressures below ~0.1 Torr. This contrasted with the behavior at higher

$$CH_3\overset{\overset{\displaystyle O}{\|}}{C}O\underset{\underset{\displaystyle CH_3}{|}}{C}HCH_2CH_3 \longrightarrow CH_3CO_2H + CH_2{=}CHCH_2CH_3 + CH_3CH{=}CHCH_3 \; (cis\text{–}trans) \tag{2.61}$$

$$(CH_3)_3CCl \longrightarrow (CH_3)_2C{=}CH_2 + HCl \tag{2.62}$$

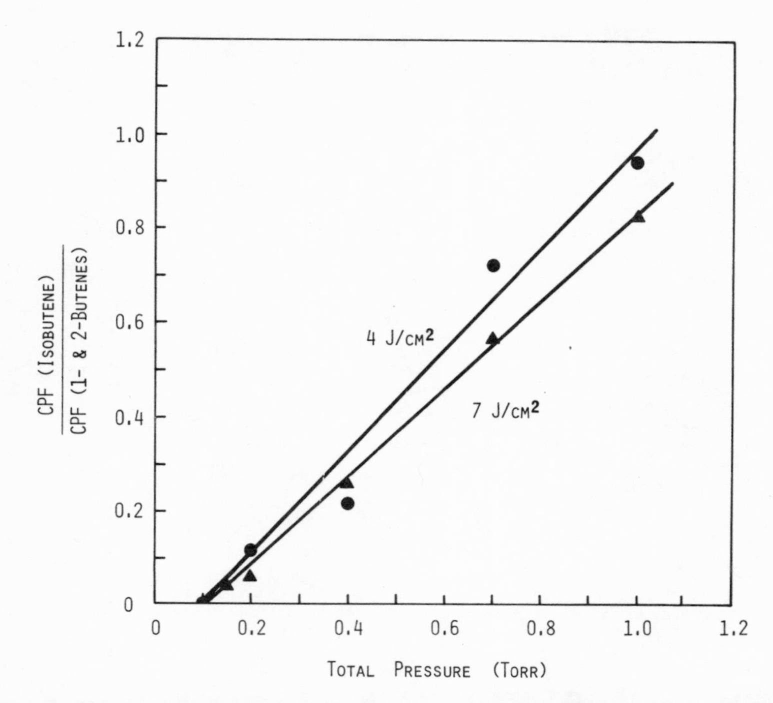

Figure 2.14. CPF(isobutene)/CPF(1-butene + *cis*- and *trans*-2-butene) vs. total reactant pressure for a 1 : 1 mixture of *sec*-butyl acetate and *tert*-butyl chloride. *Sec*-butyl acetate was selectively excited at 1029.44 cm^{-1} [$P(38)$ of 9 μ band] with fluences of 4.0 and 7.0 J/cm^2.

pressure (Figure 2.3) in which a strong dependence on fluence was observed.

Figures 2.13 and 2.14 illustrate that there is a strong effect of total pressure on the amount of the thermal component and that one must utilize pressures of less than ~0.05 Torr to avoid significant intermolecular energy relaxation. It should be noted, however, that we are observing only the onset of thermal equilibration at these low pressures. For the conditions of Figure 2.13, for example, a thermally equilibrated system would exhibit a propylene/ethylene ratio of ~7.3; total equilibration is, therefore, not approached even at ~0.4 Torr pressure.

There is significant collisional transfer of energy from laser excited ethyl acetate molecules to isopropyl bromide monitor molecules even at ~0.1-Torr total reactant pressure. Under even these relatively low pressures some of the monitor molecules must acquire in excess of ~48 kcal/mol excitation and react. A greater number certainly acquire lesser amounts of energy but an insufficient amount to transcend the threshold. This energy must be coming from the excited reactant molecules and should be manifested in lowered conversions per flash for ethyl acetate.

Table 2.2. Effect of Isopropyl Bromide on the Yield and Selectivity of 0.02-Torr Ethyl Acetate Irradiated at 1046.9 cm^{-1} with 2.0 J/cm^2. These Data are Plotted in Figure 2.15

i-PrBr (Torr)	Reaction probability	$\frac{Y_0}{Y}$ [a]	$\frac{[\text{propylene}]}{[\text{ethylene}]}$
0.00	0.19	1.0	0.00
0.03	0.10	1.7	0.0005
0.04	0.08	2.0	0.05
0.10	0.04	4.1	0.13
0.11	0.03	4.6	0.25
0.18	0.02	9.1	0.78
0.23	0.01	13.5	1.37
0.29	0.003	40.7	1.65

[a] Y_0 is reaction yield without added isopropyl bromide; Y is reduced yield at a given pressure of added isopropyl bromide.

This was, indeed, observed. Tables 2.2 and 2.3 and Figure 2.15 illustrate the effect of increasing amounts of isopropyl bromide on the laser-induced reaction of 0.02 Torr of ethyl acetate at two different laser fluences. Relatively low pressures of the monitor molecule can have a significant effect on the yield of the ethyl acetate reaction even though little of the monitor actually reacted. The effect was especially pronounced at low laser fluences. For example, with 0.02 Torr of acetate and 2.0 J/cm^2, 0.04 Torr of isopropyl bromide reduced the acetate reaction yield by half ($Y_0/Y = 2.0$) although the [propylene]/[ethylene] ratio was only 0.05. If the system was

Table 2.3. Effect of Isopropyl Bromide on the Yield and Selectivity of 0.02-Torr Ethyl Acetate Irradiated at 1046.9 cm^{-1} with 3.6 J/cm^2. These Data are Plotted in Figure 2.15

i-PrBr (Torr)	Reaction probability	$\frac{Y_0}{Y}$ [a]	$\frac{[\text{propylene}]}{[\text{ethylene}]}$
0.00	0.78	1.0	0.00
0.06	0.76	1.0	0.03
0.18	0.53	1.4	0.05
0.61	0.36	2.0	0.18
1.20	0.17	4.3	1.10
2.10	0.08	8.7	6.29
3.14	0.04	16.7	37.35

[a] Y_0 is reaction yield without added isopropyl bromide; Y is reduced yield at a given pressure of added isopropyl bromide.

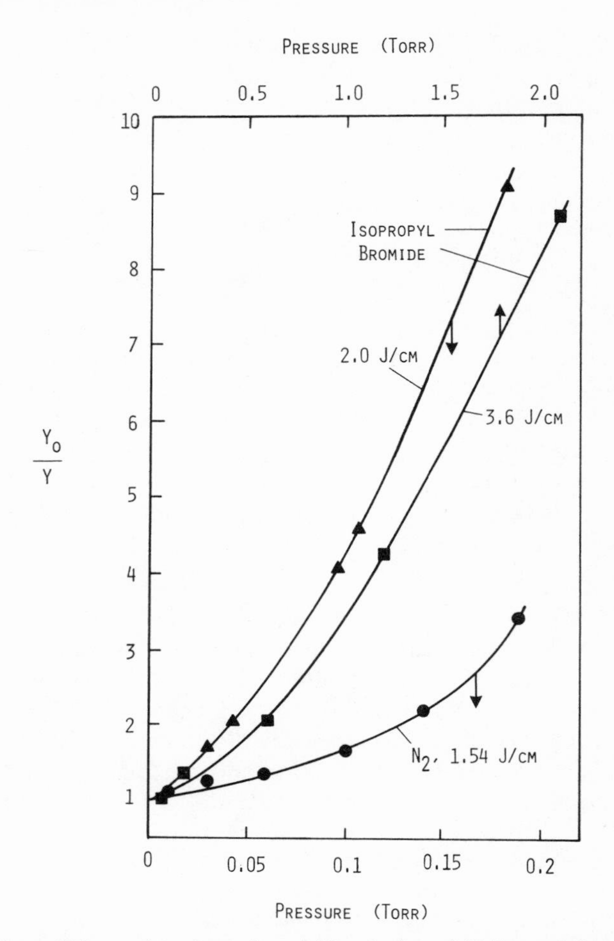

Figure 2.15. Stern–Volmer plot of reaction yield ratio from 0.02-Torr ethyl acetate without added nitrogen or isopropyl bromide (Y_0) to that with bath gas (Y) vs. pressure of added gas. ● nitrogen bath gas (bottom scale) and fluence of 1.54 J/cm^2; ▲ isopropyl bromide bath gas (bottom scale) and fluence of 2.0 J/cm^2; ■ isopropyl bromide bath gas (top scale) and fluence of 3.6 J/cm^2.

totally equilibrated, this ratio would be ~ 25. Increasing the monitor pressure to 0.23 Torr reduced the acetate yield by 13.5-fold, but the [propylene]/[ethylene] ratio was only 1.37; the equilibrium value would be ~ 140 at these relative concentrations. At higher fluences, a greater amount of isopropyl bromide was required to produce a similar effect (Table 2.3). For example, to decrease the primary reaction yield by twofold at a fluence of 3.6 J/cm^2, an isopropyl bromide pressure of 0.61 Torr was required. The [propylene]/[ethylene] ratio under these conditions was only 0.18, which compared with a totally equilibrated value of ~ 370.

Figure 2.15 displays the quenching effects graphically by means of Stern–Volmer plots in which Y_0 is the reaction yield without added isopropyl bromide or nitrogen, and Y is the reduced yield at a given pressure of the added gas. It is seen that isopropyl bromide was much more effective at quenching the reaction of ethyl acetate than was nitrogen. Moreover, there was no enhancement of reaction with addition of inert gas as is frequently observed for small molecules as a result of collisional hole filling (Quick and Wittig, 1978b; Jang and Setser, 1979).

These data are quite easily rationalized. At low laser fluences, the mean level of excitation of ethyl acetate is less than at higher fluence and the mean reaction rate is lower for those molecules excited beyond the threshold. Much of the reaction must be occurring after the laser pulse has terminated. The long mean lifetime of those molecules allows for more deactivating collisions with isopropyl bromide monitor molecules. Few individual isopropyl bromide molecules, however, ever achieve a sufficient level of excitation to react. Thus, the thermal monitor is seen to be able to affect the reaction yield rather profoundly even though it is not apparent from the monitor/reactant product yield ratio.

The above results demonstrate that collisional deactivation of ethyl acetate by the isopropyl bromide monitor is very significant even at total pressures of ~ 0.10 Torr. Figures 2.3 and 2.13 depict results utilizing mixtures comprised of 25 and 40% isopropyl bromide, respectively. Such large mole fractions of the monitor compound contribute significantly to the overall heat capacity of the reacting systems and assure considerable quenching of the ethyl acetate reaction as demonstrated. Under such conditions the isopropyl bromide realistically cannot be considered a thermal *monitor* since its presence in such large amounts exerts a significant perturbation on the system. For true monitoring of only intermolecular energy transfer only small amounts of monitor can be utilized so that the heat capacity of the system is not altered.

We have conducted several studies utilizing low pressures of a 97 : 3 ethyl acetate–isopropyl bromide mixture. Table 2.4 lists pertinent results; $P_{(i\text{–PrBr})}$ is the reaction probability per pulse for the monitor molecule. Several interesting features emerge from an analysis of the data.

1. The amount of isopropyl bromide that reacted in entries 1, 2, and 4 was far from that for a totally equilibrated system.

2. A comparison of entries (2) and (3) reveals that a fourfold increase in pressure resulted in a greater than 50-fold increase in yield of propylene from the thermal monitor. This parallels similar effects as discussed earlier.

3. Entries (1) and (2) demonstrate that an increase in laser fluence at constant pressure caused a significantly greater amount of reaction of the thermal monitor. A corresponding increase in the reaction yield of ethyl

Table 2.4. Effect of Varying Pressure, Fluence, and Laser Beam Diameter for the Irradiation of a 97 : 3 Mixture of Ethyl Acetate : Isopropyl Bromide

Entry	Pressure (Torr)	Fluence (J/cm^2)	Beam diameter (mm)	$P_{(i\text{-PrBr})}$ [a]	[propylene]/[ethylene] [b]
1	0.05	1.35	7.5	0.00008	0.0026
2	0.05	2.49	7.5	0.0011	0.0065
3	0.20	2.49	7.5	0.056	0.29
4	0.05	1.35	19.1	0.0021	0.0072

[a] Reaction probability of isopropyl bromide monitor.
[b] Ratio calculated for a totally equilibrated system ~ 0.4.

acetate was observed, however, and the propylene/ethylene ratio did not change appreciably. The increased reaction probability for isopropyl bromide merely reflects the increased energy input into the system, some of which is collisionally transferred to the monitor in competition with the cooling processes.

4. Probably the most unexpected observation was the identification of a geometric effect upon the yields from the monitor molecule. Entries (1) and (4) are identical in pressure and laser fluence but entry 1 has a much smaller irradiated volume. The diameters of the laser beams were 7.5 and 19.1 mm for entries 1 and 4, respectively. Although the photon flux within the irradiated volume was identical and the reaction probability was corrected to take into account only the irradiated volume, the yield of propylene for the smaller beam was ~ 25-fold less than for the larger beam. This effect can be rationalized by appreciating that the laser pulse generates a cylinder of vibrationally excited ethyl acetate molecules surrounded by a sea of cold molecules. The small number of monitor molecules within the hot cylinder collide with excited reactant molecules and begin to acquire the necessary energy to react but always in competition with various cooling processes. The latter are undoubtedly a complex interplay of diffusion, momentum transfer, conduction, etc., but a small, pencil-sized cylinder will dissipate energy to the surrounding cold bath more quickly than a larger activated cylinder.

The reaction probability of pure ethyl acetate (no monitor molecule) does not appear to be as seriously affected by varying the diameter of the laser beam. It might have been anticipated that cooling processes would be more effective for a small diameter beam resulting in a lower reaction probability than with a larger irradiated volume. Although there appears to be a trend toward lower reaction probability with smaller beam diameters, the effect is not nearly so pronounced as for the reaction probability of the monitor molecule as shown in Table 2.4. Of course, a fundamental difference is that nearly all the reactant ethyl acetate molecules within the irradiated cylinder are excited to some extent, and collisions following the laser pulse will more or less maintain the presumably Boltzmann-like distribution. Therefore, the lack of a geometric effect for the neat sample may result from a competition between the ubiquitous cooling processes and the possibility of an ethyl acetate molecule excited to some level near threshold acquiring sufficient additional energy to react via collision within the irradiated volume. Collisional heating of a totally cold monitor molecule with energy sufficient to transcend the threshold would not be expected to be as competitive with intermixing of molecules from the irradiated volume and the surrounding cold bath.

2.4.3.2. Possible Photodecomposition of Monitor Molecule

An obvious prerequisite for a suitable monitor molecule is that it not absorb the infrared laser radiation utilized to excite the reactant. Verification that irradiation in the absence of reactant at the frequency used to pump the reactant results in no reaction of the monitor is easy to demonstrate experimentally. However, this does not assure that the monitor molecule is photochemically unreactive during the irradiation of the reactant. Indeed, Tsang and co-workers (Tsang *et al.*, 1978) have postulated that isopropyl bromide, **11**, was photodecomposed while attempting to act as a thermal monitor for ethyl acetate, **10**, at relatively high pressures. As discussed in Section 2.3.1.2a and depicted in Figure 2.4, excitation of ethyl acetate resulted in the maintenance of a fairly high degree of specificity even at helium pressures of several hundred Torr. This result can be rationalized in terms of the helium behaving as a heat sink preventing transfer of excitation from the pumped ethyl acetate to the isopropyl bromide monitor. The inverse effect of helium at low pressures, however, was unanticipated (Figure 2.4). The authors accounted for the decrease in selectivity at low helium pressures as resulting from collisional excitation of the monitor by energy transfer from excited ethyl acetate molecules or helium to isopropyl bromide. The latter, thus, was brought into the quasicontinuum where it absorbed the laser radiation and photodecomposed.

The authors proposed a simple two-level kinetic model involving a low-level bottleneck that semiquantitatively accounted for the major observations [Scheme (2.63); **10** = ethyl acetate, **11** = isopropyl bromide, M = **11** or He]. The key feature of this model is that **11** does not decompose via a simple thermal process but through a photodecomposition ($k'_{\text{laser reaction}}$) from a collisionally generated vibrationally excited **11***. The bottleneck is created by differences in the two pumping processes, $k_{\text{laser pump}}$

$$
\begin{array}{ccc}
CH_3CO_2H + CH_2{=}CH_2 & & HBr + CH_3CH{=}CH_2 \\
\uparrow k_{\text{laser reaction}} & & \uparrow k'_{\text{laser reaction}} \\
\left[CH_3CO_2CH_2CH_3\right]^* & \underset{k_t[CH_3CO_2CH_2CH_3]}{\overset{k_t[CH_3CHBrCH_3]}{\rightleftarrows}} & \left[CH_3CH(Br)CH_3\right]^* \\
k_d[M] \downarrow \quad \updownarrow k_{\text{laser pump}} & & \updownarrow\!\!\!\!/ \; k'_{\text{laser pump}} \quad \downarrow k_d[M] \\
CH_3CO_2CH_2CH_3 & & CH_3CH(Br)CH_3
\end{array}
\tag{2.63}
$$

producing **10*** and $k_{\text{laser reaction}}$ causing further excitation of **10*** and **11*** to dissociative vibrational levels. The authors performed calculations utilizing the above model and achieved quite reasonable correlations with the experimental data although values of k_t and k_d had to be varied to achieve the best fits.

As discussed earlier, we do not believe that bottlenecks such as proposed by Tsang *et al.* (1978) exist for large molecules. Isopropyl bromide possesses a sufficiently high density of states at room temperature to qualify it as a large molecule and place it in the quasicontinuum without the need for coherent excitation through the low-lying discrete states (see Section 2.6). It is possible, however, that collisional excitation of isopropyl bromide might bring the normally transparent molecules into resonance with the laser radiation via hot-band absorption, line broadening, or some other mechanism that might account for the observations of Tsang *et al.* (1978).

In a related paper (Braun *et al.* 1978), these authors employed a laser calorimeter cell constructed of a gold-plated copper block in contact with 384 miniature thermocouples to monitor heat absorbed from laser-irradiated samples of CF_2HCl and ethyl acetate. At pressures >10 Torr, the former reactant was completely thermally equilibrated during the laser pulse. Surprisingly, however, ethyl acetate was found to be in severe thermal disequilibrium even at 20 Torr, which the authors interpreted as substantiating the concept of a two-level bottleneck mechanism. It may be noted that the authors conclude that the distribution function responsible for the dissociation of ethyl acetate cannot be characterized by a simple Boltzmann vibrational energy distribution but, instead, by a bimodal-type distribution with more molecules in levels near the dissociation limit than predicted by a Boltzmann population.

The results of Tsang *et al.* (1978), if their interpretation is correct in that the isopropyl bromide thermal monitor molecule is collisionally activated and then photochemically decomposed, complicates the use of chemical thermometry at high reactant pressures or high fluences. This is clearly not a complicating factor in the more common low-pressure studies at moderate fluences in which the great majority of reaction occurs *after* the laser pulse has terminated. The mechanism of Tsang *et al.* (1978) can be operative only at pressures and fluences sufficiently high to permit collisional excitation of the monitor *during* the laser pulse; i.e., k_t[monitor] $\gg$ laser pulselength.

The thermal method should be reliable for total reactant pressures of $\sim$ 1 Torr or less with a 100 nsec or shorter laser pulse. The maximum total pressure will be a function of the nature of the reactant molecules, fluence, and type of bath gas, if any. Collisional V–V excitation of a relatively large organic thermal monitor by an organic reactant would be more efficient than, e.g., V–T–V excitation via a monatomic bath gas such as He. More work in this area is certainly desirable.

Actually, photodecomposition of a collisionally excited monitor molecule is representative of the more pervasive problem of secondary reactions in pulsed laser-induced systems. It is frequently stated in the literature that a control reaction demonstrated that a product molecule of some laser-induced reaction was not photoreactive at the wavelength utilized. Irradiating such a product molecule and observing no reaction, however, does not assure that such a product formed from a laser-induced reaction of a suitable precursor will be unreactive. A difference in behavior may be attributable to the fact that the laser-generated product will almost certainly be formed with excess vibrational energy that might well place it in the quasicontinuum region. It is thus capable of absorbing photons *from the same laser pulse that generated it* and reacting further. It is difficult to ascertain with confidence when a phenomenon such as this is complicating a study. Varying the laser pulselength at constant fluence may be indicative of such secondary reactions, but a change in pulselength alters the laser power that might exert still a different perturbation on the reaction that may be difficult to sort out from secondary reaction complications. In the absence of high fluences, however, the majority of chemical reaction will probably occur after the termination of the laser pulse for large organic molecules, and secondary reaction from absorption of the initiating pulse will probably not be a significant complication.

2.4.4. Determining Effective Temperature and Reaction Time

A modified thermal monitor technique can be utilized to determine the magnitude and/or time duration of the transient effective temperature generated in a pulsed-infrared laser experiment for which Boltzmann conditions are thought to prevail. Instead of choosing a thermal monitor molecule to have Arrhenius parameters similar to the reactant molecule, one now picks a monitor with unimolecular values of E_a and A factor different from the reactant. Alternatively, one can utilize a bifunctional reactant, the two channels of which differ in Arrhenius parameters. In principle, one simply measures the ratio of the two products formed via the two different unimolecular reaction channels and calculates, from the respective Arrhenius parameters, the temperature required to generate that particular ratio. This procedure is demonstrated by Eqs. (2.64) and (2.65) in which R is a dual channel reactant molecule, P_a and P_b are the products from channels a and b, respectively, and T_{eff} is the effective vibrational tempera-

$$R \xrightarrow{k_a = A_a \exp(-E_a/RT)} P_a \tag{2.64a}$$

$$R \xrightarrow{k_b = A_b \exp(-E_b/RT)} P_b \tag{2.64b}$$

$$T_{\text{eff}} = \frac{E_a - E_b}{2.303R \log(A_a/A_b) - 2.303R \log(P_a/P_b)} \tag{2.65}$$

ture. This treatment applies as well to the case in which the reactant and monitor are two different compounds. There are some approximations and assumptions that must be considered in applying Equation (2.65). When the reactant and monitor are different compounds, the molecules must have a Boltzmann vibrational temperature and the pressure must be such that the reacting system is not in the unimolecular falloff region. Such use of a monitor molecule is, thus, totally different from those applications discussed earlier in which the monitor probed for a nonequilibrium energy distribution. The necessity of a total Boltzmann equilibration is dictated by the use of Arrhenius kinetic parameters that are derived under such conditions. Although it will depend somewhat upon the particular reactant and monitor, a minimum total pressure of several Torr would appear necessary to assure collisional equilibration at moderate fluences. This technique is especially applicable to sensitized reactions.

A complicating factor is that the vibrational temperature within the irradiated volume is not constant in time. The temperature will be at a maximum immediately following the laser pulse but then would begin to drop because of diffusion, momentum transfer, collisional deactivation, etc., by nonexcited molecules and, ultimately, the cell walls. There will be temperature gradients obviously generated during the diffusion–cooling process, which is further complicated by the exo- or endoergicity of any chemical reaction. Furthermore, the size of the irradiating laser beam can probably influence the maximum temperature attained and the rate of cooling as discussed in Section 2.4.3.1. This arises because a small beam will permit more rapid intermixing with molecules originally outside of the irradiated volume than will a larger irradiated volume.

Fortunately, the rates of unimolecular reactions (2.64a) and (2.64b) respond exponentially to temperature, and the great majority of reaction will occur at the maximum temperature reached in the system. Therefore, the experimental T_{eff} will probably be close to, but somewhat lower than, the maximum temperature actually generated since the ratio will also have a small contribution from reaction occurring at temperatures below the maximum.

Finally, T_{eff} refers to the effective *vibrational* temperature reached in the system since the monitor molecule or bifunctional reactant undergoes unimolecular transformation for which vibrational excitation is sufficient. T_{eff} might, therefore, be higher than the temperature anticipated for a totally vibrationally, rotationally, and translationally equilibrated system for a given absorbed energy. At high pressures and, certainly, in sensitized systems, T_{eff} will correspond to the totally equilibrated system since only a few collisions are required to Boltzmannize an ensemble of excited large molecules (see Section 2.6.2.4).

The total yield as opposed to the yield ratio is related to the length of time the reacting system remains at T_{eff}. In theory, at least, this value may be estimated from the Arrhenius relationship coupled with the reaction yield per pulse. The absolute reaction yield is related to the rate constant by the following relationship [Eq. (2.66)] where [R] and [P] are the concentration of reactant and product after time Δt, respectively, $[R]_0$ is the initial

$$\text{Reactant (R)} \longrightarrow \text{product (P)}$$

$$\begin{aligned}\text{Reaction yield} &= \frac{[P]}{[R]_0} \\ &= \frac{[R]_0 - [R]}{[R]_0} \\ &= 1 - \frac{[R]}{[R]_0} \\ &= 1 - \exp(-k_{uni}^{\infty}\,\Delta t) \qquad (2.66)\end{aligned}$$

reactant concentration, and k_{uni}^{∞} is the rate constant for T_{eff}. In a laser-induced reaction, unlike a thermal reaction, Δt is not a well-defined value but merely represents the approximate time during which the reacting system remains near T_{eff}. It is determined by the rate of cooling of the laser-excited system, which, in turn, is probably a complex function of the reaction pressure, energy absorbed, irradiation geometry, and other experimental variables. We shall see below that Δt is probably on the order of microseconds, at least for some conditions.

There are few experimental data in the literature utilizing monitor molecules to determine T_{eff} and Δt as described above. Two such reports by different groups have both involved the same bifunctional reactant, cyclobutanone. Back and Back (1979) studied the direct irradiation of this compound while Steel *et al.* (1979) utilized NH_3 as a sensitizer. The latter workers addressed directly the question of the maximum effective temperature, and their data are correspondingly more complete in this regard.

Cyclobutanone undergoes a relatively clean reaction via two channels as indicated in Reactions (2.26a) and (2.26b) (Section 2.3.1.3b); under some conditions the cyclopropane partially isomerizes to propene. Steel *et al.* (1979) measured the energy absorbed and the product ratio generated when 1 Torr of cyclobutanone and 49 Torr of NH_3 were irradiated. Figure 2.16 shows T_{eff} estimated from the ratio of products as well as calculated temperatures based on the amount of energy absorbed by the NH_3 and the heat capacity of the system; T_{eff} is seen to be $\sim$ 1000–1300°K depending upon the amount of absorbed laser energy. The heat capacity was calculated assuming both complete V, T, R equilibration ($T^{max}_{V,T,R}$) as well as equilibration of the vibrational and rotational modes only ($T^{max}_{V,R}$). Ammonia is known to undergo a rapid V–R equilibration from the pumped ν_2 mode, and it was considered possible that only V–R randomization would occur on the time scale of the experiment. It is seen that T_{eff} compared reasonably well with $T^{max}_{V,T,R}$ indicating that essentially complete equilibration of the deposited laser energy occurred before any appreciable decomposition of cyclobutanone.

These authors also determined the total amount of reaction occurring from each laser pulse. From T_{eff} and the Arrhenius parameters, it was concluded the system must remain at T_{eff} for $\sim 1 \times 10^{-5}$ sec to account for the observed reaction yield per pulse. It was estimated that momentum transfer generated by a shock wave traveling at the speed of sound in the cell ($\sim 4.3 \times 10^4$ cm sec^{-1}) would require $\sim 2 \times 10^{-5}$ sec to effect a transfer distance of 1 cm, which was considered sufficient to cool the irradiated gas effectively. This value is in very reasonable agreement with the estimated reaction time of $\sim 1 \times 10^{-5}$ sec. The authors alluded to unpublished work suggesting that Δt increased for smaller amounts of absorbed laser energy and that cooling may not occur by a shock wave under such conditions.

Back and Back's (1979) study of the direct irradiation of cyclobutanone, in part, also attempted to determine T_{eff} and Δt. Above 3 Torr the

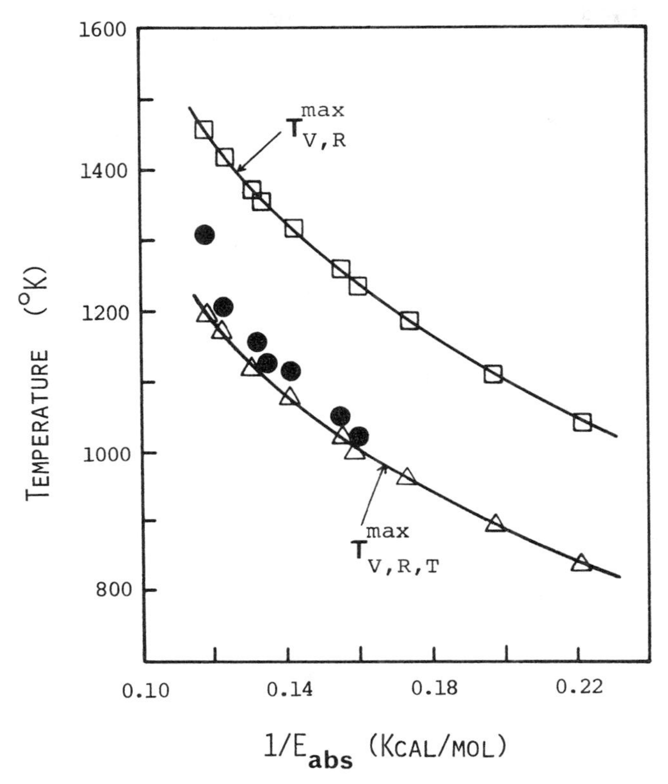

Figure 2.16. Estimated temperatures for pulsed infrared laser heating of 49 Torr of NH_3 with 1-Torr cyclobutanone. □ represents $T^{max}_{V,R}$, which denotes the vibrational temperature of the system assuming that only vibrational and rotational (not translational) equilibration has occurred; △ represents $T^{max}_{V,R,T}$, which is the calculated vibrational temperature assuming complete V,R,T equilibration prior to reaction of cyclobutanone; ● is T_{eff} estimated from the ratio of products from Reactions (2.26a) and (2.26b). T_{eff} is seen to more closely match $T^{max}_{V,R,T}$. (From Steel *et al.* 1979.)

authors assumed that the laser-initiated process was essentially a thermal decomposition and estimated T_{eff} at the highest fluence utilized as ~ 880°K from the energy absorbed and heat capacity of cyclobutanone. Although the total energy absorbed in this case was at least twice as large as any experiment of Steel *et al.* (1979), T_{eff} was lower because of the larger irradiated volume and higher heat capacity of 9.5 Torr of neat cyclobutanone as compared to a mixture of 49-Torr NH_3/1-Torr cyclobutanone. The cyclopropane/ethylene ratio was utilized to estimate crudely T_{eff} as ~ 830 ± 30°K, although the small amount of cyclopropane (< 2% of total products) produced at the relatively low temperatures achieved precluded an accurate estimate of the temperature. An effective reaction time of $\sim 5 \times 10^{-4}$ sec was also estimated from the conversions, which is ~ 50 times longer than estimated for the sensitized reaction.

In their study of ethyl vinyl ether [Reaction (2.23), Section 2.3.1.3a] Rosenfeld *et al.* (1977) observed that the ratio of products was insensitive to changes in the reactant pressure over the range 5–440 Torr. They concluded that a thermalized system resulted at these high pressures and calculated from the k_{23a}/k_{23b} ratio that T_{eff} must be ~ 1600°K. Absolute yields were not reported so no estimate of Δt can be obtained.

In another early study of this sort, Preses *et al.* (1977) attempted to determine T_{eff} and the effective reaction time for the CO_2 laser-induced single-channel decomposition of perfluorocyclobutane both neat and with added argon. Peak temperatures ~ 700–1500°K were calculated depending upon the absorbed laser energy and the heat capacity of the system. The cooling process was assumed to occur by simple heat transport, and fairly long effective reaction times were calculated: a few milliseconds without argon to 6–8×10^{-2} sec with 50 Torr of argon. These estimates appear long by several orders of magnitude as compared to other studies.

In a recent report on the infrared photolysis of cyclic $C_2F_4S_2$, Plum and Houston (1980) utilized a deactivation rate constant of 10^6 sec^{-1} in Shultz and Yablonovitch's (1978) model correlating the fractional dissociation with $\langle n \rangle$. The match of theory with experiment was reasonably good with this rate constant, indicating that Δt is on the order of microseconds.

It may be concluded that values of T_{eff} may be approximated with some degree of accuracy by means of dual channel chemical monitor molecules but that there is a rather wide range of estimated reaction times, Δt. The latter will be quite sensitive to laser fluence, reactant and bath gas pressures, etc. It is obvious that the cooling of a pulsed laser heated gas sample is of considerable complexity, and much more data must be acquired before a consistent picture can be expected to emerge. Our results with ethyl acetate and other large organic molecules would appear to require a reasonably fast cooling time, perhaps on the order of a few microseconds (see Sections 2.5 and 2.6). Simple diffusion would appear to be too slow. Since the speed of sound is ~ 1–5×10^4 cm sec^{-1} in typical low-pressure gases (Steel *et al.*, 1979; Bates *et al.*, 1970; Burak *et al.*, 1970), cooling of the irradiated volume by means of momentum transfer arising from a laser-generated shock wave would appear to provide at least one viable mechanism. For any strongly absorbing gas with rapid V–V and V–T, R energy transfer rates (which describes most large organic systems as discussed in this chapter), shock wave cooling will be much more important above ~ 1 Torr than simple diffusional cooling (Bates *et al.*, 1970). Since the thermal diffusion rate varies linearly as 1/pressure, at very low pressures and relatively low laser energy, simple thermal diffusion may become important as a cooling mechanism. The entire cooling process is obviously very complex.

It seems appropriate to conclude this section by discussing several studies that involved real-time monitoring of products generated in a pulsed, laser-induced reaction. Such studies give Δt directly but also illustrate that reaction can occur in more than one stage. The laser-induced dissociation of tetramethyldioxetane [Reaction (2.67)] is a chemi-

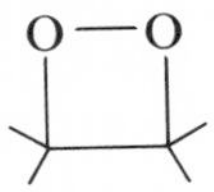

$$\text{(tetramethyldioxetane)} \xrightarrow{\text{IR laser}} 2(CH_3)_2C{=}O + h\nu \tag{2.67}$$

luminescent process generating two acetone molecules (Haas and Yahav, 1977; Yahav and Haas, 1978). The reaction was found to occur in two distinct stages: a fast, unimolecular, collision-free process and a slower contribution that the authors attributed to a collision-induced mechanism. The latter occurred after the laser pulse and accounted for $> 90\%$ of the reaction even at reactant pressures of only 0.07–0.15 Torr. The prompt reaction followed the risetime of the laser pulse, while the slow reaction peaked several microseconds after the laser pulse terminated. The fast unimolecular reaction was attributed to dioxetane molecules excited sufficiently beyond the threshold for reaction to occur during the laser pulse, while the delayed reaction was postulated to have resulted from collisional up-pumping of molecules excited to levels below threshold. This is illustrated in Reactions (2.68)–(2.70), where R** represents a reactant molecule excited to a level well beyond the reaction threshold and R* is vibrationally excited but below or, perhaps, just above threshold.

$$R \xrightarrow{\text{IR laser}} R^{**} + R^{*} \tag{2.68}$$

$$R^{**} \longrightarrow P \tag{2.69}$$

$$R^{*} + R^{*} \to\to\to\to R^{**} + R \tag{2.70}$$

Results very similar to the above have been obtained by Bialkowski and Guillory (1980) for the laser-induced decomposition of CH_3NH_2. Time-resolved monitoring of the NH_2 fragment indicated two time-dependent mechanisms: a fast, collisionless process and a slower, presumably collisional, reaction. The rate of the slower process occurred on the time scale predicted by collisional pumping [Reaction (2.70)] of translationally cold, but vibrationally excited, CH_3NH_2.

Quick *et al.* (1979) have observed similar results by monitoring the spontaneous emission from vibrationally excited HF produced in the multiphoton unimolecular dissociation of 1,1-difluoroethane, 1,1-difluoroethylene, and vinyl fluoride at pressures <0.05 Torr. Two distinct components to the HF fluorescence signal, an initial segment that rose as fast as the laser pulse and a second, more slowly rising component that persisted long after the laser pulse, were observed. The latter fluorescence was interpreted as resulting from the reaction of molecules left by the laser with sufficient energy to dissociate. The dissociating molecules were, thus, shown to have a distribution of energies, and a mean unimolecular lifetime of $\sim$ 150 nsec was estimated.

The above examples illustrate quite dramatically an important feature of the laser-induced chemistry of large molecules that has been stressed in this chapter, namely, that significant reaction can be expected to occur after the laser pulse even at pressures <0.1 Torr, the supposed "collisionless regime." We doubt, however, whether collisional up-pumping as depicted in Reaction (2.70) is of importance for highly vibrationally excited large molecules. If the initial population of excited molecules produced by the laser has a *narrow*, non-Boltzmann distribution with a mean energy less than threshold, then collisional processes will tend to generate a Boltzmann population. Under such circumstances, the fraction of molecules above the threshold might be increased by collisions as shown in Reaction (2.70). Although the distribution of molecules immediately following the laser pulse is not known for any system, we prefer to assume that a Boltzmann-like population will be produced with large molecules. Significant reaction will tend to continue occurring after the laser pulse for such molecules because of the low rate constants for excitation to just beyond the threshold. Although some additional energy may enter the system as a result of the exoergicity of reaction, various cooling processes are always competing, and, eventually, the deposited laser energy becomes so dissipated that reaction ceases.

If the laser, indeed, produces a broad, Boltzmann-like initial distribution, then collisional up-pumping as shown by Reaction (2.70) is probably not significant. Such Treanor pumping (Treanor *et al.*, 1968) is known to be important for certain small molecules at low vibrational levels. When V–V energy exchange is much more rapid than V–T relaxation, then vibrational excitation in an anharmonic molecule need not relax through a series of Boltzmann distributions, and extremely non-Boltzmann vibrational distributions can be created. This is because vibrational exchange processes between levels in a small molecule are seldom exactly resonant since the vibrational levels in a real molecule are not equally spaced, and, therefore, such energy exchange processes have a preferred direction. The features that distinguish large from small molecules as defined in this chapter, however, would appear to prohibit efficient Treanor pumping of large molecules at vibrational levels near or above threshold as depicted in Reaction (2.70). Although V–V exchange among individual molecules in a Boltzmann ensemble of highly vibrationally excited large molecules is certainly facile, such exchanges would only serve to maintain the Boltzmann population. The high density of vibrational states and relatively rapid V–T relaxation common to large molecules at high levels of excitation would prevent any non-Boltzmann favoring of high energy levels via Treanor pumping. In this regard, an information theoretic analysis of multiphoton excitation and collisional deactivation in polyatomic molecules by Jensen *et al.* (1978), likewise, did not predict a Treanor pumping process for polyatomic systems.

2.5. Experimental Data for Ethyl Acetate

We have conducted rather extensive, quantitative studies on the multiphoton excitation–reaction of a series of organic esters (Danen *et al.*, 1977; Danen and Hanh, 1980; Danen *et al.*, 1980). It appeared appropriate to present, herein, some of the main data for ethyl acetate, a prototype "large" molecule, especially for comparison with the modeling studies to be presented in Section 2.6. Most of these data are unpublished at the time of this writing, but it is planned to submit complete manuscripts to appropriate primary journals in the near future.

All experiments were conducted utilizing a Lumonics Model 103 TEA grating-tuned, multimode CO_2 laser with reactants contained in small static cells at pressures of 0.1 Torr or less for most cases. Fluences ranged from ~ 0.2 to 10 J/cm^2. The higher fluences were achieved utilizing either a long focal length BaF_2 (50 or 75 cm) lens that produced a nearly uniform fluence throughout a short (0.95 cm) reaction cell or a Galilean telescope arrangement that produced a collimated laser beam for use with longer cells. Most experiments were done with a long laser pulse that consisted of an initial spike of $\sim$ 150-nsec FWHM that contained $\sim 50\%$ of the energy followed by $\sim$ 1300-nsec tail. Short pulses were generated by removal of the N_2 from the lasing gas mix and consisted of a $\sim$ 100-nsec FWHM spike without any appreciable tail.

Some of the data for ethyl acetate have been presented already in this chapter: Table 2.4 and Figures 2.13–2.15 illustrate the effects of pressure and added bath gases on the reaction probability for ethyl acetate or *sec*-butyl acetate. These data demonstrated the necessity of working at pressures of 0.10 Torr or less to avoid thermal effects.

2.5.1. Dependence of Reaction Probability on Fluence

Figure 2.17 illustrates the fluence dependence of the reaction probability for ethyl acetate at a number of different laser frequencies. Ethyl acetate exhibits a broad absorption band centered ~ 1055 cm^{-1}; the $P(12)$

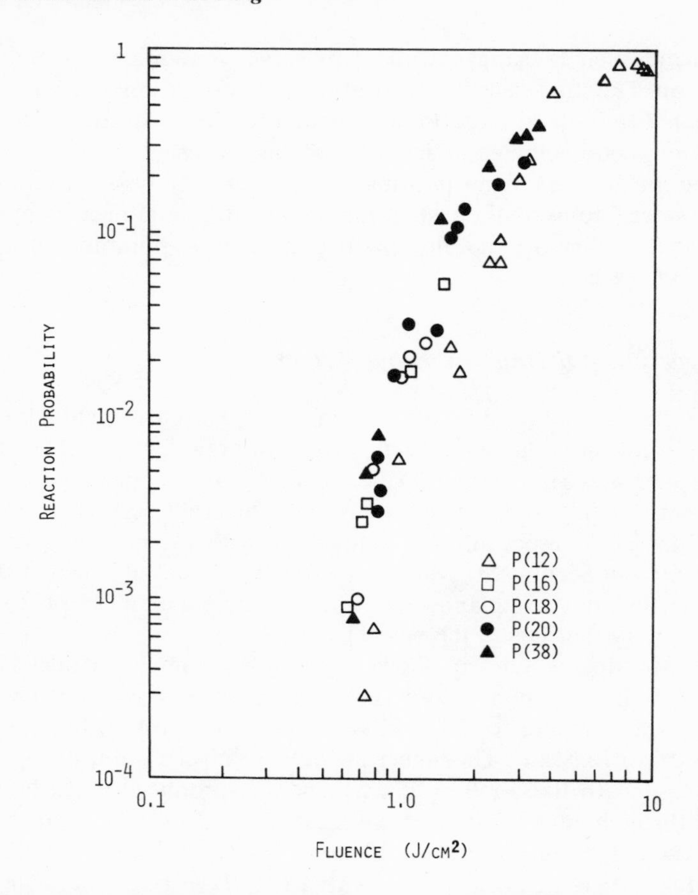

Figure 2.17. Plot of reaction probability within the irradiated volume for 0.05 Torr ethyl acetate vs. laser fluence for the indicated CO_2 laser lines: $P(12) = 1053.9$ cm^{-1}; $P(16) = 1050.4$ cm^{-1}; $P(18) = 1048.7$ cm^{-1}; $P(20) = 1046.9$ cm^{-1}; $P(38) = 1029.4$ cm^{-1}.

to $P(20)$ lines were all reasonably close to or slightly to the red of the absorption maximum while $P(38)$ was on the low-frequency shoulder. It is seen that, although there is some scatter in the data, there is only a moderate variation with laser frequency in the reaction probability vs. fluence plot. At sufficiently high fluences the reaction probability approaches 100%; i.e., all the ethyl acetate molecules within the irradiated volume undergo reaction. At intermediate fluences, there is a high-order dependence of reaction probability on fluence. The slope of the lines in the region of 10^{-2}–10^{-4} probability is ~ 4.2 indicating that the reaction yield scales as $\sim$ fourth power of the laser fluence.

2.5.2. Dependence of Cross Section on Fluence

Unlike some small molecules, the absorption of laser energy for different pressures of ethyl acetate displayed a Lambert–Beer's law dependence, i.e., σ_L was invariant with pressure [Eq. (2.71)]; F_T is the laser fluence transmittance, σ_L is the laser absorption cross-section, c is the concentration in molecules per cubic centimeters and l is the cell length in centimeters.

$$\log F_T = -\sigma_L cl/2.303 \tag{2.71}$$

Figure 2.18 illustrates the behavior of σ_L for ethyl acetate as a function of laser fluence for the $P(20)$ line near the single-photon absorption maximum and $P(38)$ which is on the low-frequency shoulder of the O–CH_2 stretching mode. It is observed that σ_L decreases at higher fluences for $P(20)$ but actually increases somewhat for $P(38)$. The latter effect can presumably be accounted for by a red shifting of the absorption band at higher fluences resulting in an enhanced σ_L. The decrease in σ_L at higher fluences for the $P(20)$ line can be attributed to such factors as anharmonic red shifting off of the absorption maximum, high percent reaction of the molecules during the early stages of the laser pulse resulting in a decreased

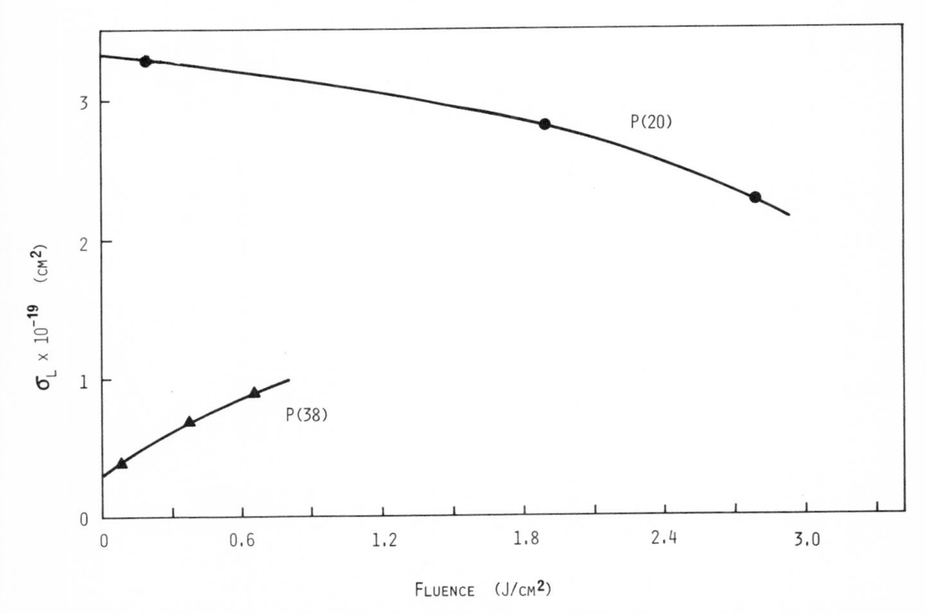

Figure 2.18. Plot of the effective laser absorption cross section, σ_L, for ethyl acetate for the $P(20)$ (1046.9 cm^{-1}) and $P(38)$ (1029.4 cm^{-1}) laser lines vs. laser fluence.

concentration of absorbing molecules for the latter part of the pulse, or saturation effects. For ethyl acetate the latter two were shown not to be important in the fluence range of Figure 2.18. This matter will be considered again in Section 2.6 dealing with the modeling studies.

Somewhat surprisingly, at low fluences σ_L extrapolates to the single-photon cross section, σ, an observation that appears to be quite general for large molecules. It is not obvious that a multiphoton absorption phenomenon should exhibit the same cross section as a single photon process. An explanation might be in the high density of states for large molecules; even though the bandwidth of the laser is < 0.1 cm^{-1}, many rotational lines lie within the bandwidth.

The observed σ_L values for large ester molecules are quite insensitive to fluence as compared to small molecules. Figure 2.19 depicts σ_L for SF_6, S_2F_{10}, and *sec*-butyl acetate ($CH_3CO_2CH(CH_3)CH_2CH_3$). It is seen that σ_L for SF_6 continually diminishes even at fluences $< 10^{-4}$ J/cm^2 while the large ester molecule is constant to $\sim 10^{-1}$ J/cm^2; S_2F_{10} exhibits intermediate behavior. These features are discussed more quantitatively by Galbraith and Ackerhatt in Chapter 1 of this volume. All of these σ_L plots

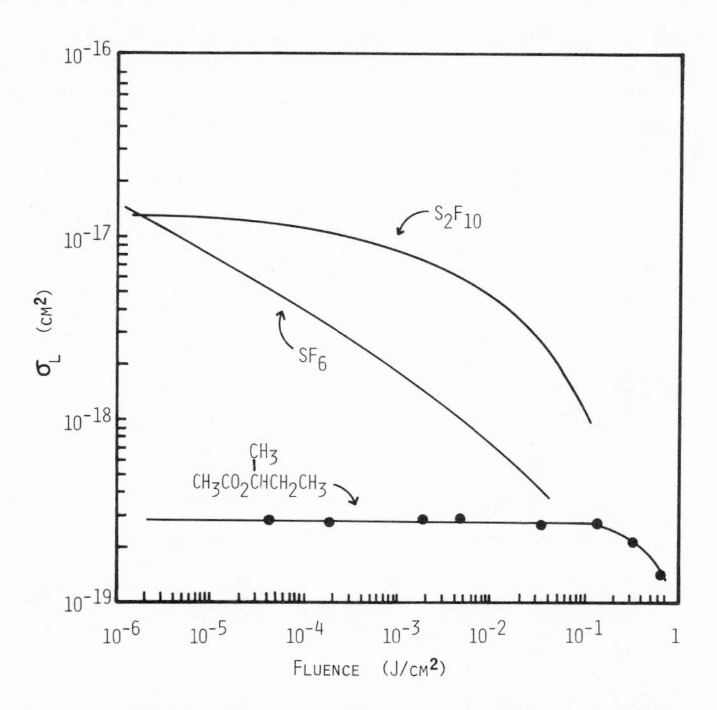

Figure 2.19. Plot of the effective laser absorption cross section, σ_L, for SF_6, S_2F_{10}, and *sec*-butyl acetate vs. fluence.

extrapolate back to the single photon σ at low fluences. The relative insensitivity of σ_L with fluence is seen to be another characteristic of large molecules, a feature presumably attributable to the large density of states possessed by such molecules and lack of an early bottleneck in the absorption process.

2.5.3. *Energy Absorption*

From the laser fluence, ϕ_0, and the laser cross section, σ_L, the total energy absorbed per molecule, E_{abs}, can be calculated from Eq. (2.72).

$$E_{abs} = \phi_0 \sigma_L \tag{2.72}$$

$$\langle n \rangle = \frac{\phi_0 \sigma_L}{hc\bar{v}} \tag{2.73}$$

Dividing E_{abs} by the photon energy gives the average number of photons absorbed per molecule, $\langle n \rangle$ [Eq. (2.73)]; h is Planck's constant, c is the speed of light and $\bar{v}$ is the reciprocal wavelength of the laser photon; the energy of a $P(20)$, 9μ band photon ($1046.9\ cm^{-1}$) is 2.99 kcal/mol. The average number of photons absorbed per reacted molecule, $\langle n_r \rangle$, is simply $\langle n \rangle$ divided by that fraction of molecules within the irradiated volume that react per pulse [Eq. (2.74)]; $P(\phi)$ is the reaction probability at

$$\langle n_r \rangle = \frac{\langle n \rangle}{P(\phi)} \tag{2.74}$$

fluence ϕ. In a sense, $\langle n_r \rangle$ is a type of quantum yield. Some preliminary, experimentally determined values of $\langle n \rangle$ and $\langle n_r \rangle$ for ethyl acetate irradiated with $1046.9\ cm^{-1}$ radiation are listed in Table 2.5 and plotted in

Table 2.5. Average Number of Photons Absorbed, $\langle n \rangle$, Average Number of Photons Absorbed Per Reacted Molecule, $\langle n_r \rangle$, and Reaction Probability, $P(\phi)$, for 0.05-Torr Ethyl Acetate Irradiated at $1046.9\ cm^{-1}$ with Various Fluences

Fluence (J/cm^2)	$\langle n \rangle$	$\langle n_r \rangle$	$P(\phi)$
2.80	31	141	0.22
1.84	25	156	0.16
1.75	24	171	0.14
1.65	23	190	0.12
1.14	17	560	0.03
1.06	16	790	0.02
0.82	13	1765	0.007

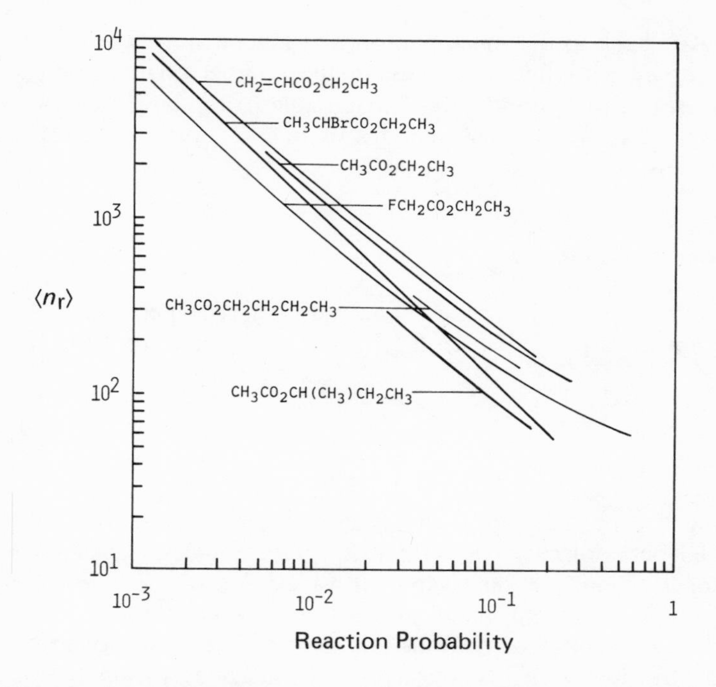

Figure 2.20. Plot of the average number of photons absorbed per reacted molecule, $\langle n_r \rangle$, vs. reaction probability for the six esters indicated.

Figure 2.20. It is to be stressed that these are preliminary data at the time of this writing and subject to possible refinement.

The data in Table 2.5 are quite informative. For example, irradiation with a laser fluence of 1.06 J/cm^2 resulted in $\langle n \rangle = 16$. The average energy absorbed at this fluence, ~ 48 kcal/mol, corresponded closely to the energy threshold for reaction of ethyl acetate and implied that about half the molecules within the irradiated volume acquired vibrational excitation to or beyond the reaction threshold. Yet, only about 2% of the molecules actually reacted. This illustrated, again, the fact that excitation to or just beyond the threshold does not assure reaction for such a large molecule. The rate constants for reaction at or a little above the threshold value are quite small, and the various cooling processes can be very competitive and effectively deactivate the majority of the excited molecules. Of course, increasing the fluence results in an exponential increase in yield. For example, increasing the fluence by less than twofold to 1.84 J/cm^2 increases $P(\phi)$ eightfold to 0.16. For these conditions, $\langle n \rangle$ is ~ 25 corresponding to a mean excitation level of ~ 75 kcal/mol indicating that the majority of the molecules within the irradiated volume are excited to above the reaction threshold.

From Table 2.5 and Figure 2.20, the $\langle n_r \rangle$ values are seen to be quite large indicating a relatively inefficient utilization of laser photons particularly at low fluences. For example, at 1.06 J/cm^2, $\sim$ 790 photons are absorbed for each molecule that ends up actually reacting. When one considers that frequently $< 10\%$ of the laser radiation is actually absorbed by the sample in these types of experiments, is seen that only one molecule is actually reacted for several thousand photons being emitted by the laser. It is to be stressed, however, that the experiments from which these $\langle n_r \rangle$ values are calculated were not designed to produce optimum, low values. The photon utilization could certainly be improved quite dramatically by using higher laser fluences, longer reaction cells, multiple reflecting optics, etc., if that were the objective.

2.6. Computer Modeling Studies

2.6.1. Literature Models

In this section we will discuss briefly, in order of increasing sophistication, some models used to interpret multiphoton excitation or multiphoton dissociation observed in small molecules and examine the implications of these theories for large organic molecules. A much more detailed discussion of the phenomenon of vibrational excitation in polyatomic molecules with particular emphasis on SF_6 and S_2F_{10} is presented by Galbraith and Ackerhalt in Chapter 1 of this book. Following this, we will present results from model calculations for such large molecules utilizing a master equation formulation. For the reader's convenience, Table 2.6 lists and defines all symbols and notations used in this section.

In the presence of strong infrared laser radiation, molecules may absorb tens of photons that may lead to dissociation or rearrangement. The process for the absorption of many photons follows a multistep mechanism. To formulate a theory for the process, one needs to solve a time-dependent Schrödinger equation for the electric dipole interactions between the molecular states and the laser field. The time-dependent Hamiltonian problem can be reduced to a time-independent Hamiltonian using the "rotating wave approximation" and "short time approximation" (Quack, 1978). Under most of the experimental conditions presently employed to study small molecules, the strongest transitions are near resonant and the transitions possess dipole interactions that are much less than the energy of a CO_2 laser photon; therefore, the "rotating wave approximation" is usually excellent. Solving the Schrödinger equation, however, is still a very difficult problem because of the large size of the transition

Table 2.6. Definitions of Symbols and Notations

E_0	reaction threshold energy (kcal/mol)
ΔE_d	deactivation energy per collision (kcal/mol)
$\langle E_{\text{vib}} \rangle_T$	average energy of vibration at temperature T
$\langle E_{\text{rot}} \rangle_T$	average energy of rotation at temperature T
$\langle E_{\text{trans}} \rangle_T$	average energy of translation at temperature T
E_{th}	thermal energy possessed by a molecule at 300°K
F_R	fraction of reaction occurring during the laser pulse
F_R^{∞}	fraction of molecules above E_0 at cessation of laser pulse (the reaction yield if totally collisionless conditions)
g_i	degeneracy of level i
ϕ	laser fluence (J/cm^2 or photons/cm^2)
I	laser intensity (photons/cm^2 sec)
K	Boltzmann constant (1.38×10^{-16} erg °K^{-1})
k_{uni}^{∞}	high pressure limit unimolecular rate constant
k_i	RRKM unimolecular dissociation rate of level i
N_i	the molecular population of level i
$\langle n \rangle$	average number of photons absorbed per molecule
P_{ij}	collisional transition probability from level j to level i
σ_{ij}	laser absorption or emission cross section from level j to level i (cm^2)
σ_i'	net absorption cross section: $(\sigma_{n+1,n} - \sigma_{n-1,n})N_n$ (cm^2)
σ_L	experimental effective laser absorption cross section (cm^2)
T_{eff}	effective (highest) transient temperature produced by a single laser pulse
Δt	effective reaction time per laser pulse
t	time (sec)
ω	collisional frequency (sec^{-1})

matrix needed in the calculations, e.g., on the order of 10^{10} for SF_6 and even larger for the organic molecules discussed in this chapter.

2.6.1.1. *Rate Equation Formulation with Incoherent Laser Excitation*

If we assume incoherent excitation,‡ the problem can be reduced to the solution of a coupled linear rate equation, and a master equation, and can be further simplified by grouping of states of the same energies into a single level with degeneracies given by the density of states at that energy. Vibrational energy levels of small molecules may be divided into at least two regions. The first region consists of several discrete states characterized by coherent excitation‡ while the second region is a quasicontinuum of

‡ A simplified distinction between coherent and incoherent excitation may be made as follows. Consider a two-level system with states a and b coupled by electric dipole radiation. The Hamiltonian can be expressed as the sum of an unperturbed Hamiltonian, $\mathscr{H}_0$, and the Hamiltonian influenced by the electric dipole radiation, $\mathscr{H}'$.

$$\mathscr{H} = \mathscr{H}_0 + \mathscr{H}' \qquad \mathscr{H} = \mu E$$

states (Black *et al.*, 1977). Most small molecules appear to reach the quasicontinuum region after absorbing a few vibrational quanta. Transitions in the quasicontinuum region are characterized by incoherent interactions. If there is no bottleneck in the coherent excitations and the incoherent excitations dominate, one can utilize a linear rate equation. We will assume that these conditions hold for most large organic molecules. The populations of molecular states during the laser pulse follow the rate equation if the collisional transition-rate and unimolecular dissociation rate are negligible during the laser pulse. This can be expressed by Eq. (2.75) in which N_i is

$$\frac{dN_i}{dt} = I \sum_j \sigma_{ij} N_j - I \sum_j \sigma_{ji} N_i \tag{2.75}$$

the population of level i, I is the laser intensity, σ_{ij} is the absorption cross section $(j \rightarrow i)$ if $i > j$ or the emission cross section if $i < j$. This rate equation, together with RRKM rate theory to describe the unimolecular reaction rate along with assumptions on the collisional probability and infrared absorption cross section, has been used by several authors to describe the dependence of dissociation yield and average energy absorbed per molecule on the energy fluence (Lyman, 1977; Grant *et al.*, 1978; Baldwin *et al.*, 1979). The distribution of molecular states predicted for an absorption cross section that decreases or is constant with increasing excitation level is narrower than a Boltzmann of the same average energy. If the absorption cross section increases with increasing level of excitation, the model gives a broad distribution that resembles a high temperature Boltzmann distribution, assuming any losses by dissociation or disturbance by collisions are negligible during the laser pulse. Although the generality of the Boltzmaann distribution has been refuted by a number of numerical calculations for certain models (Galbraith and Ackerhalt, Chapter 1 of this volume; Grant *et al.*, 1978) for the limiting case in which the dissociation yield is small during the laser pulse and the bulk absorption cross section is constant, the energy distribution is Boltzmann for the degenerate oscillator model for the vibrational density (Black *et al.*, 1979). The assumption of a

---

The quantum mechanical amplitude for these two states changes with time; i.e., $\Psi(t) = \exp(-i\mathscr{H}'t/)\Psi(t_0)$. If the two states are isolated from any other perturbation, the states retain their phase. Therefore, even with a continuous pulse, the population of the states, which is proportional to the square of amplitude, oscillates. We refer to such an oscillating frequency as the Rabi frequency, and the transition with these characteristics is termed *coherent excitation.* In the quasicontinuum, however, where intramolecular relaxation is fast between vibrational states or when collisions disturb the states, they lose their phase, and the population ceases to oscillate. We can, therefore, use linear equations to describe such *incoherent transitions.*

Boltzmann population as an initial vibrational energy distribution because of laser pumping is convenient since only the absorbed energy is needed to determine the distribution. In contrast, where this is not valid, one needs absorption cross sections for each level to solve the master equation and such data do not exist and are hard even to estimate reliably.

The two main assumptions relating to the generation of a Boltzmann distribution upon laser pumping (Black *et al.*, 1979) are the absence of an early bottleneck and a constant cross section. As compared to small molecules, the likelihood of a Boltzmann distribution is greater for large molecules that have no early bottleneck and exhibit fairly constant bulk absorption cross sections as we have discussed earlier, although there is no assurance that the level-to-level cross sections are constant with changes in level of excitation. A Boltzmann distribution is the equilibrium and broadest distribution possible as a result of detailed balance for a thermal process. But, for a radiative absorption process, the distribution can be broader than a Boltzmann distribution. In Chapter 1 of this volume, Galbraith and Ackerhalt present a model that predicts a very broad distribution for SF_6 and S_2F_{10}, the latter of which these authors consider a large molecule (see Figure 1.10, Chapter 1). Nonetheless, we will assume that a Boltzmann distribution is a reasonable approximation and will present calculations with this assumption in Section 2.6.3.

2.6.1.2. Coherent Excitation with Heat Bath Feedback Model

The incoherent excitation model neglects the coherent initial excitation through discrete levels that may be important for low-energy fluence conditions or for molecules with rather small transition interactions with the laser field even at moderate laser intensity or for molecules that have an early bottleneck. As stated above, the time-independent Schrödinger equation is practically difficult to solve because of the large size of the matrix for polyatomic molecules. There are two ways to approach the problem in the case of coherent excitation. In the first approach, a vibrational mode (or several vibrational modes) pumped by the laser field is (are) separated from the other vibrational levels (or modes) of the molecule. The latter act as a heat bath, such that only a pumped mode is treated coherently and the heat bath is coupled to the pumped mode by intra- or intermolecular relaxation. This is depicted schematically in Figure 2.21. In this model the laser energy is first deposited into the chosen pumped mode(s) and then leaks with some characteristic rate constant from this mode to the heat bath. The feedback of heat bath excitation to the pumped

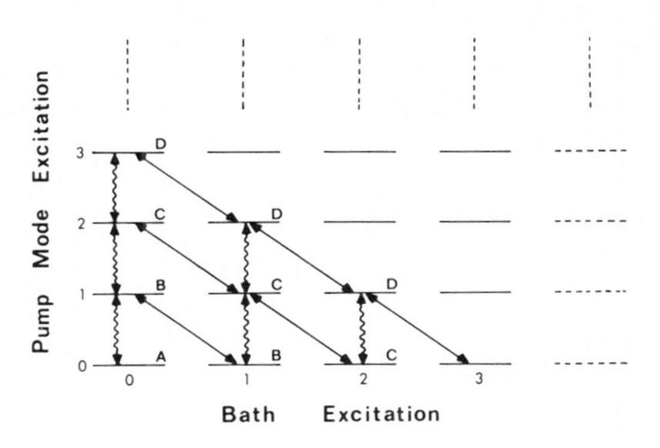

Figure 2.21. States in theoretical framework to account explicitly for both heat bath and pump mode excitation. Wavy lines denote rates of transitions induced up and down in the pump mode by the laser. Sloping solid arrows indicate intramolecular T_1 relaxation between states of the same total energy. (After Stone and Goodman, 1979.)

mode can be taken into account, although it is negligible for large molecules.

When the intra- or intermolecular relaxation rate is faster than the laser pumping rate, a linear rate equation for the quasicontinuum region can also be deduced from the heat bath model (Stone and Goodman, 1979; Stephensen *et al.*, 1979). The excitation between discrete levels is considered only up to two or three levels for small molecules. For transitions between discrete levels, a set of Bloch equations for these levels must be solved. Since large molecules may already be in the quasicontinuum at room temperature, only the linear rate equation for the quasicontinuum states need to be considered. The rate equation is the same in form as Eq. (2.75) except that the effective transition rates, $I\sigma_{ij}$ and $I\sigma_{ji}$, are based on different assumptions. The radiative transition rates can be determined from transition dipole moments and T_1 and T_2 relaxation rates in this model. These transition rates are linear in laser intensity like those in the incoherent excitation model.

According to this model, an increase in the relaxation rate results in the irreversible loss of energy from the pumped mode into the heat bath. The infrared absorption rate becomes dominant over the stimulated emission rate because of the low populations in the upper states due to the irreversible process. If the relaxation rate is slow compared with the laser pumping rate, the molecular density matrix can be divided into two uncorrelated parts describing the pumped mode and the heat bath. In this case energy leakage is small and energy remains in the pumped mode, which reduces the matrix size. But now the assumption of an RKKM decomposition rate based on the rapid intramolecular redistribution of vibration energy may fail as a result of the slow relaxation rate. There appears to be little experimental precedent for a case such as this, particularly for large molecules; however, see Section 2.3.1.4.

2.6.1.3. Two-Level Transition Models

Other approaches to multiphoton modeling group all states of similar energy and the same relevant quantum numbers with respect to the dipole selection rule into one level and deal with transitions between these levels. Quack (1978) has discussed several cases where the Schrödinger equation can be reduced to linear, or nearly linear, rate equations for coherent excitation. One groups states into such levels for coherent excition with a "rotating wave approximation" or "short time approximation." The coherent part is combined with the quasicontinuum by the effective decay rates. The rate equation for the coherent part is again the same in form as Eq. (2.75). The relationship between the effective absorption cross section and the stimulated effective emission cross section depends upon the conditions. Although this model does not explicitly contain intramolecular relaxation rates in the rate equations, the conditions under which the master equation [Eq. (2.75)] is valid take into account the relaxation rates.

Case A in Quack's model (Quack, 1978) is physically related to the transition from the discrete lower level, where intramolecular relaxation is slow between states, to the quasicontinuum upper level, where the relaxation rate between states is fast. In this case, the radiative transition rate from the upper quasicontinuum states to the lower discrete states (i.e., the emission rate from the upper level to the lower level) is negligible, while the absorption rate is finite. Thus, the populations in the lower levels exhibit an irreversible, exponential decay.

Quack's case B applies to transitions between two levels in which the intramolecular relaxation rates are fast between states in a particular level. The ratio of up- and down-transition rates is determined by detailed balance using the state densities of the two energy levels, as in the incoherent excitation model.

Quack's case C is similar to case B, but now the intramolecular relaxation rates between states in each of the two levels are slow. Case D is for the transition between two levels that have fast intramolecular relaxation rates not only between states within the levels but also between those two levels. For cases C and D, the rate equation is not linear but depends upon the initial population of the vibrational levels.

If the conditions for intramolecular relaxation (cases A and B) are satisfied this model and the heat bath model both give a linear rate equation similar to Eq. (2.75) obtained for incoherent excitation. Although the rate equation based on the three models has the same formula, the rate constants from the three models are slightly different because these models start from different assumptions. When the conditions for these models are not met, no formulated theory has been attempted to date. To understand which model is appropriate for a given molecule, high-resolution spectroscopic data is necessary from which the transition rates between closely spaced states in an energy level may be obtained.

2.6.1.4. Rate Equation Model Including Unimolecular Dissociation Rate and Collisional Effects

2.6.1.4a. Inclusion of Dissociation Rate. So far, we have considered rate equations that contain only the laser-induced transition rates. Molecules may undergo unimolecular reaction or collide with each other during and after the laser pulse, and the complete rate equation including reaction and collisional effects is given by Eq. (2.76); ω is the collisional frequency, P_{ij} is the collisional transition probability from level j to level i, and k_i is the unimolecular dissociation rate of level i. Under collisionless conditions, the rate equation [Eq. (2.76)] can be reduced to Eq. (2.77). If unimolecular dissociation is not important during the laser pulse, the collisionless rate equation becomes simply Eq. (2.75) as discussed earlier. Equation (2.75) can be rearranged as shown to produce Eq. (2.78b). Therefore, the distribution of vibrational energy states is dependent only on the laser fluence

$$\frac{dN_i}{dt} = I \sum_j \sigma_{ij} N_j - I \sum_j \sigma_{ji} N_i + \omega \sum_j P_{ij} N_j - \omega \sum_j P_{ji} N_i - k_i N_i \tag{2.76}$$

$$\frac{dN_i}{dt} = I \sum_j \sigma_{ij} N_j - I \sum_j \sigma_{ji} N_i - k_i N_i \tag{2.77}$$

$$\frac{dN_i}{I\,dt} = \sum_j \sigma_{ij} N_j - \sum_j \sigma_{ji} N_i \tag{2.78a}$$

$$\frac{dN_i}{I\,dt} = \frac{dN_i}{d\phi} \tag{2.78b}$$

($I \times dt$). Under collisionless conditions, the reaction yield depends only on the vibrational energy distribution at the termination of the laser pulse since all the molecules above the reaction threshold energy eventually react. Therefore, according to this model, the reaction yield depends only on the fluence if collisions or unimolecular reaction are not significant or do not disturb the distribution during the laser pulse. If the extent of unimolecular reaction becomes significant during the laser pulse, the distribution then depends on the laser intensity even without collisions. For more details, see Section 2.6.2 where these conditions are exemplified with model calculations.

Most model calculations have used RRKM theory (Robinson and Holbrook, 1972) for the unimolecular decomposition rates to illustrate the dependence of reaction yield on the laser fluence or intensity. RRKM theory assumes that the intramolecular redistribution of vibrational energy is rapid relative to laser pumping, which appears to be the case for most experiments reported to date. Since the three models discussed above also assume that intramolecular relaxation is rapid, there is no contradiction in using RRKM rates in the rate equations. If intramolecular relaxation is slow, both the applicability of RRKM theory and the use of the simple master equation formulation are invalid.

2.6.1.4b. Inclusion of the Collisional Effects. So far, none of the models discussed contain collisional effects. Only those experiments employing molecular beams attain true collisionless conditions. Since most experiments discussed herein have been performed in static gas cells, one needs to include collisional effects in the model. The collisional problem is especially severe for large organic molecules that have relatively long lifetimes for energies somewhat above the reaction threshold energy, E_0. To apply a full master-equation treatment, or even to follow the time evolution commencing with a Boltzmann distribution at the end of the pulse to the cessation of reaction, requires values for a large number of level-to-level vibrational energy transfer rate constants. Our understanding of such processes is incomplete. The following two sections outline some of the main features for transfer from both highly and lowly excited molecules and lay the groundwork for inclusion of collisional effects in model calculations.

Collisional energy transfer from highly vibrationally excited molecules. Knowledge of collisional vibrational energy deactivation rates for highly excited polyatomic molecules has been obtained largely from chemical activation and thermal activation experiments (Tardy and Rabinovitch, 1977). In chemical activation systems the polyatomic molecules are energized by virtue of the exoergicity of a chemical reaction producing the molecules and are all formed initially in an energized state above the dissociation threshold energy, E_0. The chemical activation distribution, with an average energy equal to the exoergicity of the formation reaction, is much narrower than a Boltzmann distribution function. After the molecules are created at high levels of vibrational excitation above E_0, they undergo unimolecular decomposition in competition with collisional deactivation. This competition can be utilized to obtain vibrational energy transfer probabilities for the molecules provided that the unimolecular rate constants are thoroughly characterized over the appropriate range of energy. A matrix of the transition probability from level j to level i (P_{ij}) for the energy levels above E_0 is defined such that the transition probabilities reproduce a given set of experimental results, i.e., $D/(D + S)$, where D is the total number of molecules decomposing per unit time and S is the total rate of stabilization. The relative probabilities of up and down transitions are determined by detailed balance and completeness conditions.

A standard approach to modeling P_{ij} assumes simple model forms for the transition probabilities. The simplest model is the stepladder form, which is an approximation for a Gaussian-type distribution:

$$P_{ij} = 1.0 - P_{ji} \qquad \text{for } i - j = \langle \Delta E_d \rangle$$

$$P_{ij} = 0 \qquad \text{for } i - j \neq \langle \Delta E_d \rangle$$

The stepladder model has the property that a particular average energy loss, $\langle \Delta E_d \rangle$, is much more probable than any other transition. This model represents efficient collisional processes.

The exponential model has down-transition probability elements of the form

$$P_{ij} = C \exp(-\Delta E / \langle \Delta E_d \rangle)$$

where ΔE is the energy difference between E_i and E_j, and C is a constant determined by normalization. The exponential model is used for inefficient collision partners.

The results from chemical activation experiments support the following general conclusions:

1. The values of $\langle \Delta E_d \rangle$ range from ~ 0.5 to ~ 12–15 kcal/mol in all systems studied.

2. The stepladder or Gaussian model, in which large energy losses are more probable than small energy losses, is more appropriate for efficient colliders, while the exponential or Poisson distribution model in which small energy losses are more probable than large energy losses is more appropriate for the less-efficient colliders.

3. The $\langle \Delta E_d \rangle$ increases for monatomic to diatomic and polyatomic molecules as collisional partner. For monatomic, diatomic, and triatomic gases with small $\langle \Delta E_d \rangle$, the exponential function is more correct. For large molecules with large $\langle \Delta E_d \rangle$, the stepladder model is satisfactory.

4. The appropriate collisional cross section is the gas kinetic cross section, $\pi \sigma_{ij}^2 \Omega^{(2,2)}(T^*)$; σ_{ij} is the Lennard-Jones diameter and $\Omega^{(2,2)}(T^*)$ is the temperature dependent reduced collision integral.

5. The $\langle \Delta E_d \rangle$ values depend, not only on the collision partner, but also the properties of the activated molecule. The relative contribution of V–V and V–R transfer, which complements V–T transfer, also depends on the individual molecule.

6. The $\langle \Delta E_d \rangle$ probably increase with an increase in total energy of the excited molecule.

7. The $\langle \Delta E_d \rangle$ is affected by temperature for less-efficient gases, while there is very little temperature effect upon the collisional models for the efficient gases.

In thermal activation (Tardy and Rabinovitch, 1966), the initial distribution of the molecules has the characteristics of a Boltzmann distribution for the given input temperature. Since the populations of the molecules above the threshold energy are very low for low-temperature experiments, the thermal and chemical activation techniques give complementary information concerning transition probabilities. Chemical activation addresses transition probabilities P_{ij} $(i, j \geq E_0)$ for energy levels well above E_0 while low-pressure thermal activation provides information concerning the probabilities, P_{jk} $(j \geq E_0, k \leq E_0)$ from energy levels below E_0 to energy levels above E_0. The general characteristics of collision efficiencies found in thermal activation systems are very similar to those observed by chemical activation techniques.

The general aspects of collisional efficiencies observed in chemically and thermally activated unimolecular systems are in contrast to the predictions of the Landau–Teller theory (Nikitin, 1974), which is appropriate for small molecules at low levels of vibrational excitation. For example, for vibrational energy transfer from polyatomic molecules at high levels of excitation He is not the most efficient of the monoatomic bath gases; the efficiency does not decrease with increasing atomic mass, and the collisional efficiency decreases, rather than increases, with a rise in temperature. Vibrational energy transfer at low levels of vibrational excitation will be discussed in the following section.

The collisional partners used in chemical activation methods are always vibrationally, translationally, and rotationally cold. Collisions between two highly vibrationally excited molecules, such as may occur in pulsed infrared laser-induced processes, have not been thoroughly explored. Information on the collisional process between two vibrationally hot but translationally and rotationally cold molecules is needed to model the variation of reaction yield on the reactant pressure in multiphoton reactions. A model involving such collisional processes will be considered in Section 2.6.2.4 along with calculations of rate equations.

Some experiments (Bado and van den Bergh, 1978; Jang and Setser, 1979) have shown that collisional deactivation in multiphoton initiated unimolecular reactions by a cold bath gas resemble chemical or thermal activation systems. The deactivation should not be confused with enhancement, by collisions, in the multiphoton excitation process that is interpreted in terms of rotational equilibration or hole-filling in the discrete energy level region of the molecule (Stone and Goodman, 1979; Jang and Setser, 1979). As discussed earlier, rotational hole-filling seems to be unimportant in large molecules.

Collisional energy transfer from low vibration levels. Vibrational relaxation by collisions for low levels of vibrational excitation (up to ~ 4000 cm^{-1}) of small molecules has been quite thoroughly explored by laser-induced vibrational fluorescence techniques or ultrasonic or shock-tube experiments (Zittel and Moore, 1973; Yardley and Moore, 1968; Weitz and Flynn, 1973a and 1973b; Knudtson and Flynn, 1973). In the former technique, a single vibrational level is populated by laser irradiation, and infrared vibrational fluorescence from several excited levels is independently observed. After molecules are pumped to the desired excited vibrational state, molecules are de-excited to nearby vibrational levels via rapid intra- or intermolecular V–V energy transfer processes followed by slow V–T,R transfer. Vibrational-to-vibrational energy transfer rates are much faster than V–T,R rates, and the probability of V–T,R energy transfer has been found to decrease exponentially as the amount of energy transformed increases.

Vibrational energy usually flows into the lowest vibrational level and deactivated directly into translational and rotational energy from that level. The fast V–V relaxation, in which a much smaller amount of energy is transferred into translation and rotation, occurs by resonant or near-resonant processes. The V–V energy transfer process has been interpreted as resulting from the interaction between the vibrational transition dipoles of the collision partners (Hovis and Moore, 1980). Figure 2.22 depicts the probability for nearly resonant V–V transfer from the asymmetric stretch of CO_2 with collision partners with comparable transition dipole moments such as isotopic CO_2, N_2O, and OCS, and with collision partners with

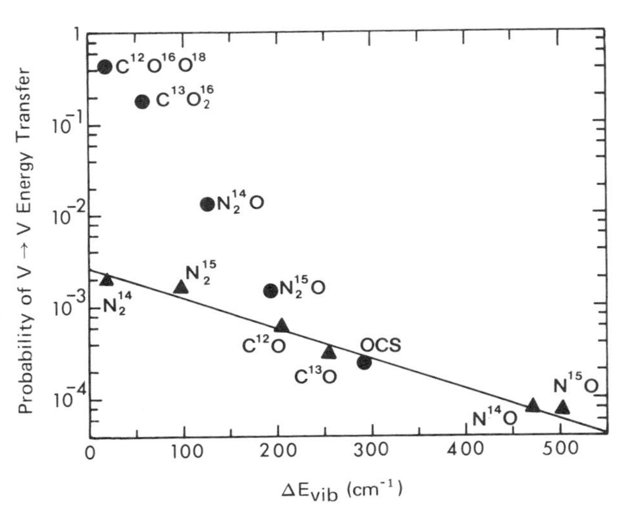

Figure 2.22. Probabilities for nearly resonant V–V transfer from the asymmetric stretch of CO_2. The isotopic CO_2 molecules, N_2O, and OCS have comparable transition dipole moments while N_2, CO, and NO have zero or small transition dipoles. (After Hovis and Moore, 1980.)

zero or small transition dipoles such as N_2, CO, and NO. Since resonant V–V transfer processes in the parent molecule produce little change in the total vibrational energy, the average vibrational quantum number of the molecules may remain the same during this process.

The slow V–T,R process (Hovis and Moore, 1980) may be primarily due to the repulsive component of the intermolecular potential, which leads to a transition probability that increases with increasing temperature. If the molecular translational velocity is much greater than the rotational velocities, the energy is transferred from vibration mostly to translation; SSH (Schwartz, Slawsky, and Herzfeld) theory is appropriate for such a process (Schwartz *et al.*, 1952; Schwartz and Herzfeld, 1954). The deactivation in SSH theory is proportional to $\exp(\mu/kT)^{1/2}$, where μ is the reduced mass. Vibrational-to-rotational transfer is more probable than V–T transfer when the rotational velocities of the atoms are greater than the molecular translational velocity and the rotational spacings are much less than kT. When V–R transfer is dominant, the probability can be obtained using a formulation similar to SSH theory (Yardley and Moore, 1968). In such a modification, the rotational angular velocity, ω, instead of translational velocity, and I/d^2 (where I is moment of inertia and d is radius of the rotor), instead of the reduced mass, are used. A combined V–T,R model formed from the above two models is appropriate for the case where rotational and translational velocities are comparable. In the V–T,R model (Zittel and Moore, 1973), the reduced mass in the probability function for the collision between two rotors is substituted by the mass parameter μ^+,

$$(\mu^+)^{-1} = m^{-1} + A_1 \mu R_1^{-1} + A_2 \mu_2^{-1}$$

where the A's are adjustable parameters depending on the asymmetry of the rotating molecule and the geometry at closest approach.

An example may best illustrate the above discussion. The ν_6 mode of CD_3H has been selectively excited and the vibrational energy redistribution, intermolecular vibrational energy transfer, and V–T,R energy transfer by several rare gases for this molecule have been mapped (Drozdoski *et al.*, 1977 and 1978). These results are summarized in Figure 2.23. The "curvature" of the CD_3I line in the plot of probability vs. reduced mass in Figure 2.23 is related to the increased importance of V $\rightarrow$ R transfer.

Since deactivation probabilities of low vibrational levels are three to four orders of magnitude lower than deactivation probabilities of vibrational levels above E_0, vibrational deactivation of low levels normally is not an important process. Vibrational deactivation in multiphoton experiments can be important in two general ways: (1) during the pulse, deactivation of intermediate energy levels may inhibit the ladder-climbing process and (2) after the pulse, the molecules with long lifetimes can be deactivated. We will consider explicitly the latter effect in the modeling studies that follow.

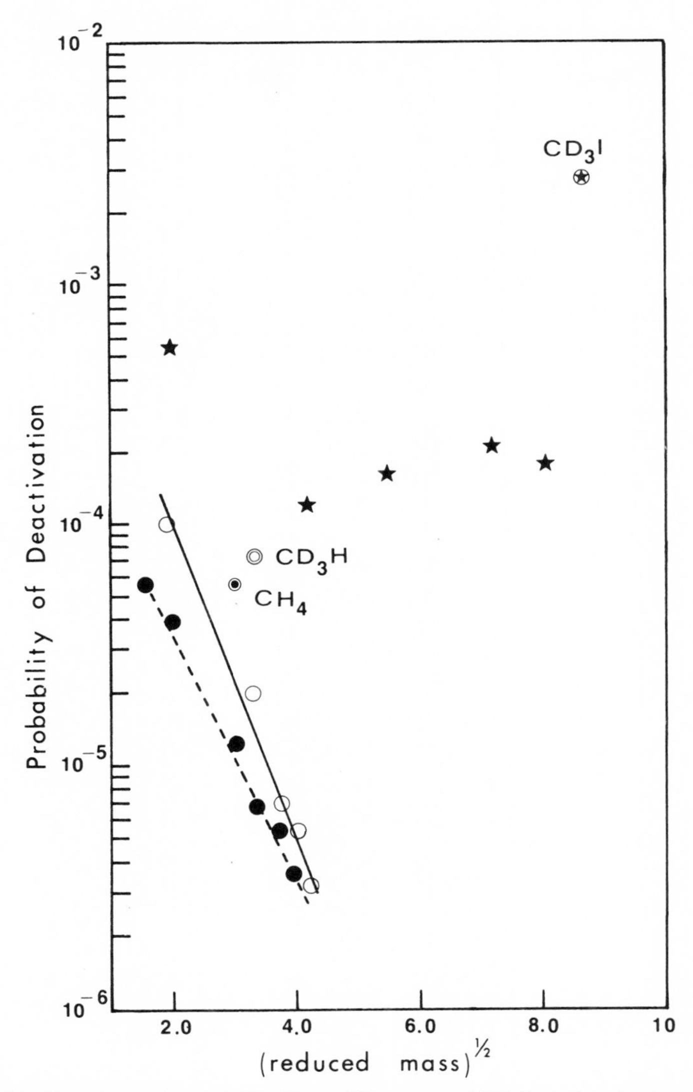

Figure 2.23. Plot of experimental CD_3H, ○; CH_4, ●; and CD_3I, ★ deactivation probabilities vs. the square root of the reduced mass of the collision partners. The circled symbols represent the self-deactivation points, whereas the uncircled symbols represent deactivation by collision with rare gas atoms. (After Drozdoski *et al.*, 1977.)

2.6.2. Model Calculations with a Master Equation Formulation for Large Organic Molecules

The unimolecular reaction models of multiphoton excitation and reaction discussed in the previous section have been used to explain phenomena observed in several small polyatomic molecules. It is necessary to examine the characteristics of large organic molecules as compared to small molecules in order to extend the models to the former.

Small polyatomic molecules are characterized by: (1) low vibrational densities at low levels resulting in coherent excitation between the first few discrete levels; (2) the importance of rotational structure and anharmonic splitting of degenerate modes for compensating anharmonicity in laser excitation; (3) high reaction threshold energies; and (4) short decomposition lifetimes.

For large organic molecules with small rotational constants and low degrees of symmetry, the rotational structure and anharmonic splitting of degenerate modes will not be as important in the multiphoton excitation process. On the other hand, the density of states in a large molecule rises so rapidly with energy and the thermal vibrational energy is so high at room temperature that the molecules may already be in the quasicontinuum state region. This may be countered to some extent, however, by strong optical selection rules that may considerably reduce the effective density of states for the multiphoton excitation process. Since the molecules may already be in the quasicontinuum region, the master equation [Eq. (2.75)] based on the three models discussed above are appropriate for large molecules. Other important characteristics of large molecules are frequently rather low reaction threshold energies but relatively slow decomposition rates, k_i, and, therefore, long lifetimes. Because of such long lifetimes, collisional processes are undoubtedly much more important for large organic molecules than for small molecules. To illustrate these differences, the RRKM rate constant $k_i(E)$ of ethyl acetate compared with $k_i(E)$ of CH_3CF_3 is shown in Figure 2.24.

We will now present and discuss model calculations for relatively large molecules with ethyl acetate as a specific example. Since, in ethyl acetate the average thermal vibrational energy at room temperature is 1080 cm^{-1} and the density of vibrational and rotational states at an energy of 1050 cm^{-1} is $> 10^3$ states/cm^{-1}, we can assume that, after the absorption of only a single photon, the molecule is already in the quasicontinuum region and that the excitation process therefore has incoherent characteristics. Such an assumption permits the use of master Eq. (2.76) based on the rate equation formulation, and we will specifically include unimolecular dissociation, collisional energy transfer, and absorption and emission of laser radiation.

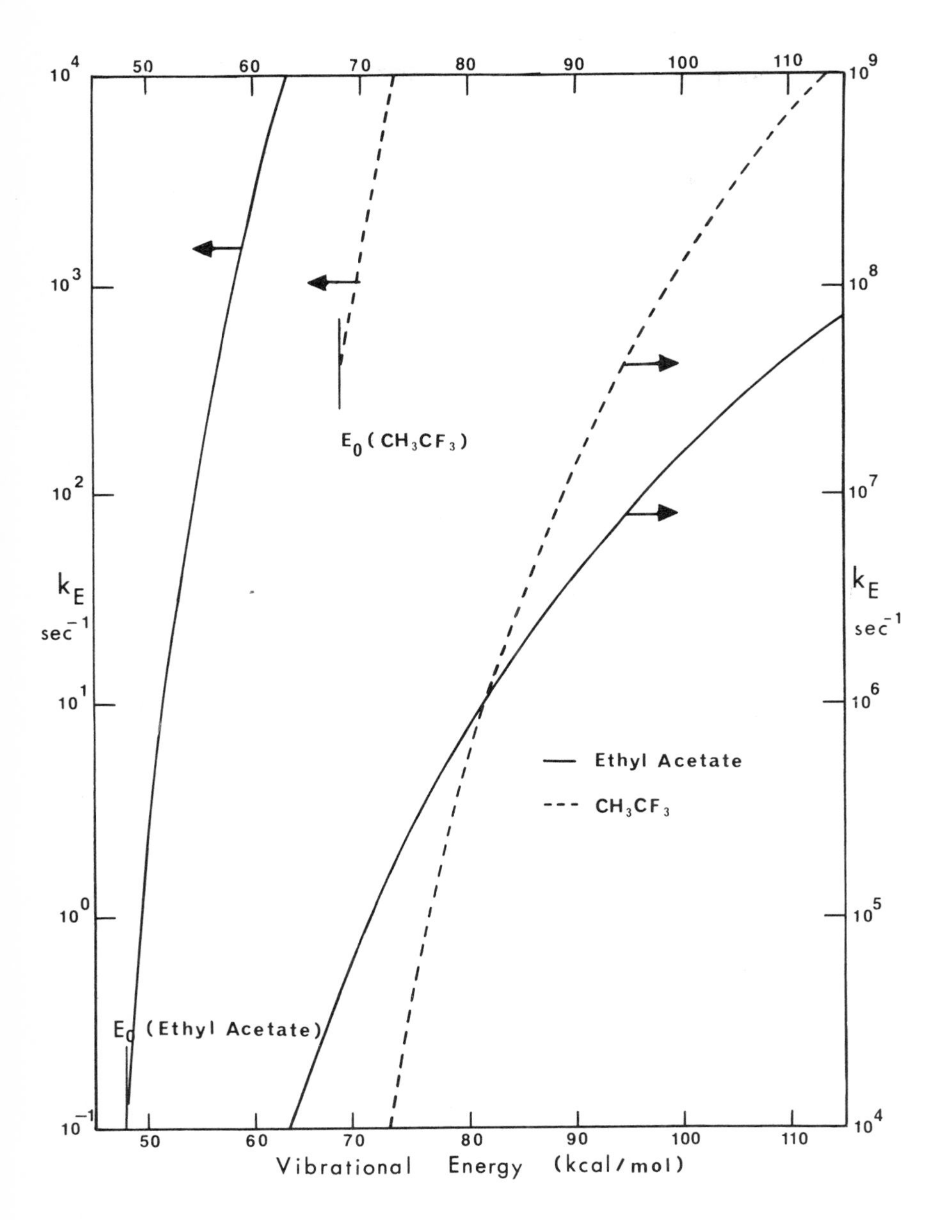

Figure 2.24. RRKM rate constants, k_E, vs. the vibrational energy of ethyl acetate (—) and CH_3CF_3 (---) as examples of a large molecule and small molecule, respectively. The reaction threshold energies, E_0, are 48 kcal/mol for ethyl acetate and 68 kcal/mol for CH_3CF_3.

As mentioned in the previous section, since the density of vibrational states increases very rapidly with energy in a large molecule, it is virtually impossible, or certainly not practical, to treat each individual vibrational state as a separate energy level. The vibrational states can be partitioned into levels of equal energy spacing as a degeneracy of the level, g_i. For most calculations, the energy level spacing, ΔE, was 1050 cm^{-1}, or the energy per CO_2 laser photon. The RRKM unimolecular rate constants were assumed for unimolecular dissociation rates, and single-photon transitions were assumed in the infrared absorption process. Two-photon or three-photon transitions were not included.

With the assumptions of incoherent absorption, the defined energy grid, RRKM rates, and single-photon transitions, we will discuss the effects of laser intensity, I, laser fluence (energy), ϕ, infrared absorption cross section, σ_{ij}, and collisional processes on the distribution function. For different values of these variables, the reaction yield, F_R^{∞}, number of photons absorbed per molecule, $\langle n \rangle$, and fraction of reaction during the laser pulse, F_R, will be compared. The input CO_2 laser pulse was assumed to resemble a square wave for most calculations.

2.6.2.1. *Effect of Laser Intensity*

Figures 2.25a–2.25c depict the vibrational energy distribution at the end of the laser pulse for pulsewidths of 100 and 300 nsec at the same fluence; i.e., the laser intensity is changed. The 100-nsec pulse is rather typical of a CO_2 TEA laser operating without N_2 in the gas mix, while the pulses of a more conventional CO_2–He–N_2 mix frequently "tail" for as long as 1–2 μsec. For all cases of Figure 2.25, a constant infrared absorption cross section was maintained; $\sigma_{ij} = \sigma_{10} = 3.3 \times 10^{-19}$ cm^2 was utilized, and induced emission was determined from detailed balance. Figures 2.25a–2.25c depict the distributions at fluences of 0.8, 1.6, and 3.0 J/cm^2 under collisionless conditions for the two different intensities.

If the collisional transition rate, $\omega P_{ij} N_j$, and unimolecular dissociation rate, $k_i N_i$, are negligible during the laser pulse, the rate equation can be written as Eq. (2.78). Recall that the distribution depends on only (intensity) $\times$ (time), $I \times dt$, which is simply the fluence, ϕ. Under collisionless conditions, perturbations on the fluence dependence are only because of unimolecular dissociation processes. At low fluences without collisions, where the fraction of reaction during the laser pulse, F_R, is negligible, the absorbed energy per molecule, $\langle n \rangle$, and the vibrational energy distributions for the two different intensities are the same and the overall reaction yields, F_R^{∞}, are, therefore, identical for the same fluence but different pulselengths (Figures 2.25a and 2.25b). For collisionless conditions, F_R^{∞} is the same as the fraction of molecules above the reaction

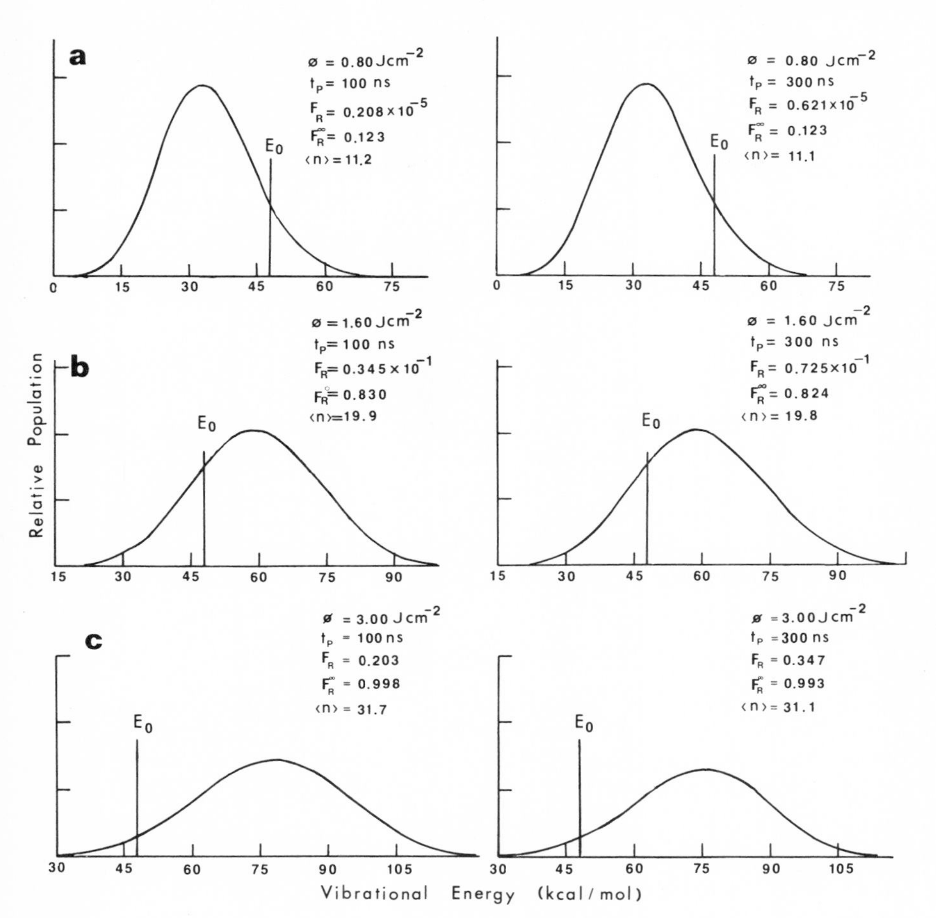

Figure 2.25. Vibrational energy distribution of ethyl acetate immediately following the laser pulse for two laser pulselengths, t_p (100 and 300 nsec) at three different fluences, ϕ, under collisionless conditions. A rectangular pulse shape for both the 100- and 300-nsec pulses and a constant absorption cross section for the infrared pumping with stimulated emission were used in the calculations. The fluences are (a) 0.8 J/cm^2, (b) 1.6 J/cm^2, and (c) 3.0 J/cm^2.

threshold energy, E_0. At high fluence, F_R becomes important and is dependent upon the laser intensity. Even at fluences such that F_R is 20 ~ 30%, however, the laser intensity dependence on $\langle n \rangle$, the distribution, and reaction yield is small. For the case of Figure 2.25c where the F_R is 20–35%, there is a threefold change in intensity, but $\langle n \rangle$ and $F_R{}^\infty$ were changed only negligibly.

Figures 2.26a and 2.26b depict cases similar to Figure 2.25 except that the pressure was 0.025 Torr for Figure 2.26a and 1.0 Torr for Figure 2.26b. Since, as discussed above, unimolecular dissociation during the laser pulse

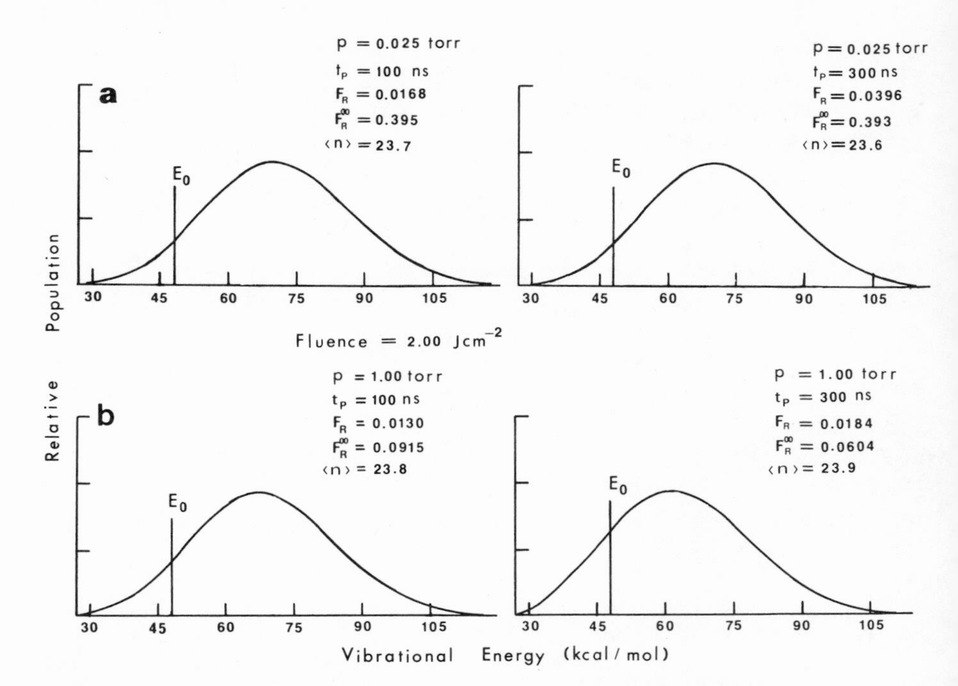

Figure 2.26. Calculated vibrational energy distribution of ethyl acetate for two different pulses (100 and 300 nsec) and two different pressures [(a) 0.025 Torr and (b) 1.0 Torr]; fluence = 2.0 J/cm^2. A constant absorption cross section with stimulated emission for the infrared pumping and a stepladder model with $\Delta E_d = 6$ kcal/mol for the collisional deactivation were used in calculations.

was virtually insensitive to laser intensity except at high fluences, the perturbation because of collisions can be shown for these two pressures at a fluence of 2.0 J/cm^2. A stepladder model was used to calculate collisional relaxation assuming a deactivation energy per collision $\Delta E_d = 6$ kcal; the collisional frequency, ω, for collisions of ethyl acetate molecules at 1 Torr was 9.6×10^6 sec^{-1}. As shown in Figures 2.26a and 2.26b the reaction yields $F_R{}^\infty$ are changed $\lesssim 0.5\%$ at 0.025 Torr in changing from a 100-nsec pulse to a 300-nsec pulse, while the $F_R{}^\infty$ are reduced about 30% at 1.0 Torr. At both pressures, modification of the distribution because of unimolecular decomposition during the pulse, F_R, are small ($< 4\%$). The collisional transition rates, ωP_{nj}, are $\sim 10^8$ sec^{-1} at 1 Torr and $\sim 2.5 \times 10^5$ sec^{-1} at 0.025 Torr, while the infrared absorption transition rate is $10^7 \sim 10^8$ sec^{-1} at 2 J/cm^2. Therefore, deactivation because of collisions during the laser pulse are expected to be important at 1 Torr and a fluence of 2 J/cm^2.

In conclusion, at low fluences where F_R is not high and at low pressures for which ω is $5 \sim 10$ times smaller than the absorption rate, the intensity dependence on the number of photons absorbed per molecule, $\langle n \rangle$, and the reaction yield, F_R^{∞}, is negligible. Therefore, a Gaussian-shaped pulse will be expected to give similar results on $\langle n \rangle$ and F_R^{∞} as the square wave pulse. Indeed, some calculations with a Gaussian pulse were performed and gave identical results. The laser intensity is unimportant in determining $\langle n \rangle$ and F_R^{∞} unless very high fluences corresponding to high reaction yields or high pressures are employed.

It is important to emphasize that the calculated F_R and F_R^{∞} values discussed in this section cannot be compared easily with any experimental quantities since important postpulse collisional phenomena are not included in the calculated quantities. In this regard, however, Lyman *et al.*, (1979) observed for SF_5NF_2 that the absorption cross section and, presumably, $\langle n \rangle$ were not affected by a > 100-fold change in laser intensity; SF_5NF_2 appears to have many of the properties of a large molecule. We have preliminary data for ethyl acetate that, likewise, suggest a minimal dependence of reaction probability at < 0.1 Torr and constant fluence for ~ 10-fold change in the laser pulselength (150 vs. ~ 1300 nsec).

2.6.2.2. *Effect of Laser Energy*

The dependence of F_R^{∞} on the laser energy (fluence) is calculated as shown in Figure 2.27 for three different absorption cross sections, σ_{ij}, under collisionless conditions. In the case of curve a of Figure 2.27, the absorption cross section increased with excitation level by the amount $\sigma_{ij} = \sigma_{10} \times i^{1/2}$. For curve b of Figure 2.27, σ_{ij} was assumed the same for all levels and in the case of curve c, σ_{ij} decreased with excitation by $\sigma_{ij} = \sigma_{10}/i^{1/2}$ with the stimulated emission determined by detailed balance, i.e., $\sigma_{ij} = (g_i/g_j)\sigma_{ji}$ in which g_i is the degeneracy of level i. The calculations show that the total reaction yields are highly dependent on the laser fluence for all three cases. The slopes of the plot of log F_R^{∞} vs. log ϕ for low fluence conditions were ~ 6–8, which is in the range of that obtained experimentally (see Figure 2.17) if F_R^{∞} is associated with yield. Since the reaction yield is so highly dependent on fluence, very accurate measurements of the laser fluence are necessary in such experiments. Figure 2.28 shows the change in the reaction yield illustrating the effect of experimental error in the measurement of fluence. The inner pair of dashed lines corresponds to an error of ± 20%, and the outer dashed lines would result when the error is ± 50% for the case of constant absorption cross section with no collisions. At low fluences, e.g., 0.5 J/cm^2, a 20% error in energy gives a change in the reaction yield by factor of ~ 5. At higher fluences, this error is not as serious according to the model calculations. Under static bulb conditions for which collisional effects are significant,

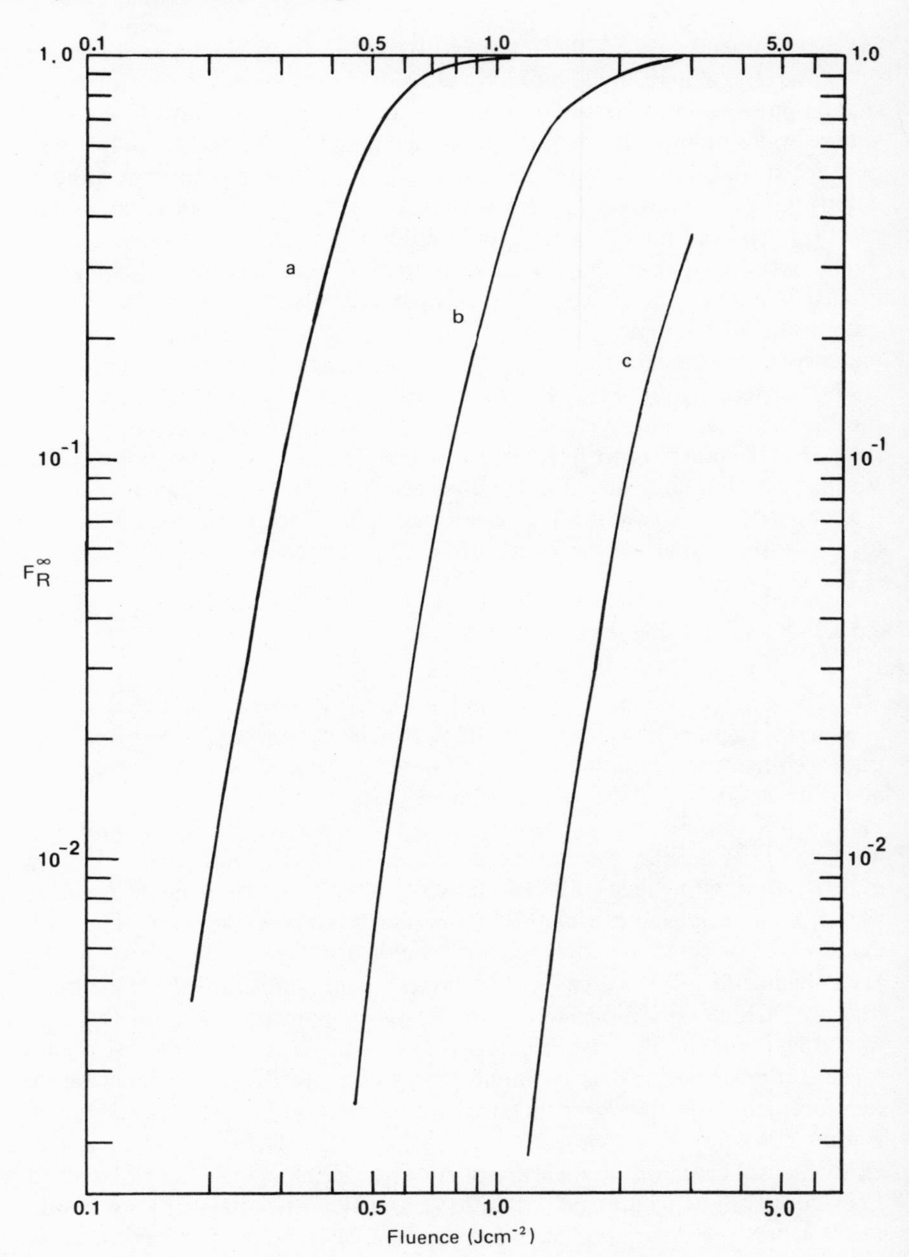

Figure 2.27. Calculated reaction yield, F_R^{∞}, vs. fluence for ethyl acetate under collisionless conditions with three different absorption cross sections: (a) increasing cross section with increasing excitation; (b) constant section section; and (c) decreasing cross section. All cases include stimulated emission determined by detailed balance.

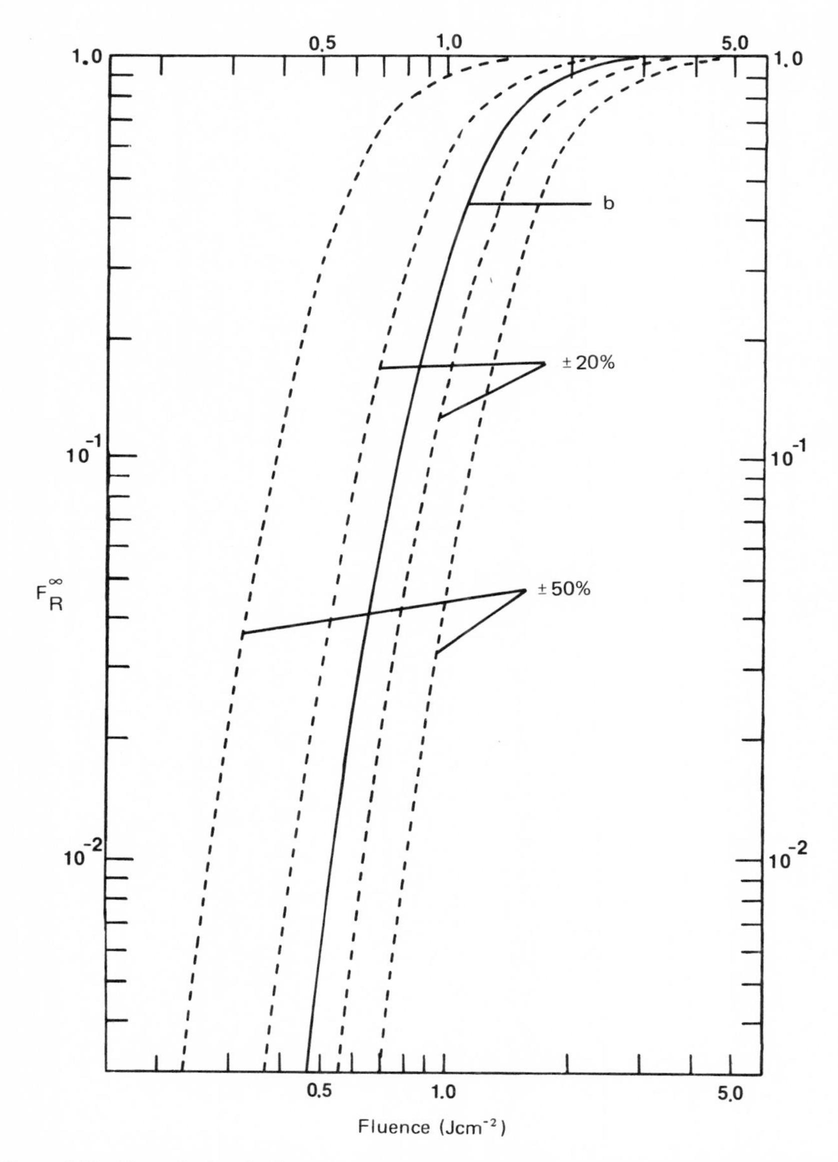

Figure 2.28. The calculated effect of uncertainty in the measurement of laser fluence on the reaction yield, F_R^{∞}, assuming collisionless conditions and a constant absorption cross section. Curve b is the same as curve b in Figure 2.27. The dotted lines indicate the calculated F_R^{∞} assuming errors in the fluence measurements of ± 20 or $\pm 50\%$.

however, the curve of log $F_R{}^\infty$ vs. log ϕ or $\langle n \rangle$ shifts towards higher ϕ (see Figure 2.33) and the errors in measurement of energy are serious even at relatively high fluences.

2.6.2.3. *Effect of Absorption Cross Section*

Four different functions were chosen for the absorption cross section, σ_{ij}, to determine the effects on the distribution function and absorbed energy. These included (a) increasing σ_{ij}, (b) maintaining a constant σ_{ij} with excitation level, (c) decreasing σ_{ij}, and (d) holding σ_{ij} constant without including stimulated emission. For (a), σ_{ij} was increased by $\sigma_{10} \times i^{1/2}$, where σ_{10} is the absorption cross section for the lowest level and σ_{ij} is the absorption cross section from level j to level i. This resulted in σ_{ij} being increased four times at the threshold energy for ethyl acetate ($E_0 = 48$ kcal/mol) and seven times at the highest level utilized in the calculations (150 kcal/mol). For case c, σ_{ij} was decreased by $\sigma_{10}/i^{1/2}$. For cases a, b, and c the emission cross section was three times smaller than the absorption cross section at E_0 and 1.5 times smaller at 150 kcal/mol.

Since we assumed a single-photon transition under collisionless conditions with small F_R, Eq. (2.78a) can be rewritten as Eq. (2.79)

$$\frac{dN_i}{I\,dt} = \sigma_{i,\,i-1} N_{i-1} + \sigma_{i,\,i+1} N_{i+1} - \sigma_{i-1,\,i} N_i - \sigma_{i+1,\,i} N_i \tag{2.79}$$

$$d\langle n \rangle = \frac{1}{N} \sum_{i=1} (i-1)\, dN_i \tag{2.80}$$

or

$$\frac{d\langle n \rangle}{dt} = \frac{1}{N} \sum_{i=1} (i-1) \frac{dN_i}{dt} \tag{2.81}$$

$$\frac{d\langle n \rangle}{I\,dt} = \frac{d\langle n \rangle}{d\phi}$$

$$= \frac{1}{N} [\sigma_{21} N_1 + (\sigma_{32} - \sigma_{12}) N_2 + \cdots (\sigma_{n+1,\,n} - \sigma_{n-1,\,n}) N_n + \cdots] \tag{2.82}$$

$$= \frac{1}{N} [\sigma_1' N_1 + \sigma_2' N_2 + \sigma_3' N_3 + \cdots \sigma_n' N_n + \cdots] \tag{2.83}$$

The number of photons absorbed per molecule, $\langle n \rangle$, can be calculated by Eqs. (2.80) or (2.81). Substituting dN_i/dt from Eq. (2.79) into Eq. (2.81) gives Eq. (2.83) where σ_i' is the net absorption cross section. If σ_i' is constant for all levels, i.e., $\sigma_i' = \text{constant} = \sigma'$, then

$$\frac{d\langle n\rangle}{d\phi} = \frac{1}{N}\sigma_i'[N_1 + N_2 + \cdots + N_n + \cdots]$$
$$= \frac{1}{N}\sigma' N$$
$$= \sigma'$$

Thus, the absorbed energy per molecule is linearly dependent on the laser fluence with the proportionally constant, σ' (see d of Figure 2.29). As anticipated, an increasing absorption cross section ($\sigma_{ij} = \sigma_{10} \times i^{1/2}$) causes the slope of $\langle n\rangle$ vs. ϕ to increase (see a of Figure 2.29). For constant σ_{ij} but

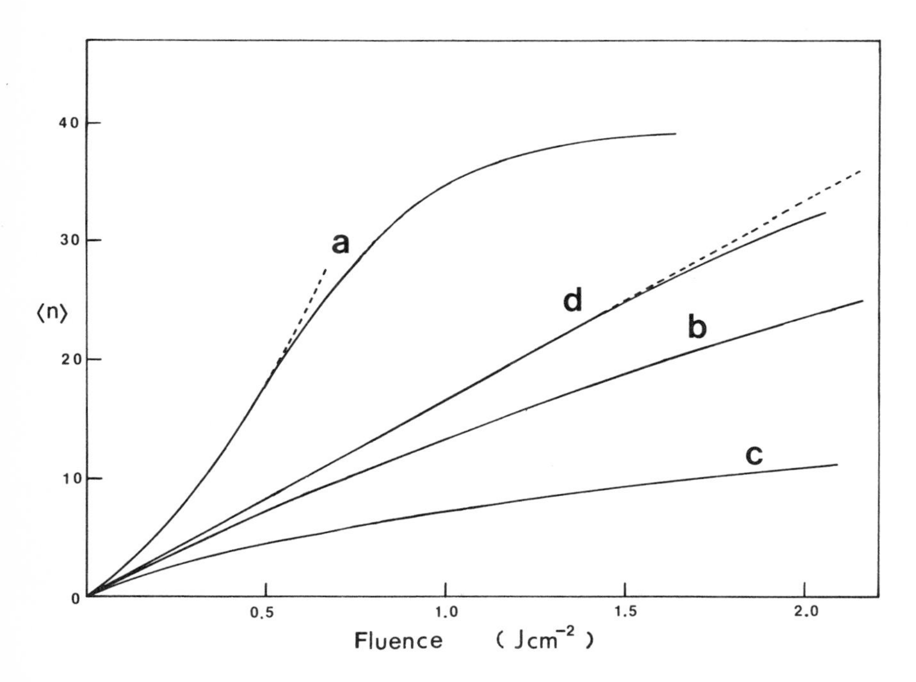

Figure 2.29. Number of photons absorbed per molecule, $\langle n\rangle$, as a function of fluence for four different absorption cross sections under collisionless conditions: (a) increasing absorption cross section with excitation level; (b) constant absorption cross section; (c) decreasing cross section; (d) constant absorption cross section without stimulated emission (i.e., constant net absorption cross section). Cases a, b, and c include stimulated emission. Dotted lines indicate cross section if unimolecular decomposition during the laser pulse, F_R, is negligible.

with stimulated emission (curve b) or decreasing σ_{ij} (curve c) the slope of $\langle n \rangle$ vs. ϕ is decreased relative to curves a and d as expected.

The bulk, effective laser absorption cross section, σ_L, is that often measured experimentally. It can be calculated simply as $\sigma_L = \langle n \rangle h\nu$ and measured experimentally according to Eq. (2.84) in which ϕ_0 is the incident laser fluence, ϕ is the fluence exiting through a sample distance of Δx, and N is the density of molecules in the irradiated volume. The results of the calculations for the four types of σ_{ij} are shown in Figure 2.30.

$$\phi/\phi_0 = \exp\left[-(\sigma_L N \, \Delta x)\right] \tag{2.84}$$

Case d of Figure 2.30 is for a constant σ_{ij} without stimulated emission, i.e., a constant net absorption cross section. It is anticipated that such a cross section would be insensitive to changes in fluence and, indeed, the cross section is constant up to a fluence of ~ 1.2 J/cm^2 after which it begins to drop slowly. The decrease at high fluence is because of uni-

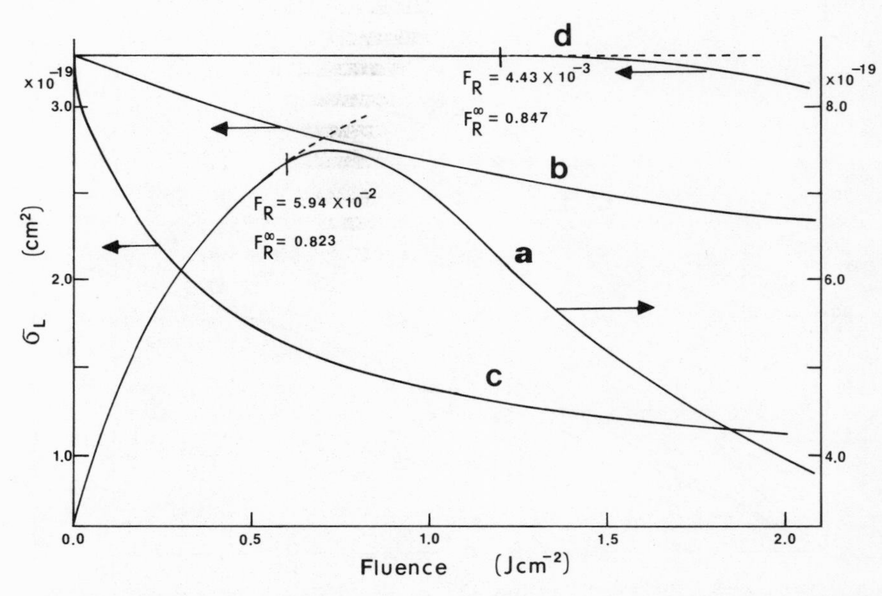

Figure 2.30. The effective laser absorption cross section, σ_L, vs. fluence for four different absorption cross sections. Curves a, b, c, d have the same conditions as defined for Figure 2.29. Dotted lines are expected if unimolecular decomposition during the laser pulse, F_R, is negligible. Deviation because of unimolecular decomposition for case a onsets at ~ 0.6 J/cm^2 ($F_R = 5.94 \times 10^{-2}$ and $F_R{}^{\infty} = 0.823$); for case d onset occurs at ~ 1.2 J/cm^2 ($F_R = 4.43 \times 10^{-3}$ and $F_R{}^{\infty} = 0.847$). For case b, $F_R = 0.396 \times 10^{-1}$ and $F_R{}^{\infty} = 0.909$ at 2.0 J/cm^2 and for case c, $F_R = 0.115 \times 10^{-6}$ and $F_R{}^{\infty} = 0.0674$ at 2.0 J/cm^2. For the latter two cases, no significant deviation because of unimolecular decomposition during the laser pulse is obtained at fluences < 2 J/cm^2.

molecular dissociation during the laser pulse. At 1.2 J/cm^2, F_R is negligible (4.43×10^{-3}) while F_R^∞ is already 0.847. At 2.0 J/cm^2, $F_R = 0.318$, which results in a slight decrease in the cross section; $F_R^\infty = 0.999$ at this fluence.

Case b of Figure 2.30 represents a constant σ_{ij} but including stimulated emission. This is the cross section utilized to generate the vibrational energy distributions at various fluences and two different pulselengths shown in Figure 2.25. As anticipated, such a cross section exhibits a significant, systematic decrease as the fluence increases. Dropoff because of reaction during the laser pulse is calculated to be insignificant for case (b) at fluences less than ~ 2.0 J/cm^2; F_R is only 3.96×10^{-2} at 2.0 J/cm^2. It may be noted that experimental plots of σ_L vs. ϕ (Figures 2.18 and 2.19) resemble more closely cases b and d of Figure 2.30 than case a (increasing σ) or case c (decreasing σ).

An increasing σ_{ij} with stimulated emission (case a of Figure 2.30) shows a rapid rise with increasing fluence to a value approximately twice the single-photon value at a fluence ~ 0.7 J/cm^2. An even higher rise would result except for depletion of molecules during the laser pulse at higher fluences. At 0.6 J/cm^2, F_R is already 5.94×10^{-2} and $F_R^\infty = 0.823$; at only 0.8 J/cm^2, $F_R = 0.309$ and $F_R^\infty = 0.970$. Therefore, it is apparent that perturbations because of unimolecular dissociation during the pulse on the absorption process become serious at fluences greater than ~ 0.6 J/cm^2 or $F_R \sim 0.1$ for an increasing σ_{ij}.

Case c of Figure 2.30 depicts a decreasing σ_{ij} including stimulated emission. This cross section falls more rapidly at low fluences than any of the other cases and results in a value for the cross section approximately one-half the single-photon value at a fluence ~ 0.5 J/cm^2.

In summary, the experimental laser cross section, σ_L, can be determined at low fluences where $F_R < 0.1$, but the unimolecular dissociation rate must be taken into account at high fluences where $F_R > 0.1$. An $F_R > 0.1$ corresponds to a quite high fraction of total reaction. Although F_R^∞ is not simply related to yield in a bulb experiment, it is probably realistic to estimate an observed reaction probability > 0.5 for $F_R > 0.1$.

Figure 2.31 shows the vibrational distributions resulting from absorption of energy equivalent to ~ 13.5 infrared photons for three different absorption cross sections compared with the Boltzmann distribution of the same average energy. Since ethyl acetate has vibrational energy equivalent to one photon at room temperature, the average energy of these four distributions are the same and equivalent to ~ 14.5 photons. The distribution gets increasingly broader in going from the case of decreasing σ_{ij} (curve c) to the case of increasing σ_{ij} (curve a). Only the distribution with increasing σ_{ij} is broader than a Boltzmann distribution. The variation of F_R^∞ with distribution (Figure 2.32), however, is not very high at moderate $\langle n \rangle$ values (i.e., $10 \lesssim \langle n \rangle \lesssim 35$), where the fraction of reaction during the

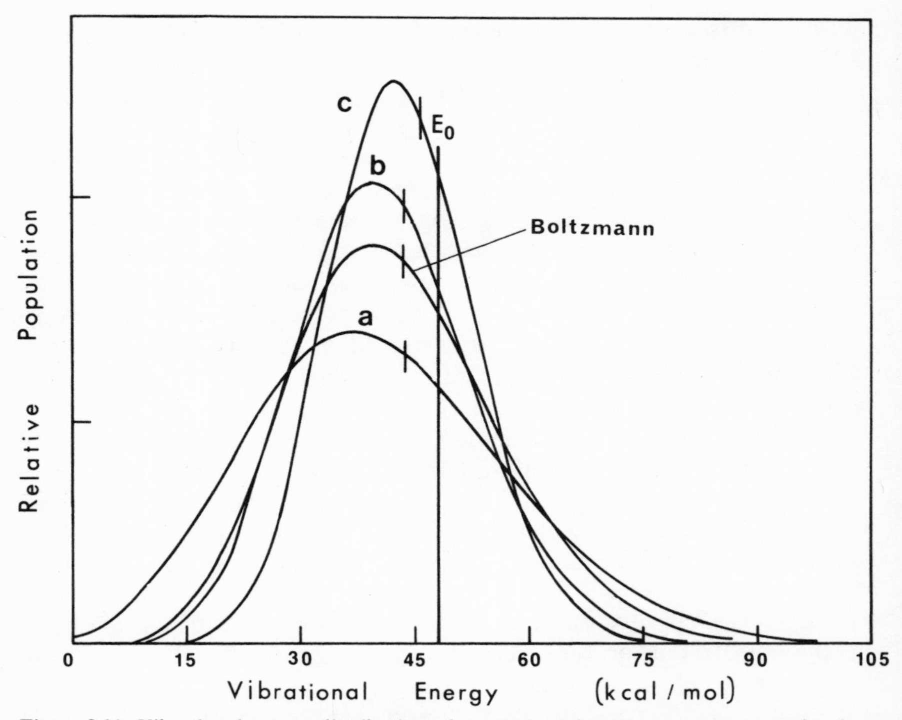

Figure 2.31. Vibrational energy distribution of approximately same average energy for three different absorption cross sections compared with the Boltzmann distribution of average energy of 14.5 photons. (a) Increasing absorption cross section, $\langle n \rangle + E_{th} = 14.3$, $\phi = 0.4$ J/cm^2, $F_R = 0.15 \times 10^{-2}$; (b) constant absorption cross section, $\langle n \rangle + E_{th} = 14.5$, $\phi = 1.0$ J/cm^2, $F_R = 0.26 \times 10^{-4}$; (c) decreasing absorption cross section with increasing excitation level, $\langle n \rangle + E_{th} = 15.2$, $\phi = 3.0$ J/cm^2, $F_R = 0.23 \times 10^{-4}$. E_{th} is the thermal energy of ethyl acetate at 300°K.

laser pulse does not disturb the distribution appreciably (e.g., at $\langle n \rangle = 25$, $F_R \simeq 0.05$). The $\langle n \rangle$ values that give $F_R{}^{\infty} = 10^{-1}$ are within ~ 9.5–11.5 for the three different cross sections and the Boltzmann distribution. Such differences in $\langle n \rangle$ value or in reaction yield would be sufficiently small to be within most experimental error limits.

We can conclude for practical purposes that for conditions such that $F_R \leq 0.10$, a Boltzmann distribution is a good approximation for the vibrational distribution existing at the termination of the laser pulse. This conclusion will need modification if collisions occur during the pulse. This will lower the mean energy and broaden (to low energy) the distribution. This realistic assumption of a Boltzmann distribution saves effort and expense by allowing one to ignore the calculation of the master equation up to the time when the laser pulse is over.

Although F_R and $F_R{}^{\infty}$ are relatively insensitive to the distributions for

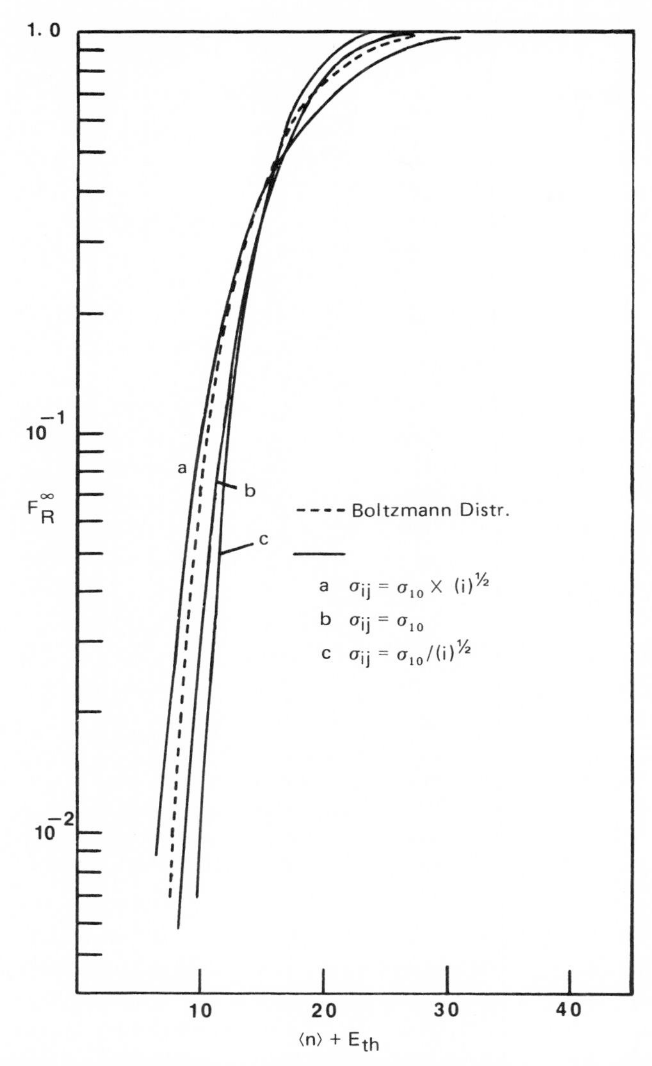

Figure 2.32. Reaction yield vs. the average energy of molecule (i.e., $\langle n \rangle + E_{th}$) for three different types of absorption cross section compared with the Boltzmann distribution under collisionless conditions. (a) Increasing absorption cross section; (b) constant absorption cross section; (c) decreasing absorption cross section.

the same average energy obtained from four different cross sections (Figure 2.32), the dependence of the reaction yield on the incident energy, ϕ, changes greatly with the absorption cross section (Figure 2.27). Highly accurate measurements of the laser fluence are important to establish the reaction yield dependence on the fluence and σ_L.

2.6.2.4. *Postpulse Collisional Effects*

Although collisions will not affect the vibrational energy distribution during the laser pulse at pressures of $\lesssim 0.1$ Torr, the reaction yield, F_R^{∞}, can be changed dramatically even at very low pressures because of deactivation of molecules with small k_E values by postpulse collisions. These slow unimolecular rates and the possibility for collisions with both cold and excited molecules make for a complex interplay of pressure effects.

A relatively simple situation involving collisional effects is the case in which the reaction cell contains large amounts of an inert bath gas and very small amounts of reactant. In such a situation, the inert gas has a large heat capacity and acts as a heat reservoir. All collisions tend to deactivate excited reactant molecules. A similar situation is that in which the reaction cell contains only reactant molecules but the irradiated volume is so small that the predominant collisional process is between a vibrationally hot reactant molecule and vibrationally cold reactant molecules. In this section we will present calculations addressing collisional problems but only for cases in which the main process is collision between an excited reactant molecule and a bath of cold molecules.

Figure 2.33 demonstrates the change in reaction yield as a function of the number of photons absorbed for 0.05 Torr of ethyl acetate with and without collisional effects. A constant infrared absorption cross section was assumed, and a stepladder model was used for the collisional probability with collisional deactivation energies $\Delta E_d = 350\ \text{cm}^{-1}$ (1.0 kcal/mol), 1500 cm^{-1} (4.3 kcal/mol), and 3000 cm^{-1} (8.6 kcal/mol); a collisional frequency $\omega = 9.6 \times 10^6\ \text{sec}^{-1}\ \text{Torr}^{-1}$, which corresponds to a polyatomic bath gas of mass $\sim$88, was utilized.

The stepladder collisional probability was used for upper vibrational levels from 10 kcal/mol below E_0 to the highest necessary in the master equation. As discussed in Section 2.6.1.4b, there is a lack of knowledge of collisional transfer probabilities for low levels of large molecules, but such probabilities are expected to range from the stepladder model to $P_{ij} \simeq 10^{-2}$ for the first few levels. Therefore, we assumed a zero probability for the lower levels. For example, collisional transition rates even with $P_{ij} \simeq 1.0$ are much slower than infrared absorption rates for these pulsed experiments. For ethyl acetate with $\sigma_{ij} = 3.3 \times 10^{-19}\ \text{cm}^2$ and $\omega = 9.6 \times 10^6\ \text{sec}^{-1}\ \text{Torr}^{-1}$, the collisional transition rate is five times slower than the infrared transition rate when the fluence is 1 J/cm^2 with a 300-nsec pulse and the pressure is 1 Torr with $P_{ij} \simeq 1.0$. Therefore, small changes in collisional probabilities for the lower levels will not change the vibrational energy distribution during the laser pulse. After the laser pulse has terminated, collisional transitions between levels below E_0 will not contribute to the overall reaction yield, F_R^{∞}, and can be ignored.

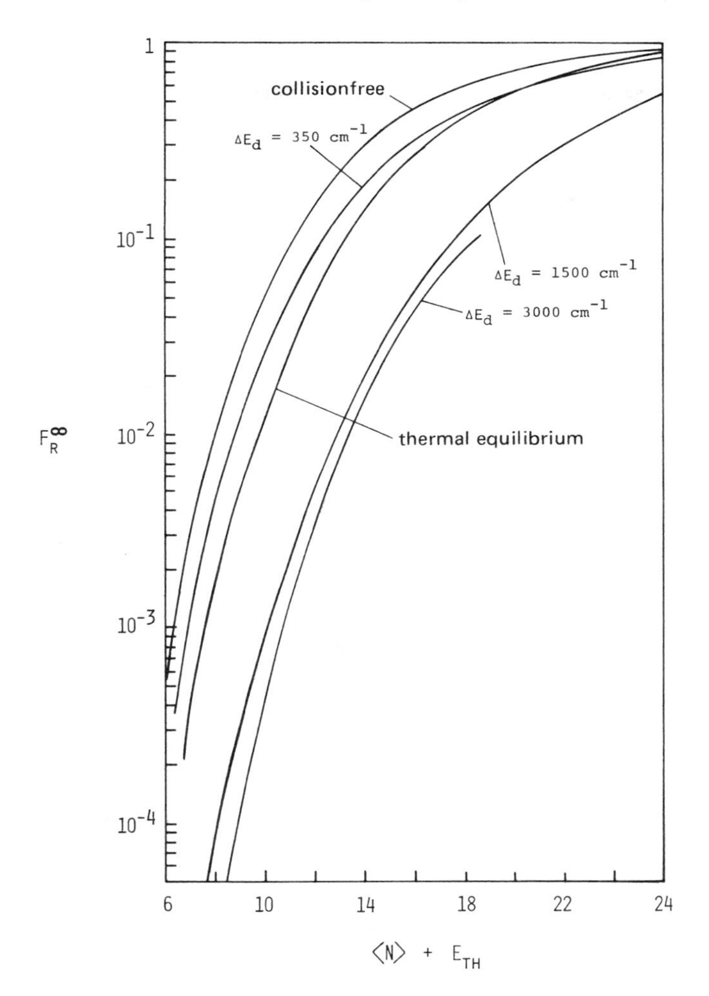

Figure 2.33. Plot of calculated reaction probability, F_R^{∞}, of ethyl acetate vs. average number of 1046.85 cm^{-1} photons absorbed, $\langle n \rangle$; one photon = 2.99 kcal/mol. Upper trace assumes no deactivating collisions with bath gas. Decreased F_R^{∞} values calculated with a stepladder model for collision with 0.05 Torr of a polyatomic bath gas of infinite heat capacity assuming $\Delta E_d = 350$ cm^{-1} (1.0 kcal/mol), 1500 cm^{-1} (4.3 kcal/mol), and 3000 cm^{-1} (8.6 kcal/mol). The thermal equilibrium curve depicts the decrease in F_R^{∞} for 0.05 Torr of vibrationally excited ethyl acetate relaxing to a thermal equilibrium (V,R,T equilibration) without collisions with bath gas.

The effect of collisions on F_R^{∞} for ethyl acetate interacting with 0.05 Torr of a polyatomic bath gas assuming various ΔE_d values is depicted in Figure 2.33 as a function of absorbed energy; similarly shaped curves would result if F_R^{∞} were plotted vs. fluence. The curve labeled "collision-free" is the yield assuming all molecules above the threshold energy react.

The effect of collisions is to reduce dramatically the yield with the extent of reduction increasing with ΔE_d and decreasing with higher average absorbed energy. For 10 absorbed photons and a $\Delta E_d = 1500\ \text{cm}^{-1}$ (4.3 kcal/mol), the yield is reduced by a factor of ~ 50 for ethyl acetate; for $\langle n \rangle = 15$, the reduction is ~ 10. For $\Delta E_d = 350\ \text{cm}^{-1}$ (1 kcal/mol), the deactivating effect is quite small and not very dependent on pressure; i.e., essentially the same result was obtained at 0.1 and 0.01 Torr. The reason the small ΔE_d is relatively ineffective is related to the large breadth of the distribution function, from which the loss of 1–2 kcal/mol does not significantly perturb F_R^{∞}.

Figure 2.34 is a Stern–Volmer plot for quenching of ethyl acetate, i.e., a plot of $(F_R^{\infty})_0/F_R^{\infty}$ vs. pressure of polyatomic bath gas where $(F_R^{\infty})_0$ is

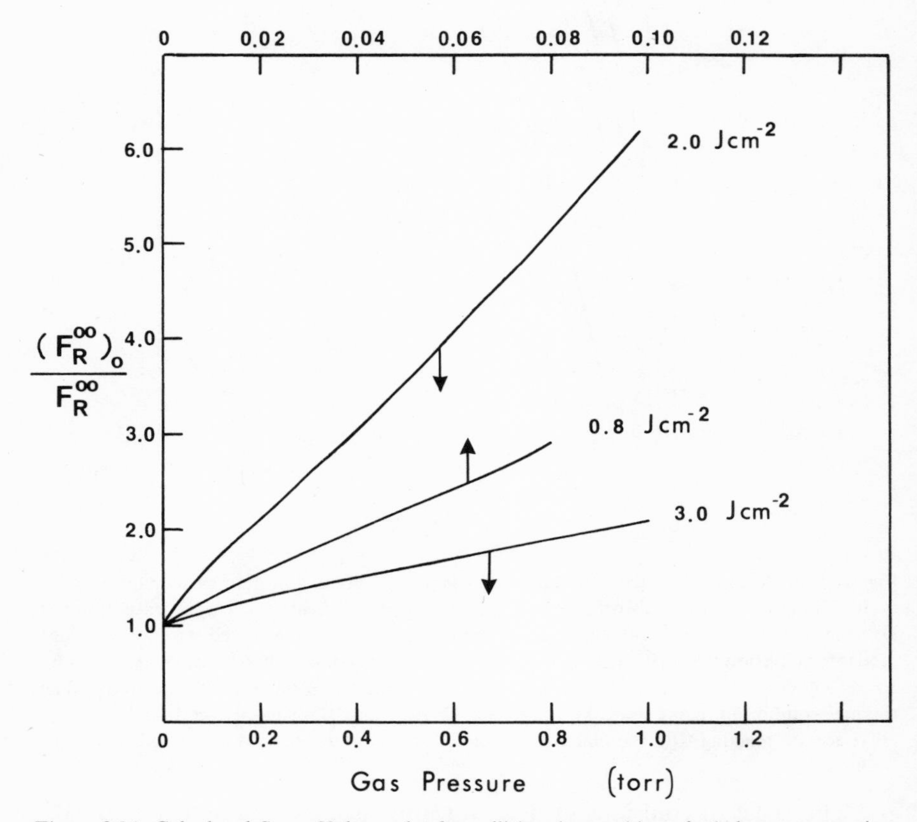

Figure 2.34. Calculated Stern–Volmer plot for collisional quenching of ethyl acetate reaction by a polyatomic bath gas. Collisional deactivation was calculated assuming a stepladder model with $\Delta E_d = 2100\ \text{cm}^{-1}$ (6.0 kcal/mol). $(F_R^{\infty})_0$ is the calculated reaction yield without collisions and F_R^{∞} is the reaction yield at the corresponding pressure of bath gas. Fluences are 0.8 J/cm², 2.0 J/cm², and 3.0 J/cm².

the reaction yield without collisions and F_R^{∞} is the yield at the indicated pressure. A $\Delta E_d = 2100$ cm^{-1} (6.0 kcal/mol) was utilized in the calculations. The collisional deactivation efficiency is highly dependent on the laser fluence with the change in reaction yield for the same pressure of bath gas greater at lower fluence. The explanation for this observation is that since the average energy of the molecules is lower at low fluence than high fluence, the effective mean unimolecular rate constant is smaller at lower fluence, and, for the same pressure of bath gas, a higher fraction of molecules will be deactivated by collisions.

The calculations in Figure 2.34 resemble the experimental conditions of Figure 2.15 in which cold, unexcited isopropyl bromide bath molecules very efficiently remove energy from excited ethyl acetate causing a pronounced detrimental effect on the reaction yield. The calculations are in qualitative agreement, at least, with experimental observations for bath gas effects.

The above discussion related to the collisional processes of a laser-excited, vibrationally hot reactant molecule interacting with a cold, inert bath gas of infinite heat capacity and could be modeled reasonably well. A discussion of collisional effects for a system comprised of only neat reactant molecules is much less tractable. We have demonstrated for pure ethyl acetate the yield per pulse is relatively insensitive to changes in reactant pressure at least for the region < 0.05 Torr and concluded that the extent of reaction must be controlled by the "cooling process" as discussed in Section 2.4.4. This phenomenon probably involves both molecular and bulk effects and is presumably very complex and too poorly understood to even begin to model. We will discuss this situation only in general terms.

Consider a static gas cell containing only reactant molecules in which a certain volume element is irradiated by the CO_2 laser. The molecules that encounter the laser radiation will absorb photons and become vibrationally excited although to different extents. If there were no collisional interactions between the vibrationally hot molecules within the irradiated volume and the surrounding cold molecules and only collisions among vibrationally excited molecules occurred, these collisions would lead to a thermal equilibrium within the irradiated volume. This equilibrium would be characterized as a steady-state distribution among all the vibrational, rotational, and translational degrees of freedom of the excited molecules. From energy conservation considerations, the redistribution of the initially deposited vibrational energy would occur as shown in Eq. (2.85).

$$\langle E_{\text{vib}} \rangle_T + \langle E_{\text{rot}} \rangle_{300^\circ\text{K}} + \langle E_{\text{trans}} \rangle_{300^\circ\text{K}} \rightarrow \langle E'_{\text{vib}} \rangle_{T'} + \langle E'_{\text{rot}} \rangle_{T'} + \langle E'_{\text{trans}} \rangle_{T'} \quad (2.85)$$

The terms on the left side of the equation are the energies of molecules immediately after excitation by the laser; i.e., the molecules possess vibrational excitation corresponding to a characteristic temperature T but are rotationally and translationally cold (room temperature or 300°K). The terms on the right side are the energies of the molecules within the irradiated volume after thermal equilibration among all vibrational, rotational, and translational degrees of freedom. This final partitioning of the initially deposited laser energy will be characterized by a Boltzmann distribution with temperature T'. It is important to realize that the heat capacity of a large molecule results primarily from the vibrational component: $\langle E_{\text{vib}} \rangle_T - \langle E'_{\text{vib}} \rangle_{T'} \simeq 5$ kcal/mol. Using ethyl acetate as an example, if we assume that a Boltzmann distribution results from the absorption of a certain average number of infrared photons, $T = 1200°\text{K}$ gives $\langle E_{\text{vib}} \rangle_{1200} = 38.0$ kcal/mol and $T' = 1075°\text{K}$. Therefore, $\langle E'_{\text{vib}} \rangle_{1075} = 31.6$ kcal/mol and $\langle E_{\text{vib}} \rangle - \langle E'_{\text{vib}} \rangle = 6.4$ kcal/mol. The decrease in F_R^{∞} for 0.05 Torr of vibrationally excited ethyl acetate relaxing to a thermal equilibrium is shown in Figure 2.33. This demonstrates that the relatively small heat capacity of the translational and rotational degrees of freedom of a large molecule is rather inefficient in quenching reaction.

Since the translationally and rotationally cold molecules become hot ($T_{\text{rot}} = T_{\text{trans}} = 300°\text{K} \rightarrow T'_{\text{rot}} = T'_{\text{trans}} = 1075°\text{K}$ for the ethyl acetate example), we may refer to this process as heating. Such a heating phenomenon can be monitored by using a mixture of a large portion of reactant and a very small amount of a thermal monitor that has a similar dependence of unimolecular rate on temperature but does not absorb the laser photons (Section 2.4). A large amount of thermal monitor must be avoided since this may result in collisional deactivation rather than a heating effect as discussed in detail in Section 2.4.3.1. Since the energy loss per collision of large molecules is $\sim$ 3–10 kcal, a few collisions will be sufficient to produce thermal equilibrium. Once the molecules within the irradiated volume are equilibrated, collisional processes will maintain this Boltzmann distribution. A consequence of this is that, as those states above the threshold are depleted by unimolecular reaction, collisions will continually repopulate those levels providing that the ΔH of reaction does not seriously affect the available energy. If the reaction is exoergic, T' may even rise somewhat, while if the reaction is endoergic, T' would be expected to drop. This idealized thermal equilibration within the irradiated volume may result in contribution to the reaction yield after the laser pulse since molecules initially produced with a vibrational energy less than threshold can be collisionally excited to react. The reaction rate constant for these conditions is the same as the high-pressure limit rate for the thermal unimolecular reaction at temperature T'. The absolute reaction yield is related to the high-pressure limit rate by Eq. (2.66) in which $T' = T_{\text{eff}}$.

But, as alluded to earlier, the real situation is not this simple. Since there is always diffusion, bulk expansion, and thermal conduction, the vibrationally hot molecules within the irradiated volume are not isolated indefinitely from cold molecules. In addition to collisions between vibrationally hot molecules and cold molecules in the irradiated volume, one must consider the relaxation of the hot irradiated volume into the cold surrounding medium. As a result of this cooling process, the effective reaction time, Δt, is no longer infinite, and the actual value gives some information on the cooling rate. The Δt depends on the cooling mechanism as well as the heating mechanism, and, if these two processes are somewhat balanced, the reaction yield may not be very dependent on reactant pressure since pressure effects may compensate. Our results with ethyl acetate and other experimental results (Jensen *et al.*, 1978; Reiser *et al.*, 1979) demonstrate that the reaction yields for large molecules are not seriously influenced by the reactant pressure when collisions between vibrationally hot molecules are important. This contrasts the situation illustrated by Figure 2.15 in which the reaction yield decreases pronouncedly as the pressure of an addded inert bath gas is increased as a result of collisional deactivation within the irradiated volume.

It was concluded in Section 2.4.4 that Δt for neat gas samples might be on the order of a few microseconds before molecular and/or bulk transfer effectively dissipated the laser energy and quenched reaction but that this parameter is probably quite dependent upon experimental variables. A consideration of Figure 2.35 suggests that such a time frame may be realistic for the laser-induced reaction of ethyl acetate at <0.10 Torr providing the population of excited molecules can be approximated by a Boltzmann distribution. From the RRKM calculated k_E values given on the right side it can be seen that a molecule must possess at least ~ 75 kcal/mol of vibrational excitation before it can react in 3 μsec ($k_E = 3 \times 10^5$ sec^{-1}). It is further evident that the absorption of ~ 12 photons (~ 36 kcal/mol) will produce an effective temperature ~ 1200°K with roughly one-fourth of the molecules excited above the threshold ($E_0 = 48$ kcal/mol). Experimentally, for 0.81 J/cm^2 fluence, $\langle n \rangle = 12$ but the reaction probability is only 0.006. This correlates with the fact that the 1200°K distribution just barely extends beyond 75 kcal/mol. That is, virtually none of the excited molecules possess sufficient energy to react in less than 3 μsec. Similarly, for the 1400°K distribution, $\langle n \rangle \sim 16$ photons, which is observed experimentally to yield $P \sim 0.02$. Thus, even though about half the molecules are excited beyond E_0, only that small fraction excited above ~ 75 kcal/mol actually react before being quenched.

In conclusion of this section, it may be noted that our modeling studies reproduce qualitatively, at least, a number of experimental observations for the laser-induced reactions of large molecules. The high dependence of reaction probability on laser fluence is predicted (Figures 2.17 and 2.27) and collisional effects with cold bath gases are reproduced reasonably well (Figures 2.15 and 2.34). The possible dependence of the

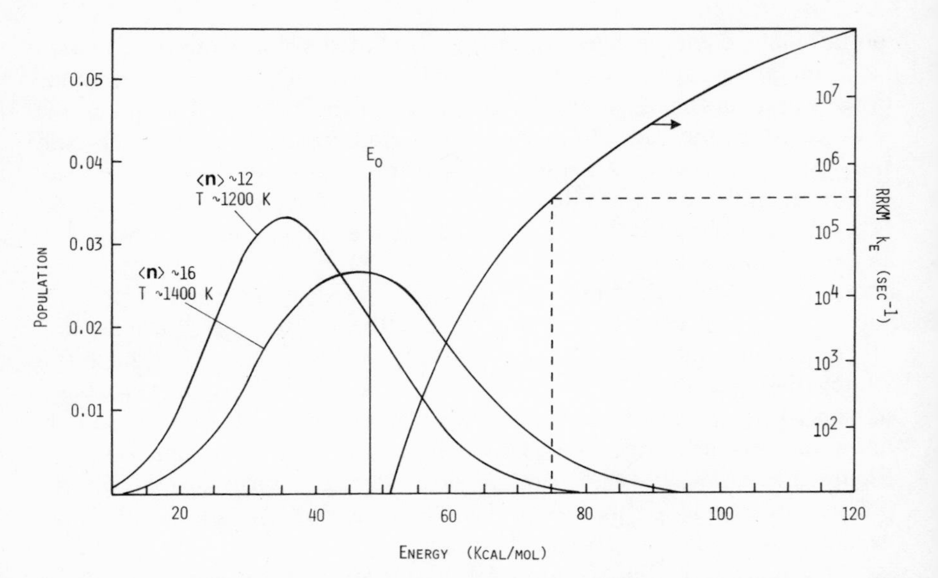

Figure 2.35. Boltzmann vibrational distributions of ethyl acetate after absorption of ~ 12 and ~ 16 infrared photons ($1046.85\ cm^{-1}$). Right-hand curve and scale designate RRKM rate constants, k_E, for unimolecular reaction. Dotted lines indicate that $k_E \sim 5 \times 10^5\ sec^{-1}$ at an excitation level of 75 kcal/mol.

level-to-level cross section, σ_{ij}, on level of excitation is not known for any system, but the modeling studies indicate that neither the vibrational energy distribution (Figure 2.31) nor the reaction yield (Figure 2.32) are seriously perturbed by assumed changes in σ_{ij} with excitation level. Therefore, the assumption of a Boltzmann vibrational distribution upon termination of the laser pulse appears reasonable for large molecules. An effect not addressed by the modeling studies is the "cooling process" by which vibrationally excited molecules of neat samples within the irradiated volume become quenched. The rate of cooling is an important factor since this determines the effective reaction time and, hence, the yield in a pulsed laser-induced reaction. It is probable that the cooling phenomenon is a complex interplay of effects, the understanding of which will challenge workers in the field of laser-induced processes for some time to come.

Acknowledgments

Acknowledgment is made to the National Science Foundation (Grants CH77-22645 and 77-21380) for support of this literary effort as well as for the experimental results reported herein. Grateful appreciation is extended to

D. W. Setser of this department for numerous helpful discussions and for a critical reading of the entire manuscript. His assistance with the model calculations was essential as these results would not have been completed, or even attempted, without his guidance. Finally, the authors thank D. Wright for expertly typing the entire manuscript.

References

Ambartzumian, R. V., and Letokhov, V. S., 1977, Multiple photon infrared laser photochemistry, in *Chemical and Biochemical Applications of Lasers, Volume III* (ed. C. B. Moore), Academic Press, New York, pp. 167–316.

Back, M. H., and Back, R. A., 1979, The decomposition of cyclobutanone vapor induced by infrared radiation from a pulsed CO_2 TEA laser, *Can. J. Chem.* **57**:1511.

Bado, P., and van den Bergh, H., 1978, Pressure dependence in the multiphoton dissociation of $^{32}SF_6$, *J. Chem. Phys.* **68**:4188.

Baldwin, A. C., Barker, J. R., Golden, D. M., Duperrex, R., and van den Bergh, H., 1979, Infrared multiphoton chemistry: Comparison of theory and experiment, solution of the master equation, *Chem. Phys. Lett.* **62**:178.

Bates, Jr., R. D., Flynn, G. W., and Knudtson, J. T., 1970, Laser-induced 16-μ fluorescence in SF_6: Acoustic effects, *J. Chem. Phys.* **53**:3621.

Bauer, S. H., 1978, How energy accumulation and disposal affect the rates of reaction, *Chem. Rev.* **78**:147.

Benson, S. W., 1978, Thermochemistry and kinetics of sulfur-containing molecules and radicals, *Chem. Rev.* **78**:23.

Benson, S. W., and O'Neal, H. E., 1970, Kinetic data on gas phase unimolecular reactions, United States Department of Commerce, NSRDS-NBS 21.

Berry, M. J., 1974, Chloroethylene photochemical lasers: Vibrational energy content of the HCl molecular elimination products, *J. Chem. Phys.* **61**:3114.

Bialkowski, S. E., and Guillory, W. A., 1980, Dynamic processes of NH_2 generated by the IR photolysis of CH_3NH_2, *J. Photochem.* in press.

Birely, J. H., and Lyman, J. L., 1975, Effect of reagent vibrational energy on measured reaction rate constants, *J. Photochem.* **4**:269.

Black, J., Yablonovitch, E., Bloembergen, N., and Mukamel, S., 1977, Collisionless multiphoton dissociation of SF_6: A statistical thermodynamics process, *Phys. Rev. Lett.* **38**:1131.

Black, J., Kolodner, P., Shultz, M., Yablonovitch, E., and Bloembergen, N., 1979, Collisionless multiphoton energy deposition and dissociation of SF_6, *Phys. Rev. A.* **19**:704.

Bloembergen, N., and Yablonovitch, E., 1978, Infrared laser induced unimolecular reactions, *Physics Today* **31**:23.

Bomse, D. S., Woodin, R. L., and Beauchamp, J. L., 1978, Multiphoton dissociation of molecules with low power CW infrared lasers, in *Advances in Laser Chemistry* (ed. A. H. Zewail), Springer, New York.

Bomse, D. S., Woodin, R. L., and Beauchamp, J. L., 1979, Molecular activation with low intensity CW infrared laser radiation. Multiphoton dissociation of ions derived from diethyl ether, *J. Am. Chem. Soc.* **101**:5503.

Braun, W., Herron, J. T., Tsang, W., and Churney, K., 1978, High intensity infrared laser irradiation calorimetry: Direct-determination of heat input to chlorodifluoromethane and ethyl acetate, *Chem. Phys. Lett.* **59**:492.

Brenner, D. M., 1978, Infrared multiphoton-induced chemistry of ethyl vinyl ether: Dependence of branching ratio on laser pulse duration, *Chem. Phys. Lett.* **57**:357.

Buechele, J. L., Weitz, E., and Lewis, F. D., 1979, Laser-induced infrared multiphoton isomerization of hexadienes, *J. Am. Chem. Soc.* **101**:3700.

Burak, I., Houston, P., Sutton, D. G., and Steinfeld, J. I., 1970, Observation of laser-induced acoustic waves in SF_6, *J. Chem. Phys.* **53**:3632.

Calvert, J. G., and Pitts, Jr., J. N., 1966, *Photochemistry*, John Wiley, New York, p. 19.

Cantrell, C. D., Freund, S. M., and Lyman, J. L., 1979, Laser-induced chemical reactions, in *Laser Handbook, Volume III* (ed. M. L. Stitch), North-Holland, Amsterdam.

Cheng, C., and Keehn, P., 1977, Organic chemistry by infrared lasers. 1. Isomerization of allene and methylacetylene in the presence of silicon tetrafluoride, *J. Am. Chem. Soc.* **99**:5808.

Colussi, A. J., Benson, S. W., Hwang, R. J., and Tiee, J. J., 1977, Intramolecular isotope effect in laser multiphoton dissociation of CH_2DCH_2Cl, *Chem. Phys. Lett.* **52**:349.

Cox, D. M., Hall, R. B., Horsley, J. A., Kramer, G. M., Rabinowitz, P., and Kaldor, A., 1979, The isotope selectivity of IR laser driven unimolecular dissociation of a volatile uranyl compound, *Science* **205**:390.

Danen, W. C., 1979, Infrared laser induced organic reactions. 2. Laser vs. thermal inducement of unimolecular and hydrogen bromide catalyzed bimolecular dehydration of alcohols, *J. Am. Chem. Soc.* **101**:1187.

Danen, W. C., 1980, Pulsed infrared laser induced organic chemical reactions, *Opt. Eng.* **19**:21.

Danen, W. C., and Hanh, N. H., 1980, unpublished results.

Danen, W. C., Munslow, W. D., and Setser, D. W., 1977, Infrared laser induced organic reactions. 1. Irradiation of ethyl acetate with a pulsed CO_2 laser. Selective inducement vs. thermal reaction, *J. Am. Chem. Soc.* **99**:6961.

Danen, W. C., Koster, D. F., and Zitter, R. N., 1979, Demonstration of Woodward–Hoffmann behavior in the pulsed, infrared laser induced reaction of *cis*-3,4-dichlorocyclobutene, *J. Am. Chem. Soc.* **101**:4281.

Danen, W. C., Rio, V. C., and Setser, D. W., 1980, unpublished results.

Dever, D. F., and Grunwald, E., 1976, Megawatt infrared laser chemistry of $CClF_3$ and CCl_3F. 1. Photochemistry, Photophysics, and Effect of H_2, *J. Am. Chem. Soc.* **98**:5055.

Douglas, D. J., and Moore, C. B., 1979, Vibrational relaxation of HF($v = 3, 4$) by HF, H_2, D_2, CO_2, and isobutene, in *Laser-Induced Processes in Molecules, Physics and Chemistry* (eds. K. L. Kompa and S. D. Smith), Springer-Verlag, New York, pp. 337–338.

Drozdoski, W. S., Fakhv, A., and Bates, Jr., R. D., 1977, Deactivation of vibrationally excited CD_3H using laser-induced fluorescence, *Chem. Phys. Lett.* **47**:309.

Drozdoski, W. S., Bates, Jr., R. D., and Siebert, D. R., 1978, Vibrational energy flow in CD_3H and CD_3H-polyatomic mixtures, *J. Chem. Phys.* **69**:863.

Frey, H. M., and Pope, B. M. 1966, Thermal unimolecular isomerization of *cis*-hexa-1,3-diene, *J. Chem. Soc. A*, 1701.

Frey, H. M., and Walsh, R., 1978, Unimolecular reactions, in *Gas Kinetics and Energy Transfer, Vol. 3* (eds. P. G. Ashmore and R. J. Donovan), The Chemical Society, London, pp. 1–41.

Fuss, W., Kompa, K. L., Proch, D., and Schmid, W. E., 1977, High power infrared laser chemistry, in *Lasers in Chemistry* (ed. M. A. West), Elsevier Publishing Co., Amsterdam, pp. 235–244.

Garcia, D., and Keehn, P. M., 1978, Organic chemistry by infrared lasers. 2. Retro-Diels-Alder reactions, *J. Am. Chem. Soc.* **100**:6111.

Glatt, I., and Yogev, A., 1976, Photochemistry in the electronic ground state. 4. Infrared laser induced isomerization of labeled compounds. A possible route for isotope separation, *J. Am. Chem. Soc.* **98**:7087.

Grant, E. R., Schulz, P. A., Sudbø, Aa. S., Shen, Y. R., and Lee, Y. T., 1978, Is multiphoton dissociation of molecules a statistical thermal process?, *Phys. Rev. Lett.* **40**:115.

Grunwald, E., Dever, D. F., and Keehn, P. M., 1978, *Megawatt Infrared Laser Chemistry*, John Wiley, New York.

Grunwald, E., Lonzetta, C. M., and Popok, S., 1979, Intermolecular energy exchange of infrared-laser excited $CHClF_2$ or SiF_4 with Br_2 at excitation energies of 70–200 kJ/mol, *J. Am. Chem. Soc.* **101**:5062.

Haas, Y., and Yahav, G., 1977, Gas phase unimolecular decomposition and chemiluminescence of tetramethyldioxetane initiated by a TEA CO_2 laser, *Chem. Phys. Lett.* **48**:63.

Hall, R. B., and Kaldor, A., 1979, Multiple IR photon laser induced reactions of cyclopropane, *J. Chem. Phys.* **70**:4027.

Hassler, J. C., and Setser, D. W., 1966, RRKM calculated unimolecular reaction rates for chemically and thermally activated C_2H_5Cl, 1,1-$C_2H_4Cl_2$, and 1,2-$C_2H_4Cl_2$, *J. Chem. Phys.* **45**:3246.

Herman, I. P., and Marling, J. B., 1979, Vibrationally stimulated addition reactions between hydrogen halides and unsaturated hydrocarbons: A negative result, *J. Chem. Phys.* **71**:643.

Holmes, B. E., and Setser, D. W., 1975, Energy disposal in unimolecular reactions. Four-centered elimination of HCl, *J. Phys. Chem.* **79**:1320.

Hovis, F. E., and Moore, C. B., 1980, Energy transfer and laser photochemistry, in press.

Hsu, D. S. Y., and Manuccia, T. J., 1978, Deuterium enrichment by CW laser-induced reaction of methane, *Appl. Phys. Lett.* **33**:915.

Hwang, W. C., Herm, R. R., Kalsch, J. F., and Gust, G. R., 1979, Multiple-photon chemistry induced by a pulsed CO_2 laser at moderate fluences, *Aerospace Report No. ATR-79(8420)-1*, May 1979.

JANAF Thermochemical Tables, 1971, 2nd ed., U.S. Bureau of Standards, Publication NSRDS-NBS 37.

Jang, J. C., and Setser, D. W., 1979, Collisional effects in infrared multiple photon induced unimolecular reactions of fluoroethane and trifluoroethane, *J. Phys. Chem.* **83**:2809.

Jensen, C. C., Steinfeld, J. I., and Levine, R. D., 1978, Information theoretic analysis of multiphoton excitation and collisional deactivation in polyatomic molecules, *J. Chem. Phys.* **69**:1432.

Johnson, R. L., and Setser, D. W., 1967, Unimolecular reactions of chemically activated C_2H_5Br, 1,2-$C_2H_4Br_2$, and 1,2-C_2H_4BrCl and the reaction of methylene with CH_2Br_2 and CH_2BrCl, *J. Phys. Chem.* **71**:4366.

Kaldor, A., Hall, R. B., Cox, D. M., Horsley, J. A., Rabinowitz, P., and Kramer, G. M., 1979, Infrared laser chemistry of large molecules, *J. Am. Chem. Soc.* **101**:4465.

Kim, K. C., and Setser, D. W., 1974, Unimolecular reactions and energy partitioning. Three- and four-centered elimination reactions of chemically activated 1,1,2-trichloroethane-d_0, -d_1, and -d_2, *J. Phys. Chem.* **78**:2166.

Knudtson, J. T., and Flynn, G., 1973, Laser fluorescence study of vibrational energy transfer in CH_3Cl^*, *J. Chem. Phys.* **58**:2684.

Kolodner, P., Winterfield, C., and Yablonovitch, E., 1977, Molecular dissociation of SF_6 by ultra-short CO_2 laser pulses, *Optics Commun.* **20**:119.

Kompa, K. L., Fuss, W., Proch, D., Schmid, W. E., Smith, S. D., and Schröder, H., 1979, Towards an understanding of infrared multiphoton absorption and dissociation, in *Non-linear Behavior of Molecules, Atoms, and Ions in Electric, Magnetic, or Electromagnetic Fields*, Elsevier Publishing Co., Amsterdam, pp. 55–63.

Letokhov, V. S., and Moore, C. B., 1977, Laser isotope separation, in *Chemical and Biochemical Applications of Lasers, Volume III* (ed. C. B. Moore), Academic Press, New York, pp. 1–165.

Lyman, J. L., 1977, A model for unimolecular reaction of sulfur hexafluoride, *J. Chem. Phys.* **69**:1868.

Lyman, J. L., Danen, W. C., Nilsson, A. C., and Nowak, A. V., 1979, Multiple-photon excitation of difluoroamino sulfur pentafluoride: A study of absorption and dissociation, *J. Chem. Phys.* **71**:1206.

Lussier, F. M., and Steinfeld, J. I., 1977, Multiple infrared photon dissociation of vinyl chloride, *Chem. Phys. Lett.* **50**:175.

Marcoux, P. J., and Setser, D. W., 1978, Vibrational energy transfer probabilities of highly vibrationally excited 1,1,1-trifluoroethane, *J. Phys. Chem.* **82**:97.

Mukamel, S., 1979, Stochastic reduction for molecular multiphoton processes, *J. Chem. Phys.* **70**:5834.

Nikitin, E. E., 1974, *Theory of elementary atomic and molecular processes in gases*, Oxford University Press, London.

Olszyna, K., J., Grunwald, E., Keehn, P. M., and Anderson, S. P., 1977, Megawatt infrared laser chemistry. II. Use of SiF_4 as an inert sensitizer, *Tetrahedron Lett.* **1977**:1609.

Plum, C. N., and Houston, P. L., 1980, Infrared photolysis of $C_2F_4S_2$: A comparison of multiphoton dissociation models, *Chem. Phys.* **45**:159.

Popok, S., Lonzetta, C. M., and Grunwald, E., 1979, Infrared laser induced bromination and chlorination of chlorodifluoromethane, *J. Org. Chem.* **44**:2377.

Preses, J. M., Weston, R. E., Jr., and Flynn, G. W., 1977, Unimolecular decomposition of *cyclo*-C_4H_8 induced by a CO_2 TEA laser, *Chem. Phys. Lett.* **46**:69.

Pritchard, H. O., Pyke, J. B., and Trotman-Dickenson, A. F., 1955, The study of chlorine atom reactions in the gas phase, *J. Am. Chem. Soc.* **77**:2629.

Quack, M., 1978, Theory of unimolecular reactions induced by the monochromatic infrared radiation, *J. Chem. Phys.* **69**:1282.

Quack, M., and Troe, J., 1977, Unimolecular reactions and energy transfer of highly excited molecules, in *Gas Kinetics and Energy Transfer, Vol. 2* (eds. P. G. Ashmore and R. J. Donovan), The Chemical Society, London, pp. 175–238.

Quick, Jr., C. R., and Wittig, C., 1978a, IR photodissociation of vinyl fluoride: time-resolved emission under collisionless conditions, *Chem. Phys.* **32**:75.

Quick, Jr., C. R., and Wittig, C., 1978b, Infrared photodissociation of fluorinated ethanes and ethylenes: Collisional effects in the multiple photon absorption process, *J. Chem. Phys.* **69**:4201.

Quick, Jr., C. R., Tiee, J. J., Fischer, T. A., and Wittig, C., 1979, A direct measurement of the unimolecular decomposition of 1,1-difluoroethane via IR laser photolysis, *Chem. Phys. Lett.* **62**:435.

Reiser, C., Lussier, F. M., Jensen, C. C., and Steinfeld, J. I., 1979, Infrared photochemistry of halogenated ethylenes, *J. Am. Chem. Soc.* **101**:350.

Richardson, T. H., and Setser, D. W., 1977, Laser induced decomposition of fluorethanes, *J. Phys. Chem.* **81**:2301.

Robinson, P. J., 1975, Unimolecular reactions, in *Reaction Kinetics, Vol. 1* (ed., P. G. Ashmore), The Chemical Society, London, pp. 93–160.

Robinson, P. J., and Holbrook, K. A., 1972, *Unimolecular Reactions*, Wiley–Interscience, New York.

Ronn, A. M., 1979, Laser chemistry, *Scientific American*, **240(5)**:114–128.

Rosenfeld, R. N., Brauman, J. I., Barker, J. R., and Golden, D. M., 1977, Infrared photodecomposition of ethyl vinyl ether. A chemical probe of multiphoton dynamics, *J. Am. Chem. Soc.* **99**:8063.

Ross, R. A., and Stimson, V. R., 1960, Catalysis by hydrogen halides in the gas phase. Part III. Isopropyl and hydrogen bromide, *J. Chem. Soc.*, 3090.

Schulz, P. A., Sudbø, Aa. S., Krajnovich, D. J., Kwok, H. S., Shen, Y. R., and Lee, Y. T., 1979, Multiphoton dissociation of polyatomic molecules, *Ann. Rev. Phys. Chem.*, **30**:379.

Schwartz, R. N., and Herzfeld, K. F., 1954, Vibrational relaxation times in gases (Three dimensional treatment), *J. Chem. Phys.* **22**:767.

Schwartz, R. N., Slawsky, Z. I., and Herzfeld, K. F., 1952, Calculation of vibrational relaxation times in gases, *J. Chem. Phys.* **20**:1591.

Shaub, W. M., and Bauer, S. H., 1975, Laser-powered homogeneous pyrolysis, *Internat. J. Chem. Kinet.* **7**:509.

Shultz, M. J., and Yablonovitch, E., 1978, A statistical theory for collisionless multiphoton dissociation of SF_6, *J. Chem. Phys.* **68**:3007.

Steel, C., Starov, V., Leo, R., John, P., and Harrison, R. G., 1979, Chemical thermometers in megawatt infrared laser chemistry: The decomposition of cyclobutanone sensitized by ammonia, *Chem. Phys. Lett.* **62**:121.

Stephensen, J. C., King, D. S., Goodman, M. F., and Stone, J., 1979, Experiment and theory for CO_2 laser-induced CF_2HCl decomposition rate dependence on pressure and intensity, *J. Chem. Phys.* **70**:4496.

Stone, J., and Goodman, M. F., 1979, A re-examination of the use of rate equations to account for fluence dependence, intramolecular relaxation, and unimolecular decay in laser driven polyatomic molecules, *J. Chem. Phys.* **71**:4068.

Sudbø, Aa. S., Schulz, P. A., Shen, Y. R., and Lee, Y. T., 1978, Three- and four-centered elimination of HCl in the multiphoton dissociation of halogenated hydrocarbons, *J. Chem. Phys.* **69**:2312.

Sudbø, Aa. S., Schulz, P. A., Grant, E. R., Shen, Y. R., and Lee, Y. T., 1979, Simple bond rupture reactions in multiphoton dissociation of molecules, *J. Chem. Phys.* **70**:912.

Tardy, D. C., and Rabinovitch, B. S., 1966, Collisional energy transfer. Thermal unimolecular systems in the low-pressure region, *J. Chem. Phys.* **45**:3720.

Tardy, D. C., and Rabinovitch, B. S., 1977, Intermolecular vibrational energy transfer in thermal unimolecular systems, *Chem. Revs.* **77**:369.

Taylor, R., 1975, The nature of the transition state in ester pyrolysis. Part II. The relative rates of pyrolysis of ethyl, isopropyl, and *t*-butyl acetates, phenylacetates, benzoates, phenyl carbonates, and N-phenylcarbomates, *J. C. S. Perkin II* **1975**:1025.

Thiele, E., Goodman, M. F., and Stone, J., 1980, Can lasers be used to break chemical bonds selectively?, *Opt. Eng.* **19**:10.

Treanor, C. E., Rich, C. W., and Rehm, R. G., 1968, Vibrational relaxation of arharmonic oscillators with exchange-dominated collisions, *J. Chem. Phys.* **48**:1798.

Tsang, W., Walker, J. A., Braun, W., and Herron, J. T., 1978, Mechanisms of decomposition of mixtures of ethyl acetate and isopropyl bromide subjected to pulsed infrared laser irradiation, *Chem. Phys. Lett.* **59**:487.

Weitz, E., and Flynn, G., 1973a, Deactivation of laser excited CH_3F in CH_3F–X mixtures, *J. Chem. Phys.* **58**:2679.

Weitz, E., and Flynn, G., 1973b, Partial Vibration energy transfer map for methyl fluoride: A laser fluorescence study, *J. Chem. Phys.* **58**:2781.

Woodin, R. L., Bomse, D. S., and Beauchamp, J. L., 1978, Multiphoton dissociation of molecules with low power continuous wave infrared laser radiation, *J. Am. Chem. Soc.* **100**:3248.

Woodin, R. L., Bomse, D. S., and Beauchamp, J. L., 1979, Multiphoton dissociation of molecules with low power CW infrared lasers: collisional enhancement of dissociation probabilities, *Chem. Phys. Lett.* **63**:630.

Yahav, G., and Haas, Y., 1978, Time dependence of multiphoton dissociation of molecules in a strong infrared field. Real-time measurement using a nanosecond laser source, *Chem. Phys.* **35**:41.

Yardley, J. T., and Moore, C. B., 1968, Vibrational energy transfer in methane, *J. Chem. Phys.* **49**:1111.

Yogev, A., and Benmair, R. M. J., 1977, Photochemistry in the electron ground state. Quantitative electrocyclic isomerization induced by multiphoton absorption of infrared laser radiation, *Chem. Phys. Lett.* **46**:290.

Yogev, A., and Loewenstein-Benmair, R. M. J., 1973, Photochemistry in the electronic ground state. II. Selective decomposition of *trans*-2-butene by pulsed carbon dioxide laser, *J. Am. Chem. Soc.* **95**:8487.

Zittel, P. F., and Moore, C. B., 1973, Model for V–T, R relaxation: CH_4 and CD_4 mixtures, *J. Chem. Phys.* **58**:2004.

Zitter, R. N., and Koster, D. F., 1976, Reaction rate difference in the laser excitation of different vibrational mode of CF_3ClCF_2Cl, *J. Am. Chem. Soc.* **98**:1613.

Zitter, R. N., and Koster, D. F., 1977, Frequency dependence of laser-initiated reaction rates of CF_2ClCF_2Cl, *J. Am. Chem. Soc.* **99**:5491.

3

Sinterable Powders from Laser-Driven Reactions

John S. Haggerty and W. Roger Cannon

3.1. Introduction

Increasingly, because of their superior properties, ceramic materials are being considered for applications where there are high stress levels and where their failure would cause a major problem. These properties include hardness, high-temperature strength, erosion, oxidation and corrosion resistance, low density, and, for some applications, specific electrical and optical properties. The use of ceramic materials in these applications can only be considered, however, if their reliability is improved. Brittle materials fail catastrophically, and the wide distribution of observed strengths specifically associated with ceramic materials forces engineers to design so conservatively that ceramics lose their intrinsic advantages relative to conventional materials. For instance, it is impossible to design for a load stress that is less than one-tenth the mean strength and retain a superior strength to weight ratio.

Strength-limiting defects are usually attributable to some specific event in the processing history that extends from powder synthesis through all the handling steps to the final consolidation into a densified part. There are many causes for strength-limiting defects; however, most defects can be avoided in fully dense parts if ceramics are made from powders with the following ideal characteristics: (1) the powder must have a small particle

John S. Haggerty and W. Roger Cannon · Energy Laboratory, Massachusetts Institute of Technology, Cambridge, Massachusetts

size, typically less than 0.5 μm; (2) the particles must be free of agglomerates; (3) the particle diameters must have a narrow range of sizes; (4) the morphology of the particles must be equiaxed, tending toward spheres; (5) the powders must have highly controlled purity with respect to contaminates and multiple polymorphic phases. A powder exhibiting these ideal characteristics should be sinterable to theoretical density without resorting to pressure or additives, and the final grain morphology should be controllable to permit achievement of useful high-temperature properties.

During the past 10 years, interest in using ceramics in heat engines has increased. The requirements for these applications are extremely demanding but if successful, will have important consequences. These applications include the all-ceramic turbine, rotors for turbochargers, caps and rings for pistons, precombustion chambers, and injector nozzles. In each case, the lack of reliable strength characteristics has precluded the use of ceramics on a large scale. Their use is highly desirable, however, because of improved performance resulting from higher temperature, more efficient operation, and reduced weight. The most promising ceramic materials candidates for these engine applications are Si_3N_4 and SiC, which combine optimum combinations of thermal conductivity, thermal shock resistance, hardness, high-temperature strength, oxidation resistance, and density.

Because existing powder synthesis techniques have not produced powders with these requisite characteristics, we have undertaken to determine whether a laser-heated, gas-phase synthesis process could in fact achieve the process conditions that we anticipated would result in uniform nucleation and growth histories. This program has emphasized synthesis of Si and Si_3N_4 powders with a lesser but increasing effort with SiC. The Si powders are used for the reaction bonding process in which densification and conversion to Si_3N_4 occur simultaneously. We have had two specific objectives: Our initial and primary objective has been to produce ceramic powders with characteristics that will permit a better understanding of the interaction between powder characteristics, densification processes, and the properties of ceramic bodies. Very specific powder characteristics are required to satisfy this objective so we have concentrated on developing detailed descriptions of time-temperature history experienced by laser-heated gases. Since it has become evident that substantial improvements in the properties of resulting ceramic parts are likely, we have considered the issues of scaling the process to production levels.

Several commercial processes are being used for synthesizing Si_3N_4 and SiC. Typically, they involve DC arcs, conventional vapor phase reactions in heated tube furnaces or nitriding or carbiding of silicon metal. The nitriding of silicon metal typically leaves a silicon core within the silicon nitride particle. Furthermore, because the process is done in the solid state, grinding and separating of particles is necessary, but this does not result in a narrow size distribution of nonagglomerated, phase pure powders. The vapor-phase methods (furnace-heated vapor and arc plasma techniques) yield a finer and more uniform powder than the nitriding of solid silicon; but these techniques have less than ideal thermal profiles and reaction zones that allow for a distribution in nucleation and growth times and the formation of agglomerates. Despite these specific process deficiencies, direct synthesis of powders from dilute gas-phase reactants is the most promising route to producing ideal powders.

The processing technique of using laser-driven gas-phase reactions, described in this chapter, offers many advantages. It is a clean process that permits cold, nonreactive chamber walls. The reaction volume is very well defined and consists only of that volume traversed by reaction gases and particles, i.e., the laser beam area. The ability to maintain steep temperature gradients in the effective thermal environment, and thus a well-defined reaction zone, appears to allow precise control of the nucleation rate, the growth rate, and exposure times, permitting the nucleation and growth of very fine particles. The available power with a CO_2 laser, the stability of the delivered power, its cost, reliability, and efficiency, allow this to be a viable process that will yield improvements in fabrication of powders for these high-performance materials.

Our ultimate objective is to develop an understanding of the interrelationships between observations and models. To reach this goal, we must be able to describe heating rates, nucleation and growth rates, and the distribution of temperature and mass flow throughout the "reaction zone." We have studied the effect of several process variables on powder characteristics, including beam intensity, gas composition (stoichiometry and dilution), gas pressure, and gas velocity. The effect of these process parameters has been correlated with both powder characteristics and process characteristics, e.g., emitted spectra, temperature, and percent conversion.

Two fundamentally different reaction types have been investigated. Most of our work has been carried out under laser intensity and gas-pressure conditions where many collisions occur between molecules during the period that the gas molecules are heated. These reactions probably proceed as normal thermal reactions. In this case, the principal attributes of the laser heat source are process control and possibly unique reaction paths because of high heating rates and resonance effects between the coherent light and the molecules. The second type of reaction is uniquely possible with laser heating. In this case, the molecules absorb sufficient energy to dissociate before colliding with the other molecules. We have investigated both types of processes.

The use of this synthesis technique as an experimental tool has tremendous potential, and its eventual use as a production tool appears increasingly probable. It is now apparent that it can be applied to elements, oxides, carbides, and nitrides. With slightly different process conditions, it can be used to deposit thin films rather than powders.

3.2. Laser-Heated Powder Synthesis

3.2.1. Process Description

3.2.1.1. General

This powder synthesis process employs an optical energy source to transfer the energy required to initiate and sustain a chemical reaction in the gas phase. In this process, the gas molecules are "self-heated" throughout the gas volume, a process that is distinct from conventional ones where heat is transmitted from a source to the gas molecules by a combination of conduction, convection, and radiative processes. The advantages of this means of heating—freedom from contamination, absence of surfaces that act as heterogeneous nucleation sites, and unusually uniform and precise process control—are discussed at length elsewhere in this volume. These attributes should permit the synthesis of powders with characteristics that are ideal for making ceramic bodies.

A laser, rather than other possible optical heat sources, has been used in this work because of the narrow spectral width of emitted light and the brightness of this type of light source. The coherence of the light is not considered an important feature for this process. Coupling between the source and absorber requires virtual coincidence between the frequency of emitted photon and the absorption lines of the reactant. If this matching occurs, the optical-to-thermal efficiency can be extremely high, and the overall process efficiency is essentially that of the laser. Overall, this efficiency is much higher than is possible with broad-band light sources because only small fractions of their light are absorbed by the gases. With CO_2 lasers, as have been used in this research, the overall process efficiency matches or exceeds that of other conventional heat sources, e.g., various types of plasmas, torches, or heated tubes. In addition to the high efficiency, the use of laser energy sources makes possible unique reaction paths that may produce powders with unusual characteristics.

Two basic ranges of laser intensity and exposure time have been investigated. The equipment and experimental procedures used for reactions carried out in the normal or thermal domain and those used for the multiphoton, unimolecular reaction domain are so different that we will discuss them separately in this part of the chapter. Section 3.2.1.2 discusses the experimental procedures and equipment used to investigate the thermal domain. Section 3.2.2 discusses the various characterizations and analyses that apply to the thermal domain. Section 3.2.1.3 discusses our experiments under conditions anticipated to cause multiphoton, unimolecular reactions. This part includes the discussion of experimental results as well as equipment and procedures because the level of effort in this area was relatively limited.

3.2.1.2. Normal, Thermal Domain Experiments

3.2.1.2a. Experimental Geometry. There are several choices of laser and gas stream geometry that can be applied to the thermal domain synthesis process. The laser and gas can intersect from opposite directions (counterflow) or orthogonally (cross flow). The effect gravity exerts on the heated gases and entrained particles can be varied by operating in horizontal or vertical directions. In varying degrees, all of these configurations have been investigated experimentally. Each has features that suggest different choices for experimental and production scale processes.

The majority of the experiments were carried out in the cross-flow configuration where the laser beam and reactant gas stream intersect orthogonally. This experimental configuration produces a very stable reaction and the reaction zone where the laser beam and gas stream intersect can be analyzed. In addition, there is very little interaction between the reaction and the cold walls of the cell.

Unfortunately, this reaction geometry has several disadvantages. Both the gas stream and the laser beam have nonuniformities that are ideally axisymmetric. Because of these nonuniformities, different volume elements within the gas stream are subjected to different laser intensity and velocity histories. In addition, the laser beam is progressively absorbed as it passes through the thickness of the gas stream, further increasing the range of exposure histories between the entering and the exiting sides of the gas stream. The second source of variable history was effectively eliminated by designing the experiment so the gas stream is optically thin. Under "reference" operating conditions, approximately 2% of the incident power is absorbed by the reactant gas stream. While this configuration is suitable for experimental purposes, it is evident that a commercial process could not permit 98% of the laser light to be wasted. With a single laser beam–single gas stream configuration, increasing optical density of the gas stream results in an improved efficiency but causes an increasingly unacceptable variance in the local beam intensity. Multiple-pass or multiple-beam optical configurations can give acceptable uniformity and efficiency.

A counterflow geometry used for some of the absorptivity and threshold experiments has certain advantages over the crossflow configuration described above. In principle, each gas element can be subjected to an identical time-intensity history as it flows into the laser beam. Also, the gas column can be made long enough so all the light is absorbed, giving maximum optical efficiency.

In practice, however, there are also disadvantages. The gas velocity profile is parabolic and the laser beam is Gaussian, so both are nonuniform. These nonuniformities can be reduced to arbitrarily low levels by manipulating the laser beam and various coaxial gas-flow velocities. By properly adjusting the laser and gas stream, it should be possible to achieve identical time-temperature histories for all gas elements, which would then result in the desired uniform particles.

Both approaches are complex, which makes experimentation difficult and requires that they be thoroughly understood before they are used to produce quantities of powder.

3.2.1.2b. Horizontal Gas-Stream Configuration. The details of the orthogonal flowing gas-cell configuration used in these experiments have been described earlier and will only be highlighted in this chapter (Haggerty and Cannon, 1978). The apparatus is shown in Figure 3.1. An either focused or unfocused CO_2-laser beam enters the reaction cell through a KCl laser window. The reactant gases, SiH_4 and NH_3, enter orthogonally to the laser beam through a nozzle into the cell at controlled pressure. The reaction produces a visible reaction flame, and the powder is carried to a microfiber filter between the cell and the vacuum pump.

A set of process conditions were defined that produced a stable reaction. These were established as a benchmark or reference condition from

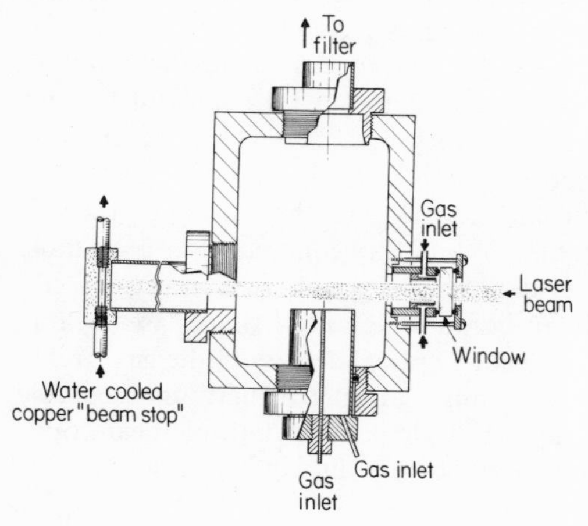

Figure 3.1. Schematic of powder synthesis cell.

which variations systematically were made. The reference conditions for Si_3N_4 synthesis (Table 3.1) are a pressure of 0.2 atm, SiH_4 and NH_3 flows of 11 cm^3/min and 110 cm^3/min, respectively, and a laser intensity of 760 W/cm^2. Approximately 2–5 W are absorbed by the optically thin gas stream. These reference conditions yield approximately 1 g/h of light brown to whitish-tan Si_3N_4 powder. Up to 5 g of powder have been produced in one experiment.

The majority of the first powder synthesis experiments had a horizontal gas stream. This configuration has several difficulties associated with it. The buoyant forces carry the heated gas and particle streams toward the KCl window and the roof of the reaction cell. This causes window breakage and also powder buildup on the cell roof, leading to possible contamination. Although 600 cm^3/min of argon across the KCl window alleviated the breakage problem, we were concerned that it would compromise our ability to analyze the laser–gas interactions and is a source of inefficiency.

The apparatus was modified so that gases flow in the upward, vertical direction to reduce or entirely eliminate these problems. Buoyant forces on the heated gas and powder act with, not against, the flowing reactant gases. This keeps the powders from depositing on the laser entry window and the cell walls for hours compared to only a few minutes with a horizontal cell, enabling prolonged powder synthesis experiments and increased powder production. Powder collection, processing and purity maintenance are all easier.

There is no indication that the powders synthesized in the vertical configuration differ from those synthesized horizontally, except in uniformity of color. Powder color varied within the cell when in the horizontal position but was more uniform in the vertical orientation.

With the cell in the vertical orientation, we have been able to reduce the argon flow rates across the KCl window and through the annular

Table 3.1. Reference Processing Conditions for Synthesis of Si_3N_4 *Powders*

Laser power density	760 W/cm^2
Cell pressure	0.2 atm
SiH_4 flow rate	11 cm^3/min
NH_3 flow rate	110 cm^3/min
Argon flow rate (to window)	600 cm^3/min
Argon flow rate (to annulus)	400 cm^3/min

sleeve by a factor of 4 without undesirable powder buildup inside the cell and on the KCl window. This, and perhaps further reductions in buffer gas-flow rates, results in a cleaner, more efficient, and more idealized reaction zone.

The increased protection to the KCl window offered by vertical operation has also allowed higher pressure reactions (0.75-atm NH_3/SiH_4) to be investigated. The increased space above the reaction zone and improved reaction stability resulting from vertical operation have allowed thermocouple measurements of reaction zone temperature. The advantages of vertical operation are sufficient so that the majority of current and future experiments will be performed in this position.

3.2.1.2c. Atmosphere Control. The reactant gases employed are electronic grade NH_3 and SiH_4. Prepurified argon is used as a buffer gas. The gas train includes a Cu oxygen getter, a molecular sieve, a dry train, and an oxygen analyzer. There are currently 10–15 ppm O_2 in the argon gas. A Ti getter may be used in order to further reduce the O_2 content in the buffer gas.

3.2.1.2d. Particle Collection. The powders are collected in a cylindrical microporous filter. Up to 5 g of powder can be collected at one time. Beyond this, the filter clogs rapidly and we are no longer able to maintain a constant cell pressure. Use of a longer filter system (about 15-g capacity) and an electrophoretic means of powder collection would increase the collection capacity.

3.2.1.2e. Powder Handling. All post-production handling of these laser synthesis powders is performed in an argon inert atmosphere glove box with <10 ppm O_2 and <10 ppm H_2O. The glove box is fitted with an extension that incorporates a press and a sintering–nitriding furnace to allow all processing to be done without exposing the powders to the atmosphere.

3.2.1.3. Unimolecular, Multiphoton Reaction Domain Experiments

A series of experiments was undertaken to determine the feasibility of inducing unimolecular, multiphoton reactions. In this type of reaction, individual gas molecules absorb sufficient energy to cause dissociation prior to colliding with other gas molecules, thus precluding a normal distribution of energies and a "thermal" reaction. The unimolecular reaction conditions were investigated to determine whether the resulting powders would exhibit unusual, advantageous characteristics.

Thermal unimolecular reactions have been well known and extensively studied for many years (see, for example, Benson, 1960; Robinson and Holbrook, 1972). With the advent of high-power, pulsed infrared lasers, it has been shown that such reactions can be induced in isolated molecules following multiple-photon excitation; in the preceding chapter, Danen and Jang have provided numerous examples. A simple calculation can be used to estimate the intensity of the laser radiation required to bring about such a process.‡ A 10.6-μm photon has approximately 0.11-eV energy, and the average energy for removal of an H atom from silane is 3.4 eV (JANAF thermochemical tables, 1971). Thus the energy of at least approximately 30 photons must be absorbed and retained to cause dissociation. The energy transfer in a collision involving highly vibrationally excited polyatomic molecules has been estimated to be $\sim$ 0.2 eV (Jensen *et al.*, 1978). Thus, the rate of photon absorption must compete effectively with this deactivation rate. At 10-Torr pressure, this implies that the reaction must occur in 10^{-7} sec or less. Assuming a photon absorption cross section of 10^{-20} cm^2 (Lyman *et al.*, 1980), an intensity of approximately 10^8 W/cm^2 is required to transmit the necessary dissociation energy/molecule in a 10^{-7}-sec pulse. Higher intensities may not produce unimolecular reactions since the gas may break down into a plasma. In fact, only molecules with appreciable absorption at the laser frequency are candidates for multiple-photon dissociation because of the breakdown limitation. With laser intensities on the order of 10^3 to 10^4 W/cm^2, and pressures near 1 atm, as in the previous section, the reactions can be expected to be exclusively thermal in nature.

The experiments discussed in this section used a tunable, pulsed CO_2 laser. The conditions that caused a reaction were determined by varying fluence and reactant pressure in the closed reaction cell. The occurrence of a reaction was determined by an irreversible pressure change, the IR spectra of the gases after irradiation, and the appearance of powder within the cell. The apparatus is shown schematically in Figure 3.2.

A Tachisto Model 215 G TEA CO_2 laser was used. Pulses contain 0.3–0.5 J with a FWHM of about 50 nsec. Three different focal length lenses were employed to vary the beam fluences at their diffraction limited waists. Peak fluences were 20, 29, and 185 J/cm^2 for 80, 67, and 28 cm focal lengths, respectively. These fluences correspond to peak intensities in the range of 4×10^8 to 4×10^9 W/cm^2.

Mixtures of reactant gases were established by successively freezing known quantities of SiH_4 and NH_3 into a liquid N_2 cooled cold finger. Experiments were undertaken after the cell had warmed to room temperature. Cell pressures were monitored with a MKS 2000 A pressure transducer before and after irradiation and both with and without condensible

‡ For very large molecules in essentially collision-free environments, it is now known that multiple-infrared-photon dissociation can be brought about by very low-power lasers. For example, the molecular ion $[(C_2H_5)_2O]_2H^+$ can be dissociated in an ion cyclotron resonance spectrometer by CW intensities of 1–0 W/cm^2 (Woodin *et al.*, 1978); the molecular complex UO_2 (hfacac)$_2$ THF, by pulsed fluences of a few mJ/cm^2 (Cox and Horsley, 1980). Silane is in a very different regime from these species, however.

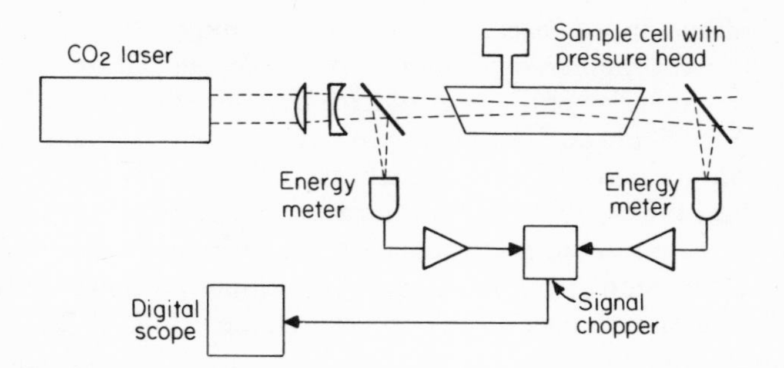

Figure 3.2. Schematic of laser, diagnostics, and reaction cell used to investigate the feasibility of unimolecular reactions.

gases frozen into the cold finger. After irradiation, the products were characterized by their infrared spectrum with a Perkin–Elmer 567 IR spectrometer.

Table 3.2 contains a summary of results for the TEA laser experiments. No reaction was noted for conditions that did not produce dielectric breakdown inside the cell, which is evidenced by a blue–white spark. The breakdown threshold is sensitive to sample pressure and peak intensity. Breakdown always occurred with fluences of 185 J/cm^2 for pure silane at pressures above 2 Torr. For fluences below 185 J/cm^2, however, breakdown did not occur for pressures up to approximately 5 Torr. For conditions where no breakdown occurred, no incondensible gas (i.e., H_2) was observed in the cell after irradiation and no change in the reagent IR spectrum occurred. In one case, Cl_2 was added to the gas mixture to "getter" any hydrogen produced and thus prevent the back reaction. For this case, however, breakdown was experienced at pressures of approximately 1 Torr using the 28-mm focal length optics. Several attempts were also made to induce reactions in mixtures of silane and ammonia. These experiments produced the same results as the silane-only experiments. For a given pressure of gas, a long focal length produced no reaction, and a short focal length caused breakdown. Lower reactant pressures might have avoided breakdown, but the quantities of products have been too small to detect with our apparatus, so this experiment was abandoned.

Although reactions were induced with both *P*(20) and *P*(24) emissions, they do not appear to have resulted from multiple-photon absorption. Reactions occurred only when a white spark was evident near the focus of the beam, an indication that dielectric breakdown occurred at these high fluences. The reactions were probably induced by plasma or arc heating rather than multiple-photon absorption.

These results are generally consistent with those published by Deutsch (1979), in which decomposition of silane was found to occur only at

Table 3.2. TEA Laser Experiments Performed on Silane

Line	p(Torr)	Lens f.l. (cm)	No. of shots	Results
Silane				
$P(20)$	4.6	80	2000	No change in pressure or spectrum; no powder
	1.21	28	2200	Slight brown powder
	5.4	28	400	Heavy powder; pressure increase; breakdown
	2.2	28	500	Light powder; pressure increase
	2.4	67	1000	No changes
	2.4	28	1500	Breakdown on first pulses
$P(24)$	2.0	28	2200	Slight powder; pressure increase; breakdown
	3.57	28	700	Powder, breakdown
	0.54	28	2000	No change
	1 + 1 Torr Cl_2	28	1500	Breakdown on first 250 shots
Silane/Ammonia				
$P(24)$	0.6/1.2	67	2300	No changes
	0.6/1.2	28	1000	No changes
	1.4/3.7	67	1000	No changes
	1.4/3.7	28	1000	Breakdown; slight powder; pressure increase

fluences of ~ 100 J/cm^2 and/or at pressures > 2–4 Torr, indicating a collision-assisted process. Recently evidence has been cited (Ronn and Earl, 1977; Lin and Ronn, 1978) that suggests that the reaction products and intermediates formed during multiple-photon unimolecular reactions and laser-induced breakdown are similar or identical to one another. The evidence is admittedly tenuous but suggests that further characterization of the products of the breakdown reaction may be of some interest. We may also mention, in this connection, the fact that Freund and Danen (1977 and 1979) have selectively removed diborane impurities from gaseous silane by decomposition following dielectric breakdown induced by high-power infrared laser pulses, without inducing decomposition of the silane itself.

3.2.2. Analyses and Characterizations

Our efforts thus far in this work have been approximately equally divided between developing descriptions of the process and of the resulting powders. The process itself is completely new and consequently has required many fundamental property measurements and new analyses to develop the most rudimentary description. Powder characterizations have followed normal practice for these types of materials. Extremely small particle size has required special handling procedures to avoid contamination. Our ultimate objective is to understand the interrelationships between process variables and powder characteristics.

3.2.2.1. Process Characterization

3.2.2.1a. Optical Absorptivities. The CO_2-laser emissions and the reported absorption peaks for SiH_4 and NH_3 are summarized in Figure 3.3. Despite the profusion of absorption peaks, which are extremely close to the emitted lines, these data cannot be used to estimate absorptivities. Unless the emission and absorption lines lie within a few Doppler widths of one another (1 Doppler width $\simeq 2\text{–}4 \times 10^{-5}$ μm), there will be virtually no coupling at low pressures. The resolution with which the absorption spectra were determined does not permit the location of the peaks to be stated with this precision. At higher temperatures and pressures, the actual peak widths are determined primarily by pressure-broadening effects. The spectra appearing in the literature were not measured with sufficient precision to allow us to calculate absorptivities. Therefore, the actual absorptivities must be measured with laser sources and with gas conditions that are very close to those of interest for the reaction. No data of this type existed for SiH_4, and only one measurement was located for ammonia. For very

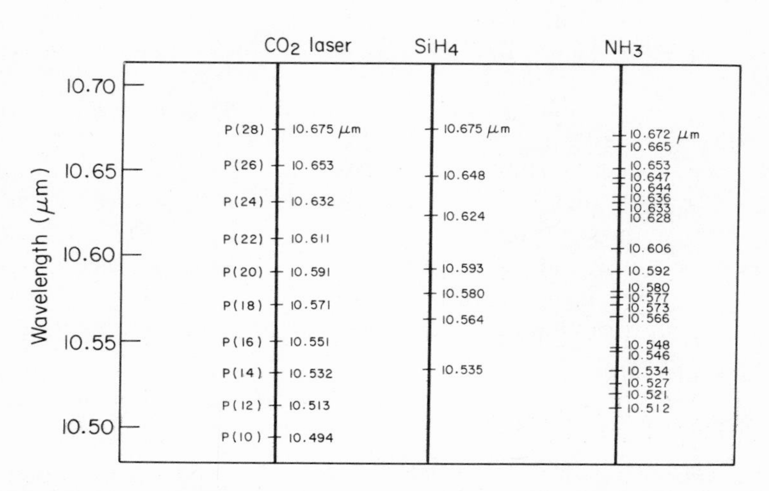

Figure 3.3. A comparison between the spectral absorption lines in SiH_4 (Tindal *et al.*, 1942) and NH_3 (Garing *et al.*, 1959) near 10.6 μm and the emission lines of CO_2 laser.

dilute ammonia in 1 atm of air, the absorptivities (Patty *et al.*, 1974) were 0.14 and 0.12 cm^{-1} atm^{-1} for the $P(18)$ and $P(20)$ emissions, respectively, of the (00°1–10°0) band, which, as noted below, are of particular interest. Absorptivity measurements were an essential part of the experimental program because their values were needed for modeling the energy absorbing processes.

3.2.2.1a.1. Absorption measurements. The absorptivity measurements were made in a manner that permitted the results to be interpreted directly in terms of the Beer–Lambert equation,

$$I = I_0 e^{-\alpha p x} \tag{3.1}$$

where, I and I_0 are the transmitted and initial laser intensities passing through a column of absorbing gas of depth x and at a pressure p. The pressure and temperature dependent absorptivity, α, is calculated directly from the experimental conditions and the observed I/I_0. The value of α is extremely sensitive to several factors that include the exact wavelength of the emitted light and both the temperature- and pressure-broadening effects that accompany absorption of light. While the theory is qualitatively useful, the quantitative measurements and their results must be viewed as largely empirical at this time. Therefore, experiments must be made very carefully over a range of conditions that approach, but do not exceed, the reaction thresholds.

We have conducted two series of absorptivity measurements. The first used an untuned Coherent Radiation Model 150 CO_2 laser. This laser was used for the majority of synthesis experiments, thus all analyses and modeling had to be based on the specific lines emitted by this laser in either pulsed or CW modes. Furthermore, the linewidths emitted from individual laser cavities vary enough from one to another to cause different effective absorptivities. The second measurement series used a line tunable CO_2 laser built at M.I.T. (Steinfeld *et al.*, 1970). This laser was used to survey the absorptivity levels for most lines emitted by a CO_2 laser in the vicinity of 10.6 μm.

3.2.2.1a.2. Measurements made with a nontunable laser

Laser emission. The coherent radiation model 150-CO_2-laser mirrors were aligned in the CW mode to emit a nominally Gaussian energy distribution, which approximately coincides with the maximum emitted power. The laser's spectrum was analyzed in CW and pulsed modes with an Optical Engineering spectrum analyzer.

In a CW mode or in long duty cycle pulsed modes (pulselength $> 50\%$ pulse period), this laser emits entirely at the $P(20)$ line of the 00°1–10°0 band (10.591 μm). In a low duty cycle pulsed mode (pulselength $\leq 10\%$ pulse period), the laser emits alternately on either the $P(20)$ or the $P(18)$ (10.571 μm) lines. The energy in individual pulses (measured with a

GenTec joule meter coupled to a storage oscilloscope) was essentially constant (within ±10%), whether the laser emitted on the $P(20)$ or $P(18)$ lines.

Absorption. The apparatus used for the absorption measurements is shown schematically in Figure 3.4. The optical path length is 10.2 cm with the O-ring sealed KCl windows in place. The cell has ports that are used for gas inlet, evacuation, and pressure-monitoring purposes. Pressure is monitored during absorptivity measurements to determine whether a reaction was induced since all of the reactions investigated produced a net increase in the number of gas molecules. Measurements were made under fixed volume conditions at predetermined initial gas pressures by alternatively measuring the pulse energy at the "transmission detector" either with or without partially absorbing gases in the cell. The ratio of these intensities yields the absorptivity directly since the effect of these absorptivities and reflections of the beam splitter and windows cancel out in the ratio. A KCl window was used as a beam splitter to reduce the pulse energy below reaction thresholds. The laser was pulsed at 1 Hz with pulselengths of 0.1 and 1.0 msec and energies of approximately 25 or 100 mJ.

The calculated absorptivities based on results of these measurements are shown in Figures 3.5 and 3.6. These energies, as we will see later, are on the order of one-sixth to one-half of that needed to raise the gas temperature to the reaction threshold. We must then conclude that during these

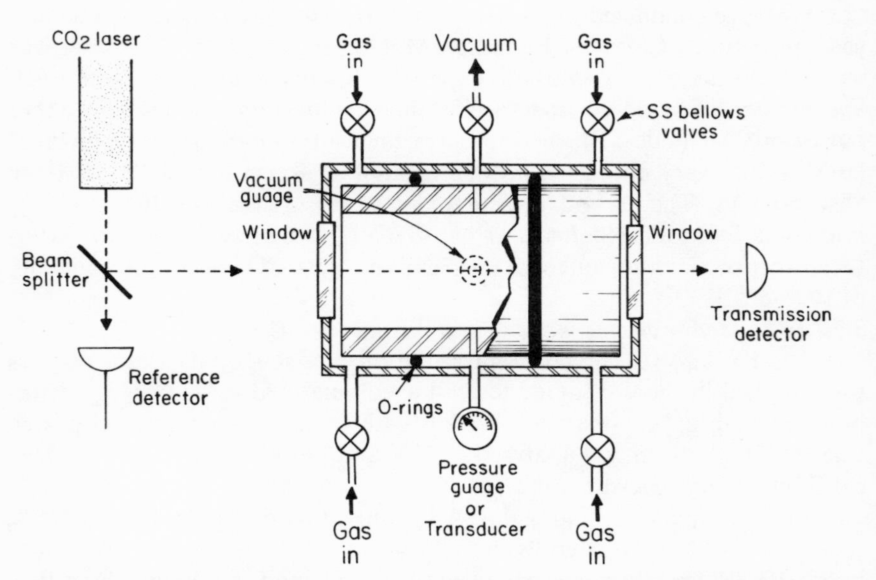

Figure 3.4. A schematic representation of the stainless-steel cell used for absorption measurements and counterflow powder synthesis experiments. The inner sleeve and O rings were removed for absorption measurements.

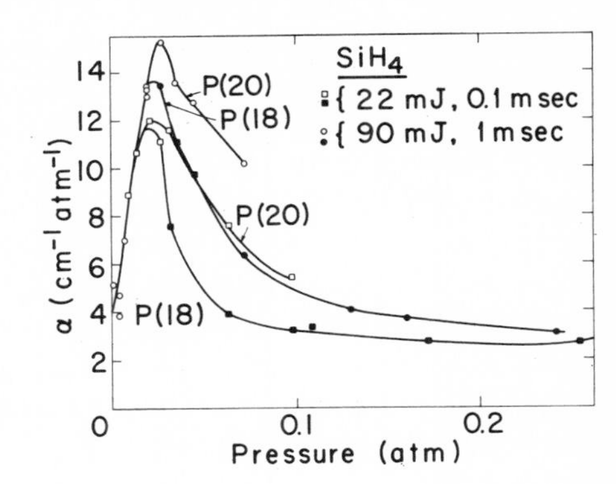

Figure 3.5. The absorption coefficients of SiH_4 as a function of pressure for the $P(18)$ and $P(20)$ CO_2 laser lines.

absorption measurements, the gas is heated by the laser pulse. Silane exhibited two distinct absorptivities that are attributed to the laser emitting on either the $P(18)$ or $P(20)$ line. The assignment of the lower absorptivity to the $P(18)$ line was based on the information from the literature (Tindale *et al.*, 1942) rather than direct correlation with spectrographic data. Ammonia exhibited equal absorptivities to the two CO_2 line that are approximately 20 to 50 times lower than SiH_4 absorptivities.

The variation in silane's absorptivity with pressure is probably because of pressure-broadening effects with closely spaced, strongly absorbing lines. With gas absorbers and laser emission sources, the apparent absorptivity is a much more sensitive function of the overlap between an absorption line and an emission line than is apparent with conventional spectroscopy. At low pressures ($P \leq 1$ Torr), the widths of the absorption lines and the laser emission lines are only approximately one Doppler width ($\Delta\lambda \simeq 2$–4×10^{-5} μm). Unless the laser emission and gas absorption lines lie within a few Doppler widths of each other, the absorptivity will be extremely low. The absorption lines broaden progressively with increasing pressure for pressure levels above a few Torr. Since the total area under the absorption peak remains constant with increasing pressure, the maximum absorption intensity decreases as the peak broadens. The variation in

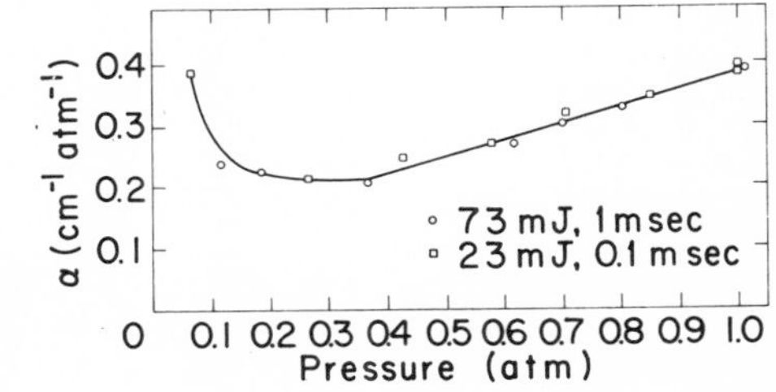

Figure 3.6. The absorption coefficients of NH_3 as a function of pressure for the $P(18)$ and $P(20)$ laser lines.

apparent absorptivity with pressure will depend on both the relative locations of the emission and (possibly many active) absorption lines as well as the details of the broadening characteristics of each active absorption line. For a single absorption line, increasing pressures, first, will cause an increased absorptivity as long as the peak broadening effects dominate and, then, may cause a decrease if the shrinking peak-height effect dominates. A continuous increase in absorptivity is expected if the emission and absorption lines are widely separated. Ascending then descending behavior is expected when they are closely spaced but are not coincident.

Silane's absorptivity, shown in Figure 3.5, follows the ascending and descending pattern. We conclude by this behavior and the high absorptivities that the active absorption line(s) is closely spaced to both the $P(18)$ and the $P(20)$ lines. It is likely that the $P(20)$ line is absorbed by the SiH_4 line at 10.593 μm, but it is difficult to determine whether the $P(18)$ line is primarily absorbed by the SiH_4 line at 10.580 or at 10.564 μm (see Figure 3.3). Ammonia's absorptivity probably results from the combined effects of many weak absorption lines that are in the vicinity of the emission lines. Thus, the absorptivities to the $P(18)$ and $P(20)$ lines are equally low and weakly dependent on pressure and are of the same order of magnitude reported previously for NH_3 (Patty *et al.*, 1974).

The high absorption coefficient of the silane is very important since it allows efficient absorption of laser energy for powder synthesis. For instance, about 70% of the laser energy would be absorbed in the first centimeter of a 1 atm 10/1 NH_3/SiH_4 mixture. Since all of the energy is absorbed directly in the reaction zone, the efficiency of the powder synthesis process is dependent only on the efficiency of the CO_2 laser. A 10–15% wall-plug-to-light efficiency is typical for a CO_2 laser.

Additional absorption measurements were made with a tunable CO_2 laser to determine if the reactant gases had higher absorption coefficients in regions other than the $P(18)$ and $P(20)$ lines of the coherent 150 laser.

3.2.2.1a.3. Measurements made with a tunable laser. Absorption measurements performed in this series differed from the others in two major ways:

1. Line selection was obtained through a tunable CO_2 laser that allowed the absorption coefficient of a large number of lines to be measured.

2. Laser intensity was sufficiently low to preclude virtually any heating of the silane or ammonia.

Two methods were used to obtain the absorption coefficients of SiH_4 and NH_3, differing slightly in technique as dictated by the availability of equipment. The apparatus in Figure 3.7 was employed for all but a few data points; it employs two choppers to modulate the sample and reference

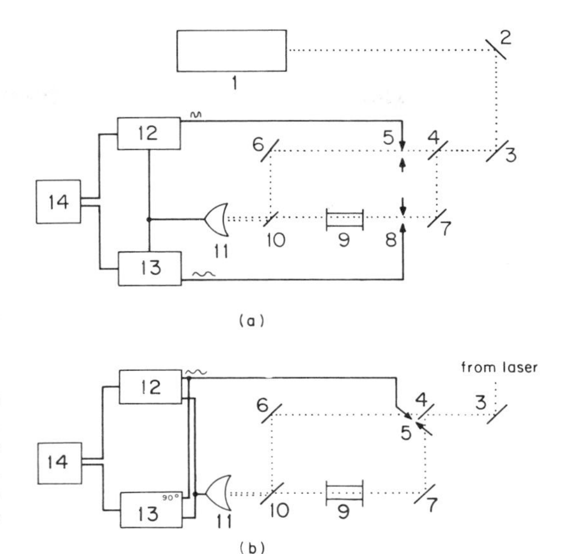

Figure 3.7. Apparatus for absorption measurements. (a) Two chopper method. 1, laser; 2,3,6,7, mirrors; 4,10, ZnSe beam splitters; 5,8, choppers; 9, sample cell, 11, pyroelectric detector; 12, reference channel lock-in; 14, PDP 8/L computer. (b) Single chopper method; lock-in 13 is set to 90° relative phase.

beams at different frequencies, allowing two lock-ins to detect the sample and reference intensities independently while using the same detector. When only one chopper was available, it was positioned to modulate the two beams at a relative phase of 90°, again allowing independent measurement with one detector (see Figure 3.7).

The laser itself consists of a 1-m gain cell with rotatable grating and a 5-m radius mirror forming the cavity; the beam emerges via a small hole in the mirror, which is mounted on a piezoelectric crystal. The stabilization network depicted in Figure 3.8 minimizes output intensity drift caused by thermal expansion of the aluminum girder supporting the optical elements. By modulating the mirror position by ± 0.5 μm, the frequency of the cavity resonance is modulated under the CO_2 gain curve. A lock-in detects the slope of the gain curve at the cavity mode frequency by demodulating the very slight ripple in the laser output and applies an error voltage to the pzt to bring the cavity mode to the gain profile.

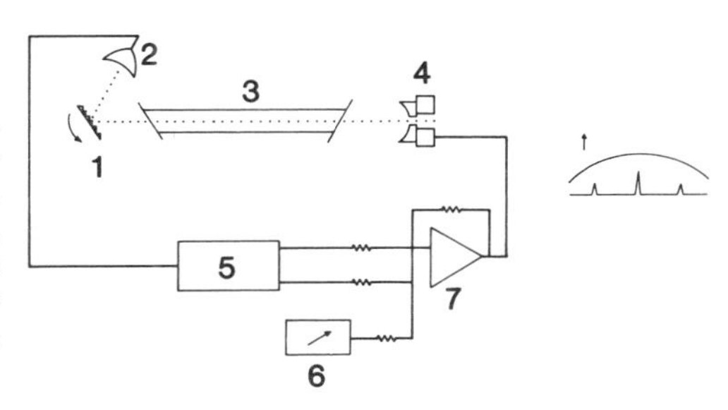

Figure 3.8. Stabilized CO_2 laser used in absorption measurements. 1, rotatable grating; 2, detector; 3, 1-m gain tube; 4, pzt mounted mirror; 5, lock-in; 6, manual dc offset; 7, high voltage op amp. Depiction of cavity resonances under a Doppler-broadened gain profile is shown at right.

Two cells were employed with 1.73- and 9.97-cm optical pathlengths fitted with perpendicular NaCl windows. When possible, sample pressures of about 130 Torr were used; for very strongly attenuated lines, lower pressures were needed. Gas pressures were read from an MKS 2000 A pressure transducer with a 0.1–1000 Torr stated range, although the head was found to be sensitive to 10^{-2} Torr. The response of the transducer was checked against a dibutylphthalate manometer. Electronic grade (Airco) silane was used from the tank after many freeze–pump–thaw cycles; anhydrous (Airco) NH_3 was treated similarly.

The procedure for measuring the attenuation of a sample is similar to the previously discussed experiments. The laser is tuned to the desired line, and the stabilization networks is activated. With the sample frozen into a cold finger in the cell, the ratio of the sample and reference lock-ins' signals was averaged for approximately 1 min (about 400 samples) by a PDP 8/L computer. The sample was then thawed and the ratio again averaged for 1 min. The laser is then tuned to the next line, etc. The transmittance, T, is given by

$$T = \frac{I/I_0 \text{ (sample unfrozen)}}{I_0 \text{ (sample frozen)}} \tag{3.2}$$

where I and I_0 are actually the voltages from the sample and reference lock-ins, respectively. Typically, the 67% confidence limits of T were $\pm 3\%$. The absorption coefficient α is calculated from T and the known sample pressure p, and cell length l, via

$$\alpha = \frac{-1}{pl} \ln T \tag{3.3}$$

All measurements were taken at room temperature.

Table 3.3 gives the absorption coefficients, α, for NH_3 and SiH_4 for laser lines in the 10.6-μm CO_2-laser band. For very strong attenuations, it was necessary to decrease the sample pressure; values for α are noted in parentheses where the uncertainty in I/I_0 (with sample) exceeded 25% because of the capacity of the gas at the particular laser line.

$P(18)$ and $P(20)$ silane and ammonia absorption coefficients measured with the untuned laser (Figure 3.5 and 3.6) agree well with those in Table 3.3. The values determined with the untuned laser were measured at comparable pressures but using higher laser fluences per pulse. Results taken at comparable pressures are given in Table 3.4. The differences between the two sets of data likely result from the different conditions and the different lasers with which the measurements were performed. A temperature change in the gas due to higher laser fluences would probably affect the values. Also the emitted linewidths are probably different for the two lasers.

Table 3.3. Absorption Coefficients (atm–cm)$^{-1}$ *for the 10.6 μm* (00°1–10°0) *Band of the* CO_2 *Laser.*

Spectral line	Wave length (μm)	Absorption coefficients[a] (atm^{-1} cm^{-1})	
		NH_3	SiH_4
R(38)	10.137	0.044 A	1.68 C
R(36)	10.148	not measured	0.27 C
R(34)	10.159	0.0072A	(small) C
R(32)	10.171	0.0065A	0.19 C
R(30)	10.182	0.047 A	0.96 C
R(28)	10.195	0.057 A	2.68 C
R(26)	10.207	0.059 A	0.70 C
R(24)	10.220	0.056 A	0.31 C
R(22)	10.233	0.071 A	1.36 C
R(20)	10.247	0.12 A	1.61 C
		0.22 B	
R(18)	10.260	7.25 B	0.87 C
R(16)	10.275	0.24 A	1.14 C
R(14)	10.289	0.47 E	0.79 C
		0.50 A	
R(12)	10.304	21.5 E	0.23 C
R(10)	10.319		0.34 C
R (8)	10.333		0.007C
P (8)	10.476	0.35 A	2.70 C
P(10)	10.494	0.16 A	0.50 C
P(12)	10.513	0.65 A	3.95 C
P(14)	10.532	0.83 A	1.94 C
P(16)	10.551	0.41 A	0.96 C
P(18)	10.571	0.18 A	11.8 C
P(20)	10.591	0.30 A	(43.0) C
			12.9 D
P(22)	10.611	0.13 A	(28.9) C
			8.17 D
P(24)	10.632	0.16 A	(34.5) C
			10.7 D
P(26)	10.653	0.42 A	7.42 C
P(28)	10.675	0.35 A	(29.4) C
			8.94 D
P(30)	10.696	0.87 A	6.46 C
P(32)	10.719	13.6 E	3.79 C
P(34)	10.741	3.5 A	5.01 C
P(36)	10.765	1.8 E	18.1 C
P(38)	10.788	1.02 E	1.05 C
P(40)	10.812	1.0 A	

[a] Conditions: A, 136 ± 2 Torr 10 cm cell; B, 130 ± 2 Torr 1.7 cm cell; C, 53 ± 1 Torr 1.7 cm cell; D, 11 ± 1 Torr 1.7 cm cell; E, 25 ± 1 Torr 1.7 cm cell; ( ) denote uncertainty in precision of results.

Table 3.4. Comparison of Optical Absorptivities Measured with Tuned and Untuned CO_2 Lasers

			Absorptivity (cm^{-1} atm^{-1})	
Gas	Pressure (Torr)	Line	Untuned	Tuned
SiH_4	53	*P*(18)	3.7–6.4	11.8
SiH_4	53	*P*(20)	7.0–10.5	43
SiH_4	11	*P*(18)	11.0	—
SiH_4	11	*P*(20)	11.0	12.9
NH_3	130	*P*(18)	0.24	0.18
NH_3	130	*P*(20)	0.24	0.30

There is also agreement with the results of Patty *et al.* (1974). They prepared ammonia samples with parts-per-million concentrations in 1 atm of air. One would expect a somewhat larger pressure-broadening effect for their experiments than for the conditions used in ours. This pressure broadening causes NH_3 absorption features to overlap the spectral bandwidth of the CO_2 laser employed. Greater pressure or laser bandwidth will generally cause greater absorption, so their values are usually slightly higher than ours. Note that for neither Patty's nor these measurements, is the laser linewidth accurately known, although the lasers are of similar construction.

The absorptivities reported in Table 3.3 have two important consequences. The first is that the highest absorptivity level is exhibited by SiH_4 for the *P*(20) emission from the CO_2 laser. Fortuitously, the highest gain, most efficient emission from the laser is most strongly absorbed by the reactant gas we selected based on other criteria. Also, there are other emission lines where SiH_4 and NH_3 have approximately equal absorptivities [e.g., *P*(34) and *P*(38) lines] or where NH_3 has a higher absorptivity than SiH_4 [e.g., *R*(18) and *P*(32) lines]. Changing the gas species that preferentially absorbs the laser radiation may influence the chemical reaction and kinetics even though the process is usually carried out in the thermal domain where a normal distribution of energies is assumed. Although we have made some absorptivity measurements, we have still not measured other data that is important for modeling the process reaction thresholds, propagation velocities, etc. Absorptivity measurements must be made at elevated temperatures as a function of pressure for unreacted, partially reacted, and fully reacted gases and entrained solids. These rather difficult experiments will be carried out in future elaboration of this work.

3.2.2.1b. Gas-Flow Modeling. For characterizing the reaction zone, it is essential to understand the reactant gas-flow characteristics. The Reynold's number for "reference" conditions is approximately 140. This Reynold's number and those calculated for other experimental conditions are well below the laminar-to-turbular transition. A computer model has been employed to characterize this laminar, streamlined gas-flow behavior (Patanker and Spalding, 1970). The results of this modeling have been qualitatively useful but have not been quantitatively consistent with observations of the reaction flame shape. We are now using a more sophisticated computer model (Spalding, 1977) that provides a more accurate description of the process.

Both models treat the gas flow as an axisymmetrical jet emerging from a circular orifice that enters into a coaxial stream. The values of the dependent variables, i.e., fluid velocity in the direction of flow, stagnation enthalpy, and mass fraction of a particular chemical species in the fluid mixture, are calculated by a finite difference numerical integration of differential equations having the form

$$\frac{\partial \phi}{\partial x} + (a + b\omega)\frac{\partial \phi}{\partial \omega} = \frac{\partial}{\partial \omega} c\left(\frac{\partial \phi}{\partial \omega}\right) + d \tag{3.4}$$

where x and ω are the coordinates in the direction of the flow and the radial coordinate, respectively; a and b are arbitrary functions of the coordinate x that are related to the mass flow rates of fluid across the internal and external fluid boundaries (The internal boundary is coincident with the symmetric axis and has zero mass flow across it.); c is a measure of the transport property that is appropriate to the variable ϕ (fluid velocity, stagnation enthalpy, mass fraction of a particular chemical species) i.e., for fluid velocity. c may have term related to the fluid viscosity and density; d represents the source of the entity ϕ within the fluid. If ϕ represents the mass fraction of a chemical species, d will be a measure of the rate of reaction or depletion of the species within the fluid by way of a chemical reaction. If ϕ stands for longitudinal velocity, the d term will contain an element that is proportional to the rate of diminution of pressure with distance x. If ϕ stands for stagnation enthalpy, d may contain a term expressing the influence of kinetic heating on the redistribution of temperature within the flow.

In the present form, the computer model is only concerned with the velocity distribution and the mass within the gas jet. It treats the fluid within the jet as a single phase containing only one chemical species with constant density and viscosity. The outer jet is assumed to have the same fluid properties as those of the inner jet. Gradients in temperature and pressure as well as the effect of an exothermic reaction are ignored.

Figures 3.9, 3.10, and 3.11 show typical computer-calculated results describing (1) the change in average velocity with distance from the inlet nozzle, (2) the boundary between the inner stream of reacting gas and the

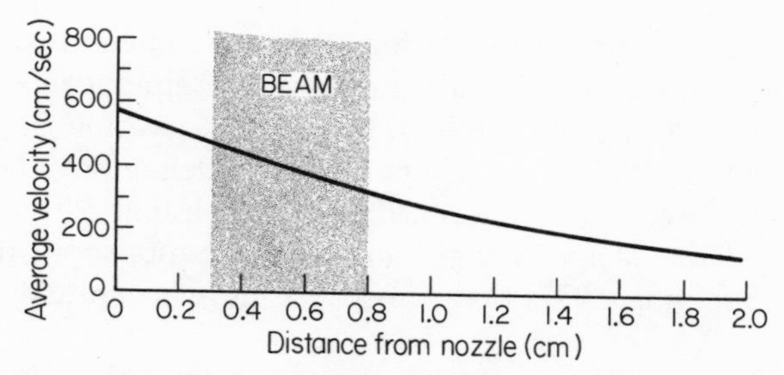

Figure 3.9. Average velocity of gas stream calculated for reference conditions.

outer flow of argon, (3) a typical velocity profile at a particular distance from the nozzle. The figures shown use reference conditions for orthogonal reactor geometry. An ammonia–silane gas mixture of ratio 10 to 1 and a volumetric flow rate of 121 cm^3/min at room temperature and 0.2 atm were used for the calculations.

Comparison of the calculated gas-flow behavior with the observed flame shape as shown in Figure 3.18 (see p. 198), shows that at the base of the flame the diameter of the flame is almost 5 times larger than the calculated gas-stream diameter. The diameter of the flame corresponding to the distance between the lower cusps, which may be indicative of the boundary between the inner and outer gas streams, is about 2.5 times larger than predicted by the computer model. Since the computer model does not consider the volumetric increase in the gas due to heating and reaction products, these effects should be added to the calculated stream width. A temperature increase from 300 to 1100°K will produce a volume increase of 3.6*x*, while the increase because of the reactions, $3SiH_4(g) + 30NH_3(g) \rightarrow Si_3N_4(s) + 26NH_3(g) + 12H_2(g)$, gives a volume increase of less than 2*x*. This corresponds to a total radial increase of about 1.6*x* if Δr is proportional to $(\Delta V)^{1/3}$, or 2.1*x* if Δr is proportional to $(\Delta V)^{1/2}$, which assumes only radial expansion. The combination of these two effects sub-

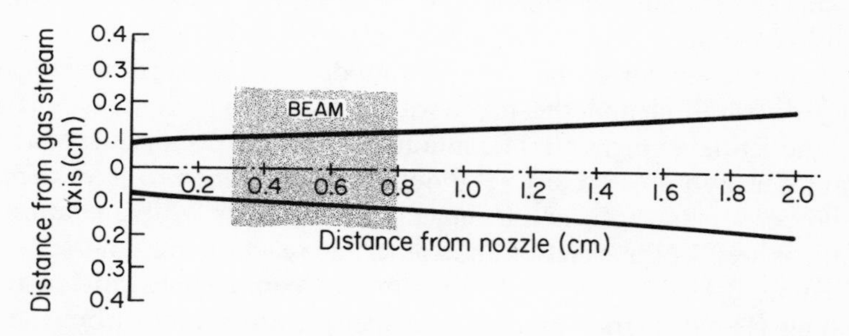

Figure 3.10. The computed width of the gas stream under reference conditions.

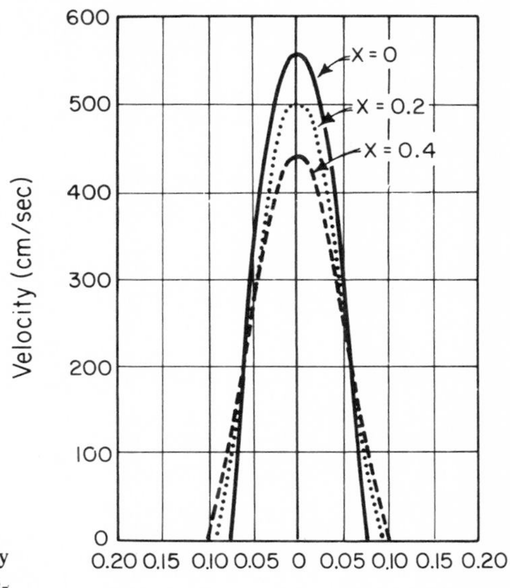

Figure 3.11. The computed velocity profiles of reactant gases at several distances (x) from the nozzle.

stantially resolves the difference between the observed and calculated stream diameters.

The new computer model will permit consideration of gradients in temperature, density, viscosity, and pressure of both the reacting gas and annular flow. A single- or multiple-step chemical reaction can be employed, and the mass fractions of the chemical species involved can be determined at various points of interest. It also has the versatility of being able to handle radiative transfer by blackbody emission and change in the size of particles carried by the gas stream because of condensation, vaporization, combustion, or other processes.

3.2.2.1c. Reaction Thresholds and Reaction Propagation Velocity. Static-and flowing-gas experiments have been carried out to map laser-intensity pulselength thresholds required to induce chemical reactions in SiH_4 and $NH_3 + SiH_4$ mixtures. These threshold determinations provided data for modeling studies and served as an empirical basis for designing a continuous synthesis process.

Threshold pulselengths were determined for various static reactant gas mixtures and pressures with fixed pulse intensities. Reaction threshold experiments were undertaken in the cell shown in Figure 3.4. A reaction was detected by a permanent pressure change in the closed volume cell. Typically, the transient pressure rise, which occurs with heating the gas, disappears within 30 msec unless a reaction occurred. Multiple pulses above threshold conditions caused visible concentrations of powder to form on the entrance window. The persistent formation of this powder caused us to abandon these static threshold determinations in favor of the flowing-gas experiments.

It was anticipated that the powder deposit on the input window would cause two serious problems. Whether formed from SiH_4 or NH_3/SiH_4 mixtures, the powders would attenuate the laser beam resulting in erroneous, higher than actual threshold determinations. In addition, the KCl windows might be broken with high-energy pulses and large quantities of powder. The only beneficial effect of powder forming on the KCl window was that the window could be used as a support for IR spectrographic analyses.

In order to eliminate deposition of powder on the windows, a stainless-steel insert was put in place to direct the flow paths of the gas streams. Gases are pumped from the cell in a radial direction at approximately the midpoint along the cell's axis. An inert, transparent gas is introduced at the laser entrance window that flows axially along the beam's propagation direction. The absorbing reactant gases are introduced at the laser exit window. They flow axially, opposite to the beam's propagation direction. They meet the inert gas at approximately midpoint in the cell where they are subjected to the full intensity of the laser beam. They are then drawn out of the cell.

The principal advantage of this cell configuration is that the reaction occurs far away from the laser windows, and reaction products do not reach the windows because of the gas-flow directions. The inert gas protects the entrance window area and the beam is fully attenuated by the time it reaches the exit window, so there is no problem with heating in that region. The exit window can be used safely for direct observation or other optical characterizations of the reaction zone. For short laser pulselengths, this experimental configuration is indistinguishable from a static synthesis experiment except for the important feature of preventing reaction products from forming on the windows. These features, combined with the good visibility of the reaction zone, proved to be a superior experiment for determining reaction thresholds than the static experiments.

Reactions were visibly evident in several ways. With long-high-intensity pulses, a weak, blood-red emission was evident near the center of the cell. Weaker or shorter pulses resulted in a momentary increase in the optical density of the gas within the cell; when sighting through the cell against a light background, laser intensity pulselength combinations above certain threshold levels caused a darkening effect. Under some light combinations, wisps of white smokelike powder were also visible. Threshold combinations were taken to be the minimum conditions that produced any visible effect. Thresholds determined in this manner agreed closely with the static thresholds identified by an irreversible pressure change.

The purpose of these experiments was to map the laser intensity pulselength thresholds that cause a reaction as a function of gas pressure and composition. The small quantities of powders that formed in pulsed experiments were not captured by any means. The results of the reaction threshold experiments with SiH_4 and NH_3/SiH_4 mixtures are shown in Figures 3.12 and 3.13.

These experimental results are reported separately on Figure 3.12 and 3.13 because the laser pulse characteristics change for pulselengths shorter than approximately 10 msec. For pulselengths longer than approximately

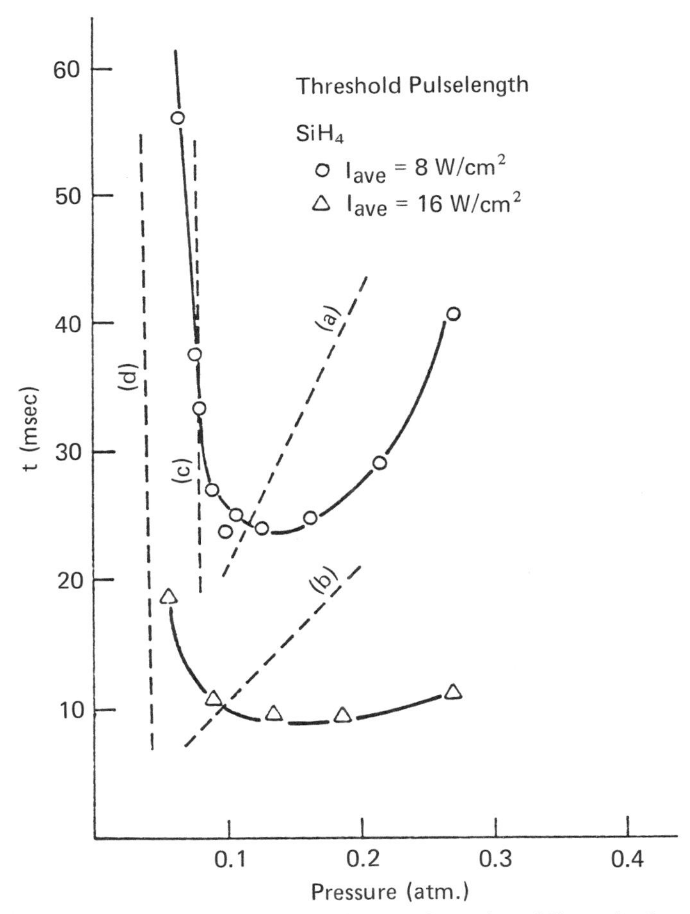

Figure 3.12. Threshold pulselengths that result in the formation of Si powder from SiH_4 as a function of pressure. Dotted lines a, b, c, and d are calculated thresholds equating the absorbed power to the sensible heat (a, c) and to the conductive losses (b, d). Lines a and c correspond to 8 W/cm^2 and b and d correspond to 16 W/cm^2.

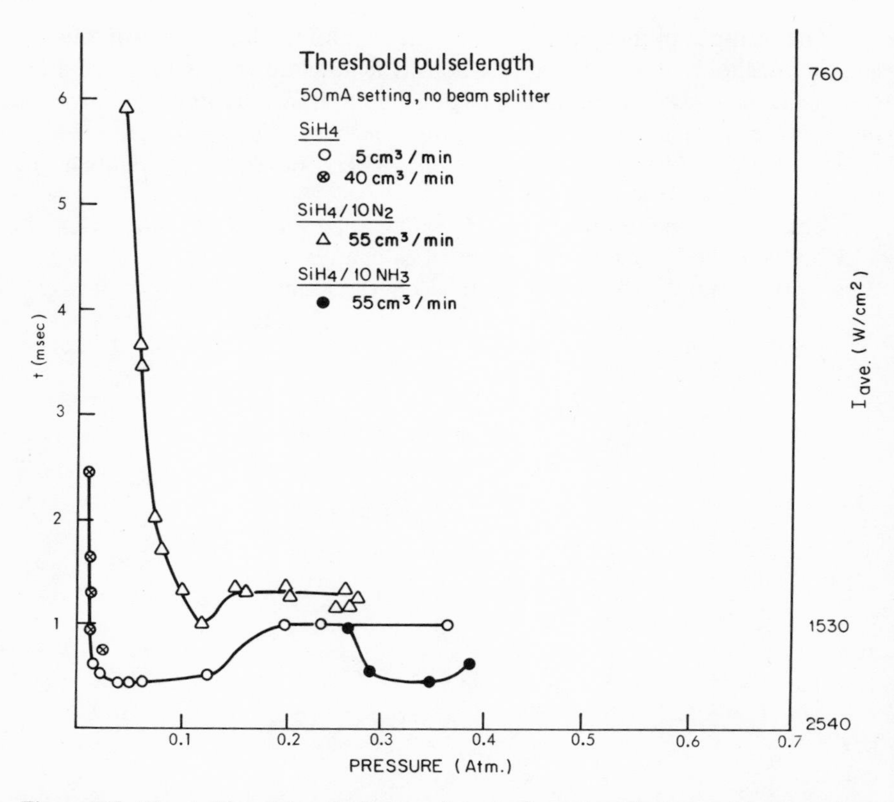

Figure 3.13. Threshold pulselengths that result in the formation of powders from several gas mixtures as a function of pressure. Pulselength dependent pulse intensities are indicated. For pulselengths greater than 10 msec, I_{avg} equals 760 W/cm^2.

10 msec and laser beam plasma current in the range of 20–50 mA, the intensity of the laser beam is independent of pulselength. Thus, the energy emitted in a pulse is directly proportional to pulselength. For pulselengths shorter than 10 msec, the average beam intensity in this particular laser becomes dependent on pulselength. Figure 3.13 indicates the average pulse intensity as a function of pulselength. The longer pulselength experiments employed a KCl beam splitter to reduce the intensity of the beam at the cell to approximately 10% of the emitted beam.

The equation that describes the absorption of laser energy and the resulting temperature rise and heat losses is straightforward to write but has not been solved in a general manner. The solution is difficult because optical characteristics of the materials are not well known and mass–heat transfer processes are intrinsically difficult. Limiting solutions have been developed.

The heat power balance in any volume element of partially absorbing gas exposed to light is given by the following equation:

$$I_0 \,\Delta A \exp(-\textstyle\sum \alpha_i p_i x)[1 - \exp(-\textstyle\sum \alpha_i p_i \,\Delta x)]$$
$$= C_p \frac{n}{V} \Delta V \frac{dT}{dt} + \Delta H \,\Delta V \frac{dn}{dt} + \text{heat transfer losses} \quad (3.5)$$

where I_0 is input intensity in W/cm^2; ΔV is the element of volume in question; ΔA is the cross section of the element ΔV; α_i is the absorption coefficient of the ith species; p_i is the partial pressure of the ith species; x is the distance of the element ΔV from the window; Δx is the thickness of the element ΔV; C_p is the heat capacity of the gas; n/v is the molar density in the volume element ΔV; dT/dt is the rate of change of temperature; ΔH is the heat of reaction in joules/mole; dn/dt is the moles of gas reacting per unit of time per unit volume. The geometrical relationships are shown in Figure 3.14. The expression on the left of the equation indicates the heat absorbed within a volume element ΔV after a beam, having an initial intensity I_0, has traveled a distance x through a partially absorbing medium. The terms on the right represent the means by which the absorbed heat can be dissipated. These are the sensible heat, the latent heat, and heat losses by conductive, convective, and radiative processes.

The details of the absorption expressions become uncertain once sufficient power has been absorbed to cause heating along the path of the beam. Sudden temperature rises cause both the α_i and the p_i values to change. The experimental results shown previously in Figures 3.5 and 3.6 show how α_i varies with moderate pressure changes under essentially constant temperature conditions. It can be anticipated that the α_i's will vary with temperature, but no experimental measurements have been made yet. Also, no quantitative α_i measurements have been made through partially or completely reacted gases. These factors, combined with uncertainties about the temperature, pressure, and density along the path of the laser beam

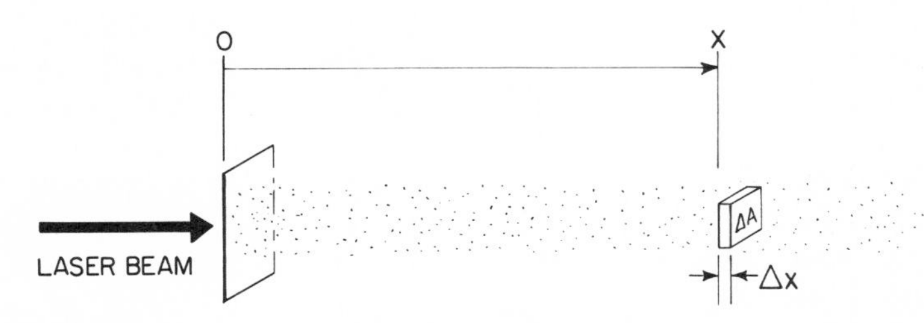

Figure 3.14. Schematic representation of the volume element (ΔV) within a column of gas that is absorbing the laser beam. At $x \leq 0$, there is no absorption of the laser beam.

preclude meaningful calculations for locations that are remote from the first interaction volume. To evaluate the terms on the right, it is necessary to develop a model for mass transfer that accompanies the intense localized heating from the laser as well as that from the exothermic reactions. Such models, and their accompanying solutions, are currently under development. Only simplified solutions have been examined at this time in order to explain the high- and low-pressure regimes of the threshold experiments shown in Figures 3.12 and 3.13.

One set of solutions considers the volume element that is first exposed to the laser beam. In the counterflow geometry, this is at the plane where the two axially flowing gas streams intersect. Since this point is subjected to the most intense illumination within the cell, it is the first to reach threshold conditions. Also, the illumination intensity is not complicated by a complex "upstream" history. For this set of solutions, we have assumed that the reaction proceeds spontaneously to completion when the premixed reactant gases reach a critical temperature. The exothermic heat, therefore, does not affect the threshold. Convective and radiative heat losses were assumed to be negligible. The assumption about convective heat losses is valid since the pulselengths are short compared to the gas residence time in the cell. Below threshold temperatures, the radiative power of the gas is very low. Conductive heat losses are considered below.

The first solution assumes that all of the absorbed heat is converted to sensible heat. Using the substitutions $\exp(-\alpha p\,\Delta x) \simeq 1 - \alpha p\,\Delta x$ and $n/V = P/RT$, and a correction for the effect of temperature on absorption $[\alpha p(T) = \alpha_0 p_0(T_0/T)]$ for a freely expanding gas gives

$$\Delta t = \frac{C(T_R - T_0)}{\alpha_0 R I_0 T_0} \tag{3.6}$$

where Δt is the threshold pulselength required to raise the reactants to the reaction temperature (T_R).

The calculated threshold pulselengths in SiH_4 for initial laser intensities equal to 8 and 16 W/cm^2 are shown in Figure 3.12 as lines a and b. The absorptivity as a function of pressure was taken as the $P(20)$ value shown on Figure 3.5. T_R is taken to be 700°C for decomposition of silane. This value was estimated for temperature ranges found in the literature (Braker and Mossman, 1971). The calculated Δt values have approximately the correct value and correct pressure dependence for pressures greater than 0.075 atm. A better fit to experimental results is obtained by choosing $T_R = 300$–$400°C$. Below a pressure of 0.05 atm, it is apparent that this simplified model does not approximate the observed behavior; threshold pulselengths increase sharply with decreasing pressure rather than continuing to decrease.

A constant mass flow was used in all these experiments, and thus the gas velocity varied inversely with cell pressure. Even at the lowest pressure, the time required for a volume element to transit the full length of the heated gas column exceeds the longest investigated pulselength by a factor of 20. Since this ratio was generally substantially larger than 20, all of the experiments fulfilled pseudostatic conditions from the criteria of pulselength and residence time. The heat flux because of the gas-flow rate at the reaction temperature is also negligible with respect to the laser beam power. Only 0.7 W is required to produce an exit temperature of 600°C at a volumetric flow rate of 25 cm^3/min. Therefore, the rapidly increasing threshold pulselengths with decreasing pressure do not result from either of these two dynamic effects.

The second solution to the heat flux equation treats the steady-state case where absorbed heat is lost to the cold cell walls by conduction. For this case, the heat transfer rate per unit length ($\dot{Q}/L$) between the inner cylindrical column of heated gas and the cell wall is

$$\frac{\dot{Q}}{L} = 2\pi k \frac{T_i - T_0}{\ln(r_0/r_i)} \tag{3.7}$$

where k is the thermal conductivity of the gas medium that is approximately independent of pressure, T_i and r_i are the inner gas column temperature and radius, and T_0 and r_0 are the outer wall temperature and radius.

The heat generated per unit length of the absorbing gas column is

$$\dot{Q}/\Delta x = I_0 \pi r_i{}^2 \alpha p \exp(-\alpha p x) \tag{3.8}$$

where the terms are the same as previously defined. Equating these two expressions and solving for pressure p with $x = 0$, gives the critical pressure where conduction heat losses with a gas temperature T_i just equal absorbed power in the first exposed volume element. For T_i equal to the reaction temperature T_R, this procedure indicates the critical minimum pressure at which a reaction can be induced,

$$P_{\text{crit}} = \frac{2k}{\alpha I_0 r_i{}^2} \frac{(T_R - T_0)}{\ln(r_0/r_i)}. \tag{3.9}$$

With $k(SiH_4) = 0.65 \times 10^{-3}$ W/cm °C, the critical pressures are 0.09 and 0.05 atm for I_0 equal to 8 and 16 W/cm^2, respectively, and are indicated as lines c and d in Figure 3.12. These critical pressures, at which the threshold pulselengths should go to infinity are in sufficient agreement with the behavior exhibited by the threshold experiments to conclude that conductive losses to the cold walls caused the departure from the assumed domination by sensible heat as used in the first solution. The time required to establish fully developed steady-state conduction in 0.1-atm SiH_4 is estimated to be approximately 40–50 × 10^{-3} sec by the equation

$$t = x^2/4\kappa \tag{3.10}$$

where κ is the thermal diffusivity.

Although highly simplified, these solutions to the heat balance equation give a useful description of the conditions required to initiate the chemical reaction. Once initiated at a critical temperature, it appears to proceed to completion in the heated volume element.

A second set of calculations was made to estimate the propagation of a reaction through a partially absorbing reactant gas. Generally, the conditions for initiating and propagating a reaction are the same as assumed in the threshold analyses. In addition, it was assumed that the gas becomes completely transparent to the laser light once the reaction occurs and also that the reaction exotherm does not affect the reaction. Computer calculations give the power absorbed in each volume element along the beam axis as a function of time and calculate the instantaneous local temperature in the same manner used for the threshold estimates in the first volume element. Reaction is assumed to occur when the local temperature reaches the reaction temperature, T_R. The extent of the reaction is followed by calculating the location of T_R for various time increments.

Representative results of these calculations are shown in Figure 3.15. The curves marked 1–5 correspond to the times indicated in the caption. Curve 1 corresponds to the condition where the threshold has just been exceeded in the first volume element. Curves 2–5 track the propagation of the reaction [at the intersection of $T(x)$ with the 973°K isotherm] through the partially absorbing reactant gas column. The velocity with which the reaction propagates through the reactants is derived directly from the penetration of the reaction as a function of time. Figure 3.15 was calculated for the SiH_4 partial pressure and the total pressure, both equal to 0.2 atm. Figure 3.16 summarizes the results of similar calculations for pressures ranging from 0.04 to 0.2 atm. The time intercept (at distance = 0) is the threshold time required to induce the initial reaction. A plot of these intercept times as a function of pressure will give the equivalent of curve a in Figure 3.12, except, in this case, a constant αp product was assumed rather than using experimentally measured values for $\alpha(p)$, as used in Figure 3.12. The reciprocal of the slopes of the time–distance curves are equal to the penetration velocities (V_p). The calculated values of V_p are indicated for each pressure. Similar calculations have been made for a wide range of laser intensities, gas mixtures, and ambient pressures.

Results such as these are extremely important for designing process experiments. In the orthogonal synthesis configuration, the gas must reside in the laser beam long enough for the reaction to penetrate through the

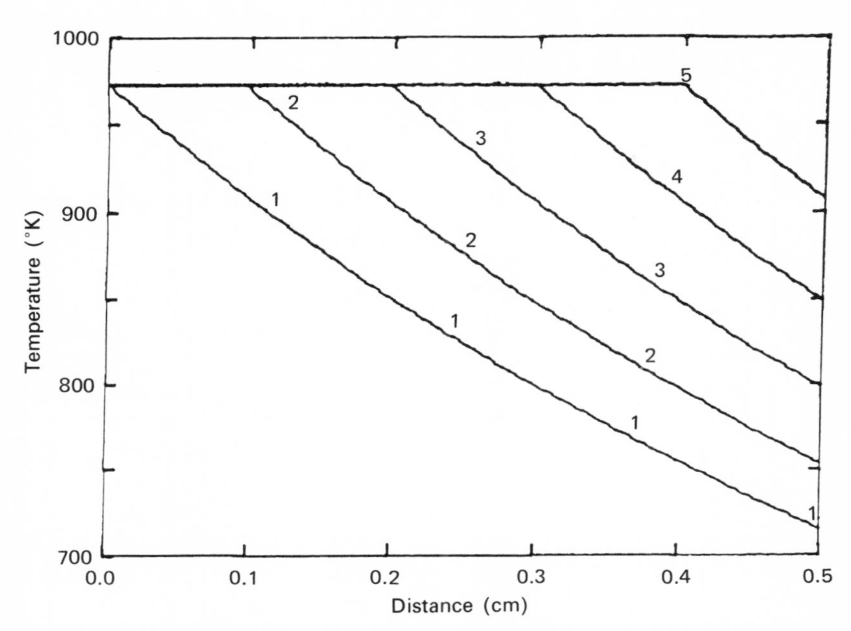

Figure 3.15. The temperature profile within 0.2-atm SiH_4 at various times after exposure to the laser beam. The times are (1) 41 msec, (2) 43.2 msec, (3) 45.7 msec, (4) 48.1 msec, (5) 50.6 msec, after the laser beam is switched on. The beam intensity is 8 W/cm^2, and the product αp is assumed to be constant and equal to 0.6 cm^{-1}.

entire gas stream. With a thin reactant gas stream in this experimental configuration, the residence time needs to be only slightly greater than the threshold time since the entire stream is subjected to an approximately uniform laser intensity. Achieving a stable reaction position in the CW counterflow configuration requires that the reactant gas velocity and the reaction penetration velocity be equal and opposite to each other. If the gas velocity is too low, the reaction will penetrate to the back of the cell; if it is too high, it will prevent the reaction. In pulsed, counterflow experiments, the pulselength reaction–velocity product cannot exceed the length of the reaction chamber.

Although these analyses have not been verified quantitatively by experiment, observations have qualitatively agreed with predicted behavior. In the orthogonal synthesis experiments, residence times were generally made long enough to insure full penetration of gas stream. In specific experiments the reaction was disrupted by raising the reactant gas velocity to a level where the residence time approximately equaled the calculated threshold time. In pulsed, counterflow experiments, pulselengths were shorter than times calculated for the reaction to penetrate to the end of the cell. No evidence of reaction was found at the end of the cell.

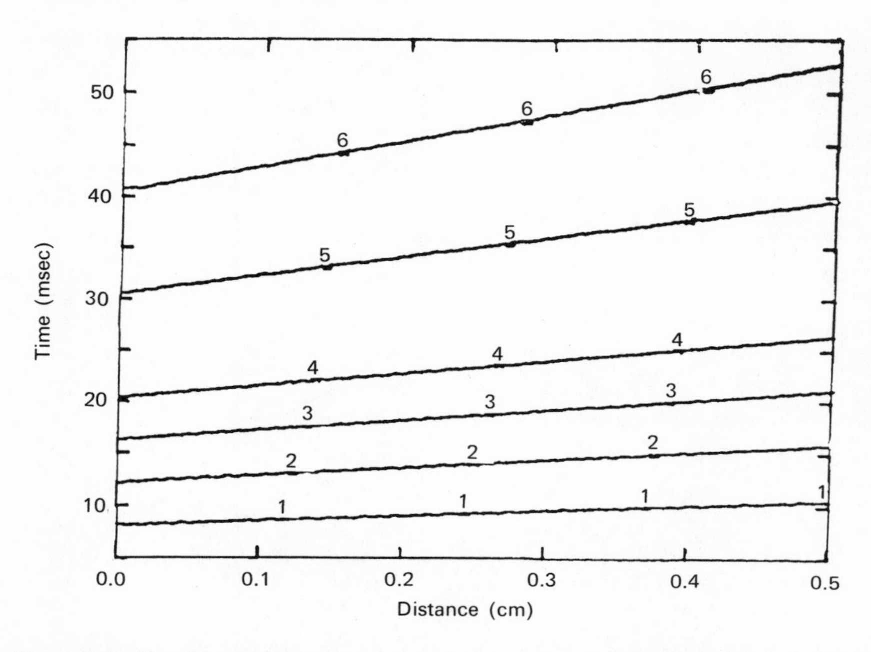

Figure 3.16. The time necessary to propagate the reaction zone to a specific distance for several SiH_4 pressures. The pressures are (1) 0.04 atm, (2) 0.06 atm, (3) 0.08 atm, (4) 0.10 atm, (5) 0.15 atm, (6) 0.20 atm. The calculated propagation velocities are (1) 233 cm/sec, (2) 135 cm/sec, (3) 104 cm/sec, (4) 86.2 cm/sec, (5) 56.8 cm/sec, (6) 42.5 cm/sec.

Other more detailed analyses of the orthogonal configurations are recounted later in this chapter. Even though they are fairly complete, they require additional experimental data to permit an accurate description of the process. The least understood optical property is the effect of temperature on the absorptivities of the reactants of the gas from the time a reaction has been initiated until it is completed. The detailed gas flow paths and rates that result from laser heating and exothermic reactions require further understanding. With this base, it should be possible to develop a fundamental description of the nucleation and growth kinetics. Besides satisfying the obvious scientific interest, this knowledge is required to optimize the process variables and cell geometry.

3.2.2.1d. Reaction Zone Mapping. An accurate empirical description of the physical boundaries and regions of the reaction zone is useful and important, since it provides a reference against which the accuracy of analytical models, such as gas heat transfer, chemistry, kinetics, and fluid flow, can be evaluated. It can also be used to provide good direct estimates of many experimental parameters such as heating rates and optical absorptivities of reacted and unreacted regions of the stream. This section will describe the physical and thermal characteristics of the laser-heated reaction zone.

Description and analysis of the spectral features emitted from the reaction zone as well as our analysis of various parameters such as gas velocities, gas mixtures, and laser intensities, are discussed in later sections.

Many features of the laser-heated process are visible to the eye and can be recorded on conventional films. Figure 3.17 shows a photograph of the reaction flame taken under reference process conditions and in a cross flow configuration. Figure 3.18 is a graphic, scaled representation of the reaction. It locates the flame and particulate products with respect to the inlet gas tip, the calculated gas-stream boundary, and the CO_2-laser beam. The calculation of the gas-stream boundary was discussed previously. The locations of various features were recorded photographically, by cathetometer and by sighting directly against graph paper positioned behind

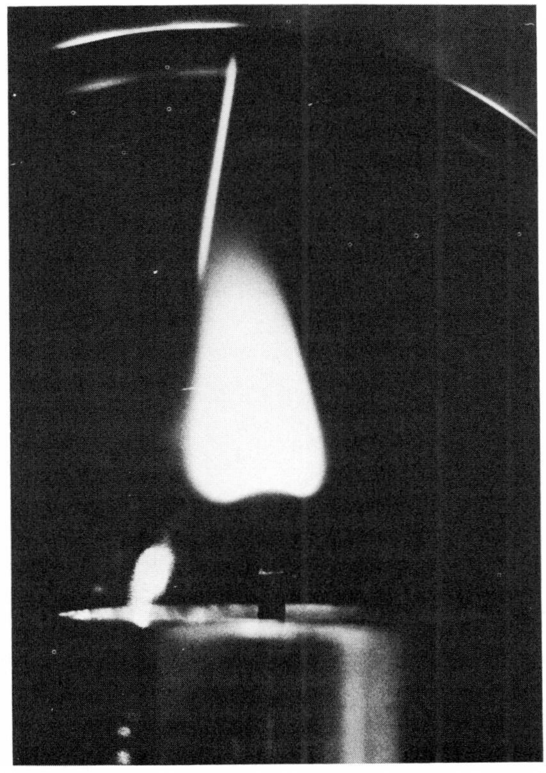

Figure 3.17. Photograph of reaction flame under reference conditions. The bar of light above the flame results from light (He–Ne laser source) scattered from the particulate reaction product.

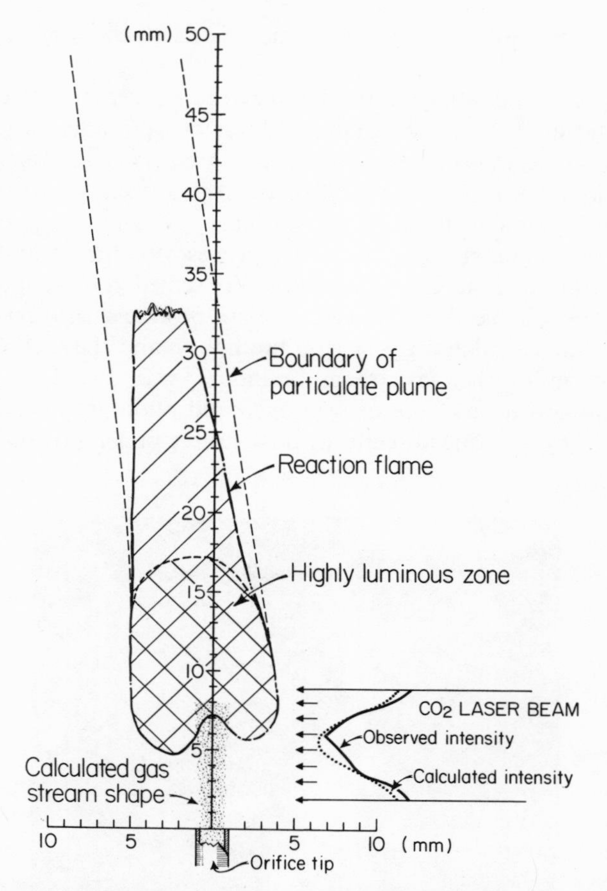

Figure 3.18. Schematic presentation of reaction photographed in Figure 3.17. Particulate plume boundary was mapped with He–Ne laser. Scattering from particles was evident to lowest point in luminous zone. Shown at lower right is the laser beam profile determined from a burn pattern in PMMA.

the flame. The latter proved simplest and provided adequate precision. The time variance of the flame position negated any advantage in precision gained by using the cathetometer. The flame positions are generally reproducible to within approximately 0.5 mm for any set of process conditions.

The distribution of laser intensity observed in the CO_2 beam is also shown in Figure 3.18. It was recorded by burning into a PMMA block where the depth of penetration is proportional to intensity. For comparison, the Gaussian distribution of an ideal TEM_{00} mode is also shown. The actual intensity is sensitive to the alignment or "tuning" of the optical cavity and is found to vary slightly with the power level and time for any specific alignment. Although a difference between ideal and actual intensity distributions is evident, these results show that the total exposure to which a volume of gas is subjected as it passes through the laser beam is reasonably represented by the Gaussian function.

The position, size, shape, and distribution of colors are easily reproduced for this reaction flame. There are distinct zones within the flame with the reference process conditions used for synthesizing Si_3N_4. There is a lower highly luminous zone and an upper transparent zone that terminates with a "feathery," nondistinct boundary. All other boundaries are sharp and have constant positions. Altering the input laser intensity, the reactant gas velocity, stoichiometry, or ambient pressure causes the flame to change character. The effect of gas velocity, stoichiometry, and dilution are treated analytically in a later section.

Two apparently anomalous features exhibited by the reaction flame are explained by the process conditions. The deflection of the flame and plume to the left of the vertical axis is caused by the Ar flow from the laser entrance window at the right side. The deflection angle is influenced by the reactant gas and/or argon flow rates. As will be shown analytically, the bottom contour of the flame (one upper and two lower cusps) is determined by the parabolic velocity profile in the premixed reactants, rather than mass transport effects that cause similar shapes in diffusional flames (Gaydon and Wolfhard, 1970). The two lower cusps should be at the same level; but the relative positions of the right and left lower cusps is caused by dilution of the energy absorbing gas by Ar from the right side.

It has been possible to determine the existence of a particulate reaction product at any point by light scattering using a He–Ne laser source ($\lambda = 6328$ Å). In this manner we have been able to identify where a reaction proceeded to any extent within SiH_4 and $NH_3 + SiH_4$ flames as well as where and how the reaction products leave the reaction zone. As the photograph in Figure 3.17 shows, the scattered light is readily visible either above the flame or within the transparent upper region of the flame. Within the luminous zone, it was necessary to make observations through a narrow bandpass filter to suppress the background light.

Scattering from particulate reaction products was evident throughout the flame. At the bottom side, scattering was first evident at a point coincident with the lower boundary of the luminous zone. Above that point, the intensity of scattered light appeared constant within the sensitivity limits of the eye. Particles travel upward in a cylindrical plume as denoted by dotted lines in Figure 3.18. Within the accuracy of the measurements, the plume diameter is equal to the maximum diameter of the luminous zone. We could not observe any distortion from a circular cross section in the plume.

The abrupt emergence and essentially constant intensity of scattered light supports our previous assumption that the reaction occurs spontaneously and goes to completion when a threshold temperature is reached.

Obviously, the terms "spontaneously" and "to completion" are meant to be references against our ability to make faster or higher resolution observations since the reaction requires a finite time to go to completion and occurs while traversing over a corresponding distance in the flame.

Measurements of CO_2-laser power transmitted through the reaction zone support another assumption regarding the transparency of the reacted gases and particulate product to the 10.6-μm light. Burn patterns have been made to map the laser intensity distribution both with and without reactant gases in its path. These results indicated that the reaction products were no more absorbing than the reactant gases. More precise statements cannot be made because of the long optical pathlength in the cell ($l \simeq 22$ cm) that tends to accumulate unreacted NH_3 during a run. With the long pathlength, the absorption caused by the ammonia became the same order of magnitude as that absorbed by the reactant stream. It was evident, however, that there was neither a sharp boundary nor any systematic difference across the beam corresponding to the reaction that occurs at approximately the center of the beam (Figure 3.18).

3.2.2.1e. Emissions from the Reaction Zone. The relatively bright and stable visible emission of light from the reaction flame suggested that this type of nonintrusive analyses could be used to characterize the laser-induced reactions as has been done with other types of more conventional flames (Gaydon and Wolfhard, 1970). In principle, the emitted spectra should provide a spacial description of chemical species and temperature in the reaction zone. This type of characterization should complete the description of the reaction that occurs within the flame boundaries discussed in the previous section.

Synthesis of Si by pyrolysis of SiH_4 and synthesis of Si_3N_4 from various reactants have been studied by many scientists (Purnell and Walsh, 1966; Haas and Ring, 1975; Galasso *et al.*, 1978; Billy *et al.*, 1975; Cochet *et al.*, 1975; Greskovich *et al.*, 1976; Lin, 1977), yet no definitive model for these reactions has emerged for conventional process conditions. Most authors suggest that the pyrolysis reaction occurs by a step-by-step polymerization reaction (Purnell and Walsh, 1966; Galasso *et al.*, 1978) involving a Si_2H_6 radical. High molecular weight hydrides ultimately condense to pure silicon, simultaneously giving off H_2. It is also suggested (Cochet *et al.*, 1975) that Si_3N_4 is formed by Si- and N-bearing polymers condensing to form the nitride.

The reaction path for this specific process must be studied further. There are several reasons to believe that the descriptions in the existing literature do not apply directly to this reaction. The most obvious reason is the lack of an accepted description of even, slow, highly controlled reactions. In hydrocarbon flames, it has been found that reaction schemes deduced from measurements on slow reaction (combustion) processes cannot be applied to propagating flames with any sense of certainty. The slow reactions are usually dominated by catalytic and other surface effects. Also, the reaction temperature in the flame is usually much higher than those measured in slow or spontaneous ignition experiments (Gaydon and Wolfhard, 1970). The laser heat source itself might also be suspected of affecting the reaction by various resonance effects in the molecules. For these reasons, the prior work on the chemistry of the reaction is useful background, but it is not directly applicable to the laser-heated synthesis process. The temperature distribution is obviously unique to this process.

Two sources of emissions can be expected. Banded or line emissions can be expected from reaction intermediates and reaction products. Of the possibilities, SiH_2 (Dubois, 1968), SiH_1 (Pearse and Gaydon, 1976), NH_2 (Gaydon and Wolfhard, 1970; Dressler and Ramsay, 1959), and H_2 (Pearse and Gaydon, 1976) have transitions that emit in the visible and ultra violet. The optical characteristics of the intermediate SiH_3 have not been studied because it is apparently too reactive. Continuous thermal emission can also be expected from hot Si and Si_3N_4 particles whose wavelength dependence is determined by the high-temperature real and imaginary indices of the particles, as well as size effects. The determination of local chemistry and temperature requires the identification, characterization, and modeling of each source of emitted light.

Three spectrometers have been employed to characterize the emission from the reaction flame for wavelengths between 4000 and 8500 Å. Moderately high-resolution (5 Å) spectra were measured with a 0.25-m Jarrell–Ash monochromator coupled to a RCA C53050 photomultiplier. This apparatus was calibrated against a tungsten lamp to permit absolute intensities to be measured over the wavelength range. High resolution measurements (0.5 Å) were made with a 1-m SPEX double monochromator. Both instruments employed a PAR lock-in amplifier that yields a dynamic range of 10^4. Lower-resolution (30 Å) measurements were also made with an optical multichannel analyzer (OMA) instrument that analyzes the entire spectrum simultaneously. The time-variant intensities of the flame make it extremely tedious to characterize the spectra using either of the monochromator instruments with a confidence level that is superior to that of the lower resolution multichannel analyzer. The high-resolution instrument was used to determine whether the broad spectrographic features consisted of a "forest" of lines that were spaced closer than resolution limits of the other instruments.

Figure 3.19 illustrates spectra emitted from CO_2-laser-heated SiH_4, NH_3, and NH_3/SiH_4 gases under process conditions that are close to the reference conditions. These specific results were obtained with the 0.25-m Jarrell–Ash monochromator at the denoted wavelengths. Equivalent results were also obtained with the multichannel analyzer instrument. At pressure

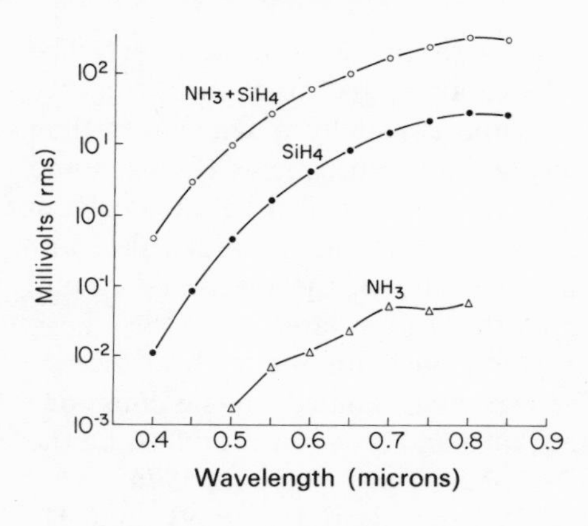

Figure 3.19. Emission spectra from reaction flame measured with a 0.25-m Jarrell–Ash monochromator. Cell pressure was 0.2 atm and gas-flow rates for SiH_4 were 11 cm^3/min and for NH_3 were 110 cm^3/min. Measurements were made only at the discrete wavelengths indicated in the figure. Millivolt output from PM tube amplifier was corrected for sensitivity of PM tube at a particular wavelength. Thus values are proportional to intensity.

levels at approximately 0.2 atm or greater, the spectra emitted no resolved features. Overall, the intensity of the $NH_3 + SiH_4$ flame is approximately 1000 times higher than the NH_3 flame and 20–50 times higher than the SiH_4 flame. The NH_3 flame is fundamentally different from the other two since the latter produce hot particulate reaction products that contribute to the emission. Products from the NH_3 flame result only from gaseous reactant, intermediate, and product species.

At low pressure ($P \leq 0.07$ atm), band emissions were evident in SiH_4 and $NH_3 + SiH_4$ flames with the multichannel instrument for wavelengths between 0.4 and 0.5 μm. Typical flame emissions are shown in Figures 3.20 and 3.21. These peaks disappear completely into the continuous thermal emission by a pressure level of 80 Torr (0.1 atm). Broad bands are also evident in the NH_3 flame at these low pressures. They persist to the higher pressures with progressive broadening and diminishing height relative to the background, to the point where the emission appears effectively continuous.

The weak peaks observed with the SiH_4 and $SiH_4 + NH_3$ flames have not been identified with any certainty. The high-resolution spectrometer has shown that, within a resolution of 0.5 Å, these peaks do not consist of a multitude of closely spaced lines. The band structure observed with the NH_3 flame is probably the ammonia α bands reported for NH_3–O_2 and CH_4–NO_2 flames (Braker and Mossman, 1971). No NH emission was detected although the multichannel analyzer is less sensitive in this region ($\lambda = 3360$ Å) and a strong thermally emitted background may have obscured it.

These results indicated that it will be very difficult to extract information about the chemical species in the flame near the reference conditions because of the dominance of the emissions by the hot particles. Pressure- and temperature-broadening effects also contribute to the masking of this information. Low-pressure reactions can be studied to identify species that

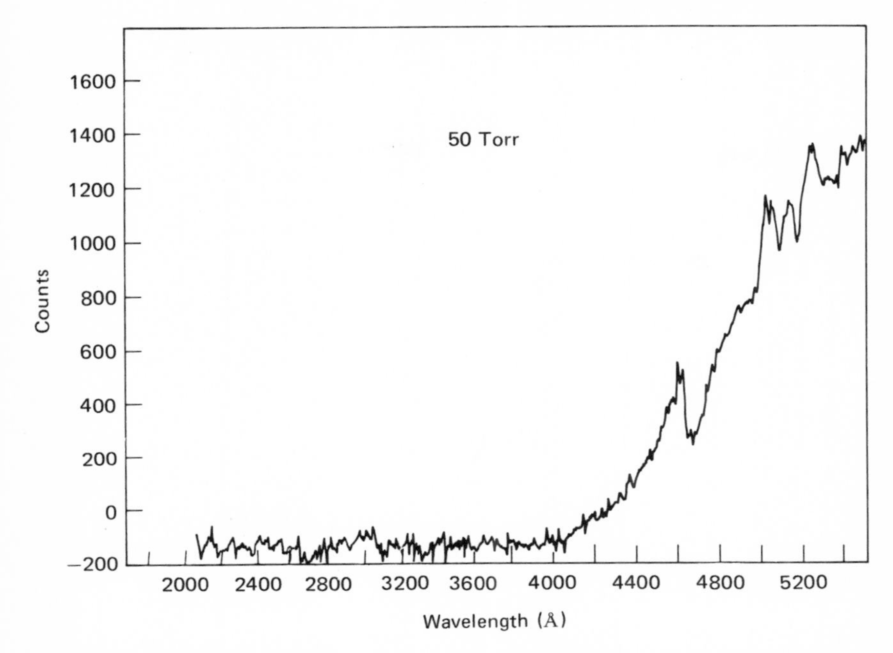

Figure 3.20. The spectrum emitted from the SiH_4 reaction flame at 50 Torr pressure. The spectrum was measured with a PAR OMA II multichannel analyzer. The abscissa is given as the difference in counts between a scan of a flame and background scan.

are present under those conditions. Their existence at the higher pressures must be verified by appropriate modeling and analyses.

3.2.2.1f. Temperature Measurements. We attempted to make temperature measurements by analyzing the spectrum and by direct measurement techniques, including thermocouple and brightness pyrometer observations. The thermal emission can also be used to determine particle temperatures. This analysis must be based on the optical properties of the hot particles that will change throughout the flame with progressive reaction, varying stoichiometry, size, and temperature. We have made initial calculations of particle temperature based on room temperature properties of the particles. Assuming Wien's approximation of Planck's law and Kirchhoff's law, it can be shown that the relative thermal intensity is

$$I_\lambda = \frac{C_1 \alpha_\lambda \exp(-C_2/\lambda T)}{\lambda^5} \tag{3.11}$$

Using measured absorptivities (α_λ) for the particles, least-square curve-fitting techniques yield an optimum fit to the experimental data for a

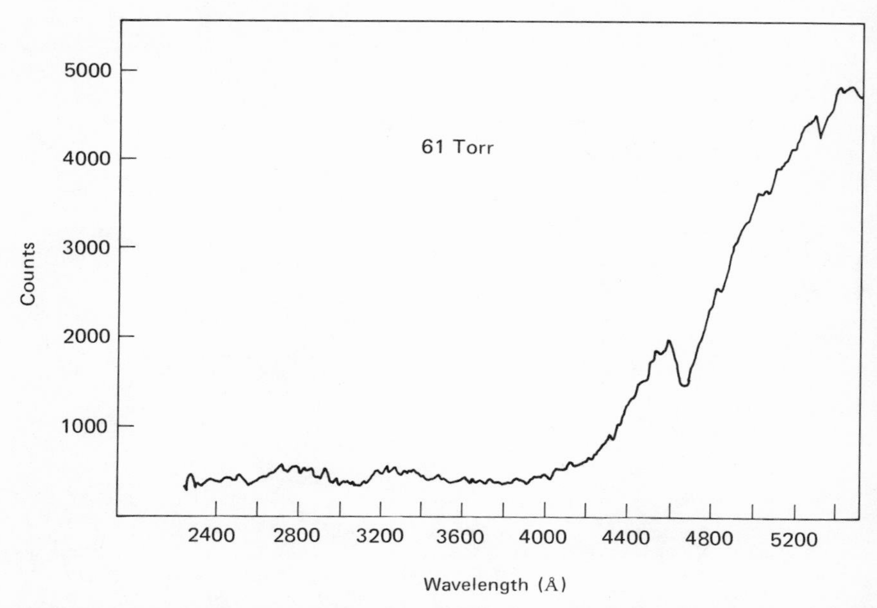

Figure 3.21. The spectra emitted from the SiH_4 reaction flame at 61 Torr pressure. The conditions are the same as Figure 3.20.

temperature of 1400°C. The agreement between observed and calculated results is very good at short wavelengths as shown in Figure 3.22.

The value of the "optimum temperature" determined by this technique represents a more serious issue than the poor fit between observed and calculated values of long wavelengths. This temperature is higher than those measured by two direct observation techniques and is also higher than accepted reaction temperatures. As indicated below, measurements of flame temperature with thermocouples indicated 709°C and optical pyrometry using a brightness pyrometer indicated 867°C (Table 3.7). The SiH_4 pyrolysis reaction (Braker and Mossman, 1971) and the reaction of SiH_4 with NH_3 (Greskovich *et al*., 1976) are known to occur rapidly above 700°C. Thus, other indications show that the flame temperature is substantially lower than 1400°C.

In the first direct measurement technique, a chromel–alumel thermocouple was inserted in the cell at the expected flame position. First, only NH_3 was introduced into the cell, and the thermocouple measured the temperature directly above the short luminous area. The temperature varied from 331 to 794°C as the pressure varied between 0.08 atm and 0.75°C. When the SiH_4 was introduced, the temperature immediately began to decrease steadily. Powder collecting on the thermocouple bead was thought to be responsible for the temperature drop.

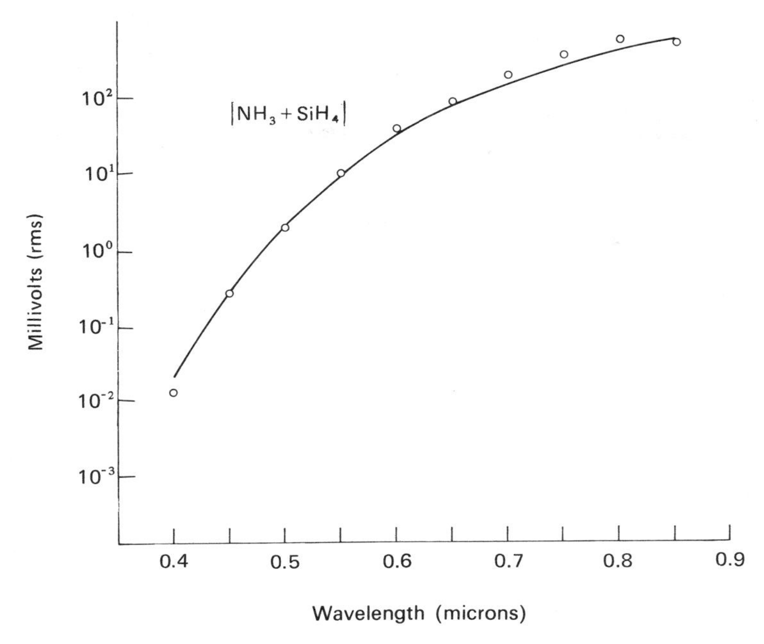

Figure 3.22. A least-squares curve fit made by varying C_1 and T as referenced in the text, for data shown in Figure 3.19. The best fit temperature is 1400°C.

The second technique, used more as a relative temperature measurement, was a brightness pyrometer sighted directly on the flame. The use of this technique rests on the flame radiation being a blackbody radiation from particles in the flame. The emissivity of such fine particles is expected to be much less than unity, so measured temperatures are expected to be low. A second effect that lowers the measured temperature is the low density of particles in the flame. If the area of radiation is not solid with particles, then its brightness cannot be compared with the brightness of a solid that is the basis of a brightness pyrometer.

Although, based on the above comments, the measured pyrometric temperature may be low, the relative temperatures from run to run may be correct since the emissivity of the approximately constant-sized particles is nearly constant. Rather large differences in emissivity are necessary to affect the temperature appreciably. Therefore, these relative temperatures may also be nearly accurate. Throughout the remainder of this chapter, therefore, optical pyrometric measurements are reported. Although the absolute temperature of the flames is still uncertain, it appears to be in the vicinity of 800°C.

3.2.2.1g. Analysis of the Cross-Flow Process. Process variables have been manipulated to determine their effect on both the process characteristics and the particle characteristics. This section discusses our analysis of the process. The following section considers their effects on powder characteristics.

Analyses have been developed that intend to explain empirical observations and infer what is occurring in the reaction zone. These analyses are becoming more complex and involve increasingly more thorough fluid flow and heat transfer considerations. The results are so sensitive to many extrapolated properties and assumptions that seemingly reasonable calculations produce results that range from being extremely plausible to clearly nonsensical. Simplified calculations can also be made that are so free of assumptions that they serve to test the accuracy of the more complex analyses. Examples of the latter are calculated heating rates, reaction times, and depletion volumes.

3.2.2.1g.1. Order of magnitude rate estimates. We have made the following order of magnitude estimates for the reference process conditions used to synthesize Si_3N_4 powders. With a volumetric flow rate of 120 cm^3/min, the average gas stream velocity decreases from 570 cm/sec at the nozzle to approximately 400 cm/sec at the center line of the laser beam axis. Mapping the reaction zone showed that reaction products were evident within 3–5 mm penetration into the laser beam. Although the reaction temperature has not been determined accurately, we can state that it occurs at 800 ± 200°C. Combined, these parameters and observations indicate that the average heating rate is approximately 10^6 °C/sec from the time the gas enters the beam to the point where reaction product is first evident. The exposure time required to initiate a reaction is approximately 10^{-3} sec. The time required to complete the reaction is less than 7.5×10^{-3} sec assuming that, at most, the reaction occurs throughout the highly luminous zone discussed in the mapping section (Figure 3.18).

3.2.2.1g.2. Depletion volume. Once the reactant gases reach the reaction temperature, particles begin to nucleate and grow within the gas stream. As far as we can determine, this point coincides with the beginning of the visible flame. We have shown by mass balance measurements that the silane is almost completely depleted from the reactant gases, so particle growth is not limited by elapsed growth time but by the depletion of silane between neighboring particles. This depletion volume is estimated by comparing the mass of silicon in a particle with the mass density of silicon atoms in the gas stream.

With the reference NH_3/SiH_4 conditions, the density of silicon in the gas stream is 5.6×10^{-6} g of Si/cm^3 and the mass of silicon in a 175-Å Si_3N_4 particle is 5.5×10^{-18} g. If the density of silicon in the gas stream is adjusted to account for the conversion efficiency, the resulting depletion volume is 1.1×10^{-12} cm^3, which corresponds to a sphere of gas with a diameter 1.2×10^{-4} cm and an entrained particle density in the gas of 9.3×10^{11} particles/cm^3. If the gas volume increases by a factor of 1.15 because of the reaction $3SiH_4 + 30NH_3 \rightarrow Si_3N_4 + 26NH_3 + 12H_2$, the resulting particle density is 8.1×10^{11} particles/cm^3 after the reaction is completed.

Similar calculations have been made for Si_3N_4 powders synthesized at different pressures and laser intensities. These results are given in Table 3.5. Within the precision of the calculation, the diameter of the depleted volume of gas is not affected significantly by these variables.

For the Si reference condition, the mass of silicon in the 435-Å particles is approximately 9.9×10^{-17} g of Si/particle, and the density of silicon in the gas stream is 6.8×10^{-5} g of Si/cm^3. This corresponds to a depletion volume with a diameter of 1.38×10^{-4} cm. This approximately equals the diameter calculated for Si_3N_4.

The time required, t, to deplete a sphere of radius, r, by a diffusion rate limited process can be estimated by

$$t \simeq r^2/6D \tag{3.12}$$

where D is the diffusivity in the gas and is calculated from kinetic theory by

$$D = 1/3\lambda\bar{c} \tag{3.13}$$

$$\lambda = \frac{1}{1.414\pi d^2}\frac{kT}{P} \tag{3.14}$$

and

$$\bar{c} = \left(\frac{N_0 kT}{\pi M}\right)^{1/2} \tag{3.15}$$

where λ = mean free path; $\bar{c}$ = average molecular velocity; d = molecular diameter (4.2 Å); k = Boltzmann gas constant; N_0 = Avogadro's number; P = pressure (0.2 atm); M = molecular weight (32 g/mole); T = absolute temperature (1100°K). The values of parameters given in parenthesis were used for sample calculations.

Table 3.5. Diameters of Depletion Volumes for Various Si_3N_4 *Synthesis Conditions*

P (atm)	I (W/cm^2)	Particle diameter (cm)	Depletion volume diameter (cm)
0.2	760	1.75×10^{-6}	1.23×10^{-4}
0.5	760	2.05×10^{-6}	1.03×10^{-4}
0.75	1020	2.21×10^{-6}	0.99×10^{-4}

Substitution of these representative values into these expressions indicated that a 1.2×10^{-4} cm diam sphere can be depleted in approximately 2.2×10^{-10} sec if diffusion is rate controlling. This time is a factor of approximately 10^7 shorter than the estimated residence time in the reaction zone. This large difference does not determine that the reaction does not occur throughout a major portion of the reaction flame, but it clearly shows that if it does, the process is not rate controlled by gas-phase diffusion. The extremely short times required to cause termination of the growth process by overlapping depletion volumes and the uniformity of the calculated depletion volumes suggests that the observed particle size is controlled by the nucleation rate.

3.2.2.1g.3. Observed effects of process variables on reaction zone characteristics. Several process variables have been manipulated during development of this synthesis process, and all had smooth, progressive effects on the reaction zone characteristics. Since no abrupt changes were observed, which would suggest that the process changed in any fundamental way, we can discuss the effects as trends. Investigated process variables include gas-stream velocity, gas mixture, laser intensity, and chamber pressure. The techniques used to characterize the shape of the reaction flame and its position relative to the laser beam as well as some other general characteristics were discussed in the section on reaction zone mapping. The observations cited in this discussion were made in this manner.

Under most processing conditions, the shape of the reaction flame was concave into the flame on the side facing the nozzle. The contour between the two cusps nearest the nozzle is defined by the parabolic velocity distribution in the gas stream. Increased gas velocity causes the concave shape to increase in its penetration into the body of the flame. This follows predicted behavior because the outer boundaries of the stream move at the velocity of the argon carrier stream that flows through the annular orifice. The maximum velocity must therefore increase in proportion to the volumetric flow rate. A gas flow rate that is greater than two times the reference flow rate causes the central cusp to pass beyond the laser beam. Once this occurs, the reactant at the stream's center axis cannot be heated to the reaction temperature, and the flame goes out with progressively increasing velocity. Lower than reference velocities cause the concave shape to disappear progressively. With a volumetric flow rate equal to or less than half of the reference conditions, the flame flattens out on this side. Qualitatively, this behavior follows that expected on the basis of velocity distributions in the gas stream.

With constant laser power, beam intensity, total pressure, and volumetric flow rate, any dilution of the silane gas with a lower absorptivity gas causes the reaction threshold boundary to penetrate further into the laser beam. The reason for the effect is obvious. These observations were made with both Ar and NH_3 diluents in the vicinity of reference conditions. With extreme dilution by NH_3, the overall absorptivity should be dominated by the NH_3; so, the effect should disappear progressively. This dilution effect also accounts for the relative position of the two cusps nearest the nozzle tip. The cusp nearest the laser entrance window penetrated further into the beam than the other despite the higher laser intensity at this side. Dilution by the Ar gas stream used to protect the window causes this shift.

With constant laser power, beam intensity, volumetric flow rate, and gas-stream composition, increased pressure caused reduced penetration of the reaction threshold boundary into the laser beam. This is primarily a gas-stream velocity effect because gas velocity decreases with increasing pressure under these conditions. To a first approximation, pressure should have no effect on the time required to heat the gas to the reaction threshold because the mass per unit volume and energy absorbed per unit volume both increase proportionally with pressure.

3.2.2.1g.4. Analysis of the effects process variables have on reaction zone characteristics. Since the laser beam intensity and gas-flow velocity are both spacially variant, the analysis of the temperature–time history experienced by a voume element of reactive gas must reflect these variables. The analytical problem involving a gas stream having a parabolic velocity distribution that intersects a laser beam having a Gaussian-shaped intensity distribution is not easily solvable. The dimensionality of the problem cannot be reduced since both the laser beam and gas stream have circular cross sections. This problem can be simplified by considering a small gas volume element at some particular position in the gas stream and performing a stepwise calculation to determine the temperature-time history of the volume element as it passes through the laser beam.

Before elaborating on this analysis, it is beneficial to examine the possible intensity–time profiles a volume element in various positions in the stream may experience. If the gas stream and the laser beam are aligned so that their central axes intersect (see Figure 3.23), the gas volume element in the center of the gas stream, which will have a maximum velocity, will be subjected to the full diameter and the maximum power intensity of the laser beam. Any other gas element in the plane m, shown in Figure 3.23, will also intersect the maximum beam intensity distribution but will have a velocity related to its distance from the center line, which is less than maximum. Volume elements within the plane m should reach reaction threshold temperatures in a shorter penetration distance than volume elements traveling at the same velocity in any other region of the gas stream. It is these gas volume elements that probably form the lower boundary of the flame, as exemplified in Figure 3.18, and this boundary should be indicative of the velocity profile within the beam. The following analytical

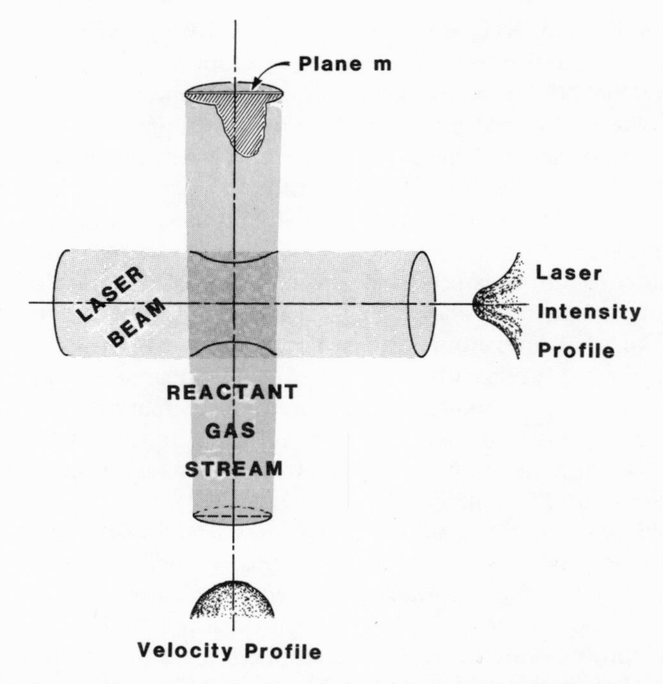

Figure 3.23. A schematic of the reactant gas stream intersecting the laser beam.

procedure was employed to calculate the temperature-time history for any particular reactive gas volume element.

The path of a gas element through the laser beam is divided into a finite number of intervals. In each interval, the gas element is assumed to have a constant velocity and is subjected to a constant laser beam intensity. The laser beam intensity in each interval is calculated from a Gaussian-shaped distribution. The gas velocity at any point is determined by the computer model previously described. The volume element is assumed to have a square cross section of width Δx along the edge and a length l determined by the number of intervals in the path. It is assumed that there is no dilution of the gas stream by the annular Ar stream. Thus, the gas volume consists of only NH_3 and SiH_4 in proportions defined by the volumetric flow rate. After reaction occurs, it is assumed that the residual gas stream and particles produced by the reaction are transparent to the laser beam.

The power absorbed (ΔP) by a specific gas element in a particular spacial interval can be calculated from

$$\Delta P = I(\Delta x l)[1 - \exp(-\sum \alpha_i p_i \, \Delta x)] \tag{3.16}$$

where I is the average laser intensity in the interval; $(\Delta x l)$ is the cross section of the element exposed to the beam; α_i is the absorption coefficient of the ith chemical species; p_i is the partial pressure of the ith species; Δx is the thickness of gas element exposed to the beam. Using the substitutions $x_i p_{\text{tot}} = p_i$, $\sum x_i \alpha_i = \alpha_{\text{ave}}$ and $1 - \exp(-a) = a$, we obtain

$$\Delta P = I \, \Delta x^2 l \alpha_{\text{ave}} p_{\text{tot}} \tag{3.17}$$

Assuming all absorbed power is converted to sensible heat, the temperature increase, $T_2 - T_1$, in the spacial interval can be calculated from

$$T_2 - T_1 = \frac{\Delta t \, \Delta P}{C_p \dfrac{n}{v} V}. \tag{3.18}$$

Substitutions of $n/V = p_{\text{tot}}/RT_{\text{ave}} = 2p_{\text{tot}}/R(T_2 + T_1)$ and $\Delta t = l/\bar{v}$, where $\bar{v}$ is the computer calculated velocity within the interval and simplification yield

$$\frac{T_2 - T_1}{T_2 + T_1} = \frac{RlI\alpha_{\text{ave}}}{2\bar{v}C_p}. \tag{3.19}$$

The gas element temperature when it leaves the interval T_2 can be calculated with the knowledge of T_1, the temperature of the element when it enters the interval. The heating rate within the interval as well as a temperature-time history for the entire path of this volume element through the laser beam can also be calculated.

Figure 3.24 shows the calculated time-temperature histories of gas volume elements traveling along the center line of the gas stream and at distances $1/4r_0$ and $1/2r_0$ from the central axis, where r_0 is the radius of the reactant gas stream at a particular distance from the nozzle tip, corresponding to the calculated boundary between the reactive gas stream and annular stream of Ar. These three volume elements lie in a plane corresponding to the plane m shown in Figure 3.23 and are exposed to the maximum power intensity distribution. Also shown in Figure 3.24 are the reaction times calculated from the positions where the flame was first observed in the mapping exercise.

The calculated temperatures at the reaction times are relatively constant for the three gas elements. Also, the average heating rates between inlet temperature and reaction temperatures are essentially equal for all three gas elements. They range from 5.5×10^5 to 6.2×10^5 °C/sec. These are about a factor of 2 lower than we calculated as an order of magnitude estimate. In contrast, the instantaneous heating rates experienced at the observed reaction times and temperatures do show a wide range of variation. The instantaneous heating rate calculated for the central axis gas

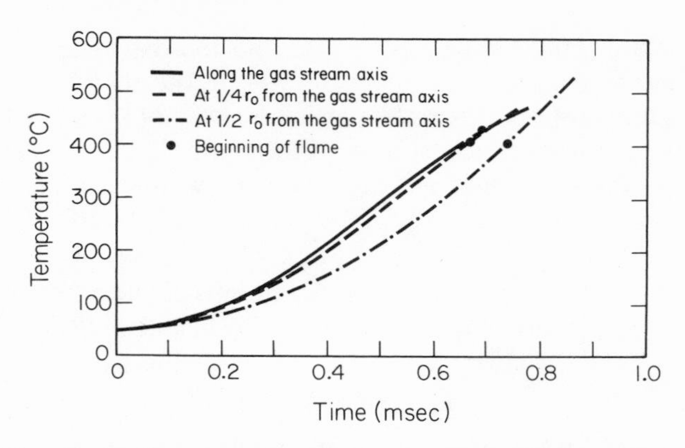

Figure 3.24. The calculated temperatures of volume elements of reactant gas after they enter the laser beam at time = 0. All volume elements lie in plane *m* of Figure 3.23, and r_0 is the radius of the gas stream. The times corresponding to the points designated as the beginning of the flame were calculated from the measured distance to the flame as given in Figure 3.18 and the calculated velocity of the particular elements in the reactant gas stream. Reference flow conditions are assumed.

element is 3.8×10^5 °C/sec. The rates increase progressively in the outward direction to 6.9×10^5 °C/sec at $1/4r_0$ and 9.6×10^5 °C/sec at $1/2r_0$. The instantaneous heating rate at the reaction threshold temperature is more important than the average heating rate since it influences nucleation rate, superheat, stoichiometry, etc. It is probable that achievement of more uniform laser intensities and plug flow gas streams will produce more uniform particle size distributions.

It should be noted that the temperatures calculated by the preceding analysis do not agree with either measured estimates of the reaction temperature for this process or spontaneous reaction temperatures reported in the literature for these gases (Braker and Mossman, 1971). It appears that the error is systematic because all of these and other calculated reaction temperatures are close to 400°C. Probably, the source of the error is the assumed constant optical absorptivity. High-temperature absorptivity measurements, such as those presented for low temperatures, are required to resolve this discrepancy.

This analysis shows that the heating rate is not explicitly related to pressure. This follows the assumption that $\exp(-\alpha p\,\Delta x) \simeq \alpha p\,\Delta x$, where $\alpha p\,\Delta x$ is small. Since Δx can be chosen arbitrarily small, the assumption always holds. At high pressures it can no longer be assumed that the gas is optically thin. We anticipate a pressure effect that causes a systematic variation in heating rates across the gas stream resulting from progressive absorption of the laser beam. These analyses also assume that α_i is independent of p_i and p_{tot}. It should be re-emphasized that the pressure dependence of the absorption coefficients of SiH_4 and NH_3 are not known. The conclusion regarding a lack of pressure dependence is clearly an approximation. Its limits of validity require further definition.

Figure 3.25 shows the heating curves for gas elements traveling through the center of the gas stream into the maximum intensity distribution of laser beams with average power intensities of 765, 2×10^4, and 1×10^5 W/cm^2. The NH_3/SiH_4 ratio is 10/1, and the flow characteristics at each intensity are the same. These calculated curves were extended to temperatures well above the reaction temperature to show qualitatively the shape of the entire heating curve and to indicate the extremely large effect the intensity has on gas heating rates. The time equivalent of the beam diameters are indicated for the intensities of 2×10^4 and 1×10^5 W/cm^2. The figure shows that with average intensities of 2×10^4 and 1×10^5 W/cm^2, the gases reach reaction temperatures so soon after entering the beam that the reaction is initiated before the maximum intensity and maximum heating rates are experienced. The average heating rates and instantaneous heating rates at the reaction threshold both increase with beam

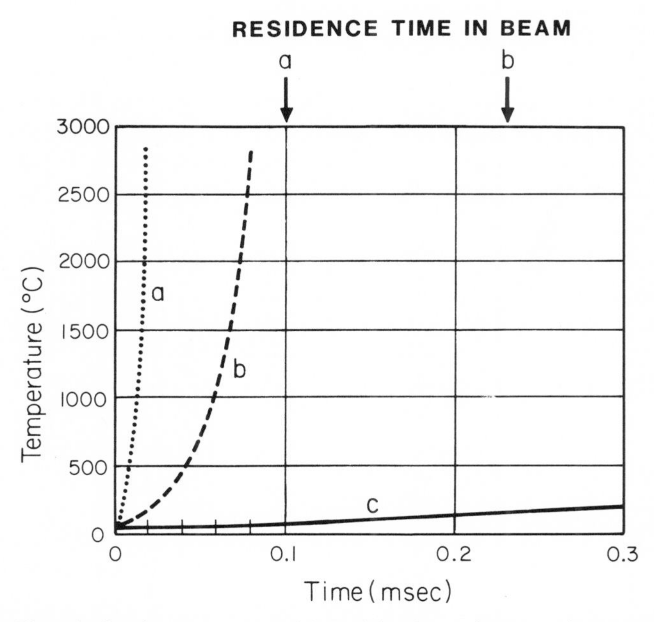

Figure 3.25. The calculated temperatures of the axial volume elements of reactant gases after they enter the laser beam at time = 0. Curves a and b refer to focused beams with intensities of 1×10^5 and 2×10^4 W/cm^2, respectively. Residence times a and b were calculated from the diameter of the focused laser beam at two different intensities and from the calculated velocities of the reactant gas stream. Curve c refers to the volume element of gas at the reference condition using an unfocused laser beam (760 /cm^2).

intensity. The average heating rates for intensities of 2×10^4 and 1×10^5 W/cm^2 may be as high as 2×10^7 and 1×10^8 °C/sec, respectively. Beam intensity should have a major effect on reaction kinetics and, consequently, particle characteristics.

Figure 3.26 shows the calculated effect of varied NH_3/SiH_4 ratio on the gas heating behavior for the central axis gas element that travels through the diameter of the laser beam. The NH_3/SiH_4 ratio was varied from 5/1 to 10/1 to 20/1. In each case, the average power was 760 W. The time equivalent of the position where the flame was first visible is indicated. The average heating rate increases with increasing silane content from 3.7×10^5 °C/sec for the 20/1 NH_3/SiH_4 ratio to 7.5×10^5 °C/sec for the 5/1 NH_3/SiH_4 ratio. The instantaneous heating rates at the observed reaction point increase more dramatically with increased silane content. They change from 1.5×10^5 to 1.3×10^6 °C/sec as the NH_3/SiH_4 ratio changes from 20/1 to 5/1. The calculations suggest that the temperature at the reaction point increases slightly with increasing silane content in the reactant gas stream. Because of uncertainty in the values of observed and calculated reaction temperatures, we have less confidence in this conclusion than those regarding heating rates at the reaction threshold.

The effect of the gas stream velocity through the laser beam is shown in Figure 3.27. The velocities of the gas elements as they entered the beam were 375, 500, and 625 cm/sec. These velocities do not correspond to specific run conditions but were selected to exemplify their effect on heating rate at the reaction temperature. A cell pressure of 0.2 atm and an average Gaussian distribution laser beam intensity of 760 W/cm^2 were

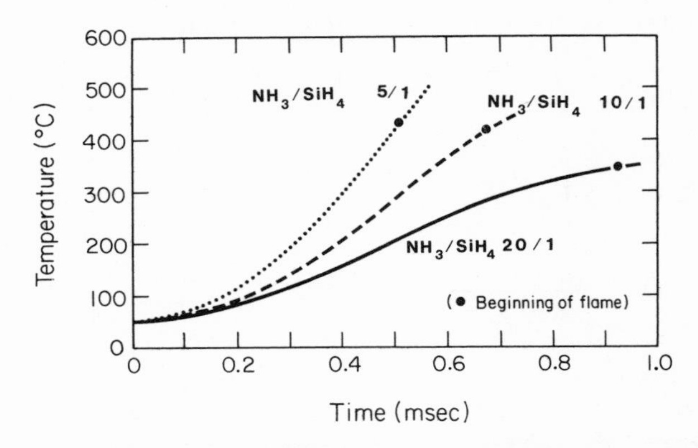

Figure 3.26. The calculated temperatures of the axial volume elements of reactant gas after entering the laser beam at time = 0. The times corresponding to the points designated as the beginning of the flame were calculated from the measured distance to the beginning of the flame and the calculated velocities of the axial volume elements.

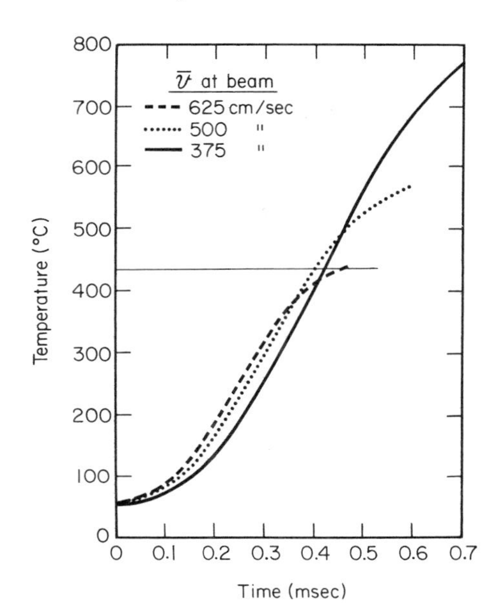

Figure 3.27. The calculated temperature of axial volume elements of reactant gas after entering the laser beam at time = 0. The velocities refer to average velocities of the reactant gas stream.

assumed. The calculated average heating rates to reach an assumed reaction temperature of 440°C were essentially constant, i.e., 1.0×10^6, 1.1×10^6, and 9.4×10^5 °C/sec for velocities of 375, 500, and 625 cm/sec, respectively. The instantaneous heating rates at the temperature exhibit a stronger dependence on velocity. They decrease with increasing gas velocity, e.g., from 1.3×10^6 °C/sec at a velocity of 375 cm/sec to 2.9×10^5 °C/sec at a velocity of 625 cm/sec. While this conclusion is similar to that made regarding lower velocity elements within a gas stream, this calculation illustrates the effect of a gas element passing beyond the maximum laser intensity prior to reaching the reaction threshold temperature. At the point where the flame is nearly "blown out," the instantaneous heating rates are very low when the reaction is finally initiated.

These results illustrate that achieving an understanding of the effects of process variables on rate kinetics and powder characteristics will require precise control of these variables. It is also probable that more uniform powder characteristics will be achieved only with either very uniform conditions or carefully manipulated laser intensity and gas-flow conditions, which are not intuitively evident. Clearly, because the reaction happens quickly in a very small region, it is not necessarily uniform in character.

3.2.2.1g.5. Effect of latent heats. The previous analyses have not included the effects of latent heats associated with the chemical reactions. They have, in fact, been thermal analyses of the process up to the point the reaction just begins. The following simple calculations do permit conclusions to be made regarding the possibility of a self-sustaining reaction without the input of energy from the laser and the anticipated maximum temperature rise in the reaction products.

To a first approximation, a reaction can be self-sustaining only if the latent heat released exothermically during the reaction exceeds the sensible heat required to raise the reactants to a temperature level where the reaction proceeds rapidly. For the Si_3N_4 synthesis reaction carried out under reference condition, the mass balance and heat balance equations are as follows. The overall reaction is

$$3SiH_4 + 30NH_3 \longrightarrow Si_3N_4 + 26NH_3 + 12H_2 \qquad (3.20)$$

assuming the excess ammonia does not crack. The heat of reaction is

$$\Delta H = -3\Delta H_{SiH_4,\,T} - 4\Delta H_{NH_3,\,T} + \Delta H_{Si_3N_4,\,T} \qquad (3.21)$$

At 1100°K (JANAF, 1971)

$$\Delta H_{SiH_4} = 2.14 \times 10^4 \text{ J/mol}$$

$$\Delta H_{NH_3} = -5.56 \times 10^4 \text{ J/mol}$$

$$\Delta H_{Si_3N_4} = 7.45 \times 10^5 \text{ J/mol}$$

Thus, at 1100°K

$$\Delta H = -1.95 \times 10^5 \text{ J/mol } SiH_4$$

With 92% conversion (see Table 3.7), the actual heat released is approximately 1.80×10^5 J/mol SiH_4. The sensible heat required to raise the 10/1 NH_3/SiH_4 reactant gas mixture at 1100°K is 4.4×10^5 J/mol SiH_4. Thus, under reference conditions, the Si_3N_4 synthesis reaction cannot be self-sustaining because the sensible heat requirement far exceeds the latent heat released. This conclusion may not hold for all NH_3/SiH_4 ratios since the latent exceeds the sensible heat with stoichiometric gas mixtures. For the SiH_4 pyrolysis reaction,

$$SiH_4 \longrightarrow Si(s) + 2H_2 \qquad (3.22)$$

at 1100°K, the latent heat is

$$\Delta H = 2.14 \times 10^4 \text{ J/mol}$$

and the sensible heat required to raise the gas to this temperature is

$$\Delta H_{\text{sensible}} = 5.76 \times 10^4 \text{ J/mol}$$

With 65% conversion (see Table 3.7), the effective latent heat (1.60×10^4 J/mol SiH_4) is less than the sensible heat requirement. This reaction cannot be self-sustaining unless the reaction is induced at a much lower temperature.

The temperatures we have measured have been at points within the reaction flame, so they correspond to a location where the reaction has proceeded to an unknown extent. The calculated temperatures at points where the reaction products were just observed are lower than measured temperatures by 400–600°C. These calculated temperatures must be lower than the actual temperatures, because they do not take the exothermic heat into account. With the same conversion efficiencies used to compare sensible and latent heats, we estimate that the adiabatic temperature rise in the excess reactants and reaction products would be 260°C for Si_3N_4 and 181°C for Si synthesis processes. It is evident that this temperature rise can account for a part of the difference between observed and calculated temperatures.

3.2.2.2. Powder Characterization

The characteristics of the covalent ceramic powders used to fabricate high-performance ceramic parts ultimately determine the properties of the fabricated bodies. We have attempted to characterize fully the physical, chemical, and crystalline characteristics of the laser synthesized Si, Si_3N_4, and SiC powders. The results of this analysis will allow us to develop a more accurate model of the powder synthesis process and indicate how close these powders are to the ideal characteristics described earlier. Our major synthesis efforts have been to produce Si powder for reaction bonding and Si_3N_4 powder for sintering. We have undertaken initial synthesis experiments with SiC. Powder evaluation has employed a number of techniques to determine the various physical, chemical, and crystalline properties of these powders.

Specifically, the particle size and distribution have been measured using SEM, TEM, BET (equivalent diameter), and x-ray line broadening. The particle shapes have been studied by TEM and SEM. The surface areas of the powders were measured by single-point BET analysis, and the powder density was measured by He pycnometry. Chemical analyses were performed using neutron activation analysis for O, wet chemical analysis for Si, H, N, O, and C, and emission spectrographic procedures for other impurities. Infrared spectroscopy was also employed to determine the nature of the elemental bonding in selected powders. The crystalline characteristics of the powders were evaluated using Debye–Scherrer x-ray diffraction, electron diffraction, and dark field electron microscopy.

The results of the powder evaluation for Si, Si_3N_4, and SiC are presented below. A comprehensive tabular listing of the individual run conditions and the resultant powder characteristics appear in Tables 3.6 and 3.7, respectively.

3.2.2.2a. Si Powders. Several runs have been made with the goal of

Table 3.6. Summary of Run Conditions for Powder Synthesis of Silicon, Silicon Nitride, and Silicon Carbide

Run number	Power density (W/cm^2)	Pressure (atm)	SiH_4 (cm^3/min)	NH_3 (cm^3/min)	CH_4 (cm^3/min)	C_2H_4 (cm^3/min)	Argon (cm^3/min)	Gas velocity at nozzle max (cm/sec)	Gas velocity at laser beam max (cm/sec)	Reaction zone temperature (°C) (pyrometer)
209S	760	0.2	20	0			1000	188	106	—[a]
213S	1020	0.2	20	0			1000	104	36.1	—
214S	760	0.2	11	0			1000	194	36.1	—
603SN	760	0.2	11	110			1000	1140	949	867 thermocouple 709
212SN	760	0.2	11	110			1000	1140	949	—
609SN	2×10^4	0.2	4.5	45			1000	468	287	985–1020
612SN	1×10^5	0.5	5	45			1000	424	355	—
021SN	1020	0.75	11	110			1000	304	231	—
023SN	891	0.75	11	110			1000	304	231	1050–1125
611SN	760	0.08	5	44			1000	1160	660	—
610SN	760	0.5	11	110			1000	456	337	1060–1080
605SN	760	0.2	17.8	178.2			1000	1850	1650	—
020SN	760	0.2	11	55			385	622	382	—
614SN	760	0.2	20.2	101.8			1000	1140	954	1025
615SN	760	0.2	5.8	115			1000	1140	949	675–700
607SN	760	0.2	6.1	60.5			1000 (dilution 54.4)	1502	934	990
022SC	865	0.2	11			9	1000	170	86	865
613SC	760	0.2	11		7		1000	188	—	710

[a] Denotes measurement not made.

Table 3.7. Summary of Powder Characteristics

Run number	Powder color	Analysis wt% Si	N	C	O	H	Free Si	BET Surface area (m^2/g)	BET Equiv. dia. (Å)	TEM dia. (Å)	Crystal structure	Conversion (%)	Infrared spectroscopy
209S	Brown	99	0.02		2.92	0.15	100	59	436	490	Cryst. Si	65–75	
213S	Brown	—	—		1.0	—	100	48	543	630	Cryst. Si	80–90	
214S	Brown	—	—		3.4	—	100	56.9	458	470	Cryst. Si	70–75	
603SN	Light brown–tan	72	26		1.25	—	35	117	176	168	Amorph. Si_3N_4 + Cryst. Si	—	
212SN	Light brown–tan	—	—		—	—	—	152	137	—	Amorph. Si_3N_4 + Cryst. Si	87	
609SN	Tan–white	60.1	—		—	—	2	165	114	150	Amorph. Si_3N_4	—	
612SN	White	—	—		—	—	Very low	190	98	120	Amorph. Si_3N_4	—	
021SN	Light tan	64	33		0.96	—	15.6	88.5	221	—	—	87	
023SN	Tan	66.6	29.7		0.3	—	23	91	220	—	—	—	
611SN	Dark brown	82.6	—		1.8	—	60	134	167	—	—	—	
610SN	Light tan	63.9	—		2.7	—	14	92	211	150	Amorph. Si_3N_4 + little Cryst. Si	—	
605SN	Brown	—	—		—	—	—	124	151	—	Amorph. Si_3N_4 + Cryst. Si	—	
020SN	Light brown	69	33		2.2	0.13	18.5	94	210	—	Amorph. Si_3N_4 + Cryst. Si	—	
614SN	Brown	68	23.7		0.5	—	35.6	130	155	—	Amorph. Si_3N_4 + poorly Cryst. Si	78	
615SN	Brown	70	23.5		—	—	38.1	119	155	—	Amorph. Si_3N_4 no Si peaks	108	
607SN	Brown	71.4	—		3.5 5.6[a]	—	36	120	173	—	—	—	
022SC	Brown to dark brown	80		14.1	1.37	—	49	97.5	247	230	Some Cryst. but not indexed	51	Si–C bonding
613SC	Brown	96.9		0.2	0.33	—	99	84.3	305	—	Cryst. Si	58	No Si–C bonding

[a] Second analysis.

producing Si powders for characterization to evaluate the reproducibility from run to run and to supply powder for reaction-bonding studies. The reaction to form Si powders is

$$SiH_4(g) \xrightarrow{h\nu} Si(s) + 2H_2(g)$$

3.2.2.2a.1. Physical characteristics. The Si production runs 209S, 213S, and 214S were made at 0.2 atm with SiH_4 flow rates of 20 and 11 cm^3/min and a laser intensity of 760 W/cm^2. These powders, light to dark brown in color, were prepared in several gram lots. We will present the data for 209S in detail, and then compare 209S with 213S and 214S to evaluate the reproducibility from run to run.

The Si powders were spherical in shape and had a wider range of particle sizes than did the Si_3N_4 powders. Figure 3.28 shows the Si particles that range in size from 100 to 1000 Å. The average size of the Si particles was 490 Å with a standard deviation of 194 Å by TEM and had equivalent spherical diameters of 436 Å as calculated from a BET surface area of 59 m^2/g. Figure 3.29a shows the particle size distribution for powder lot 209S as measured by TEM. It should be noted that the range

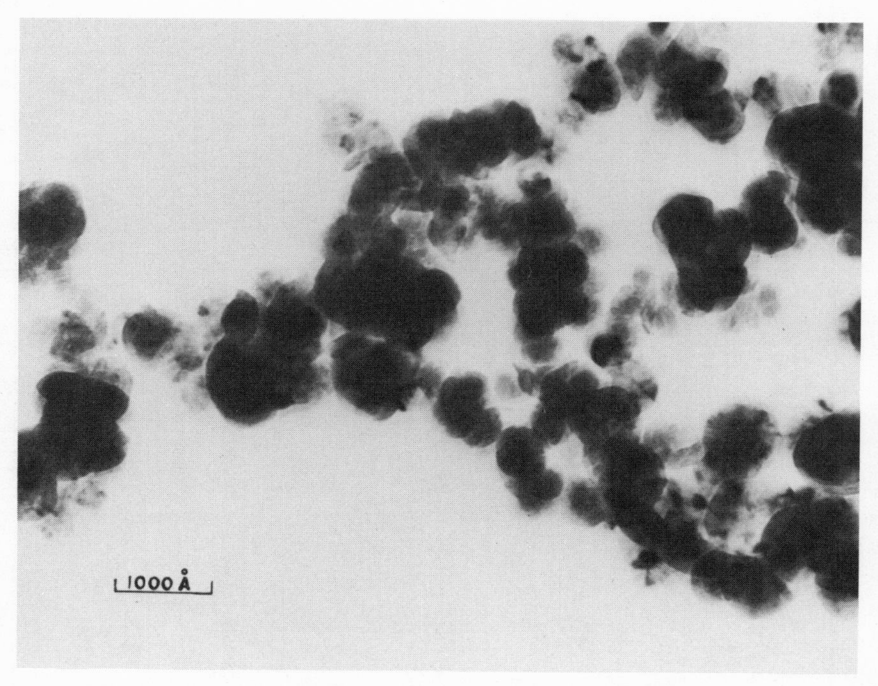

Figure 3.28. TEM photograph of Si powders from lot 209S.

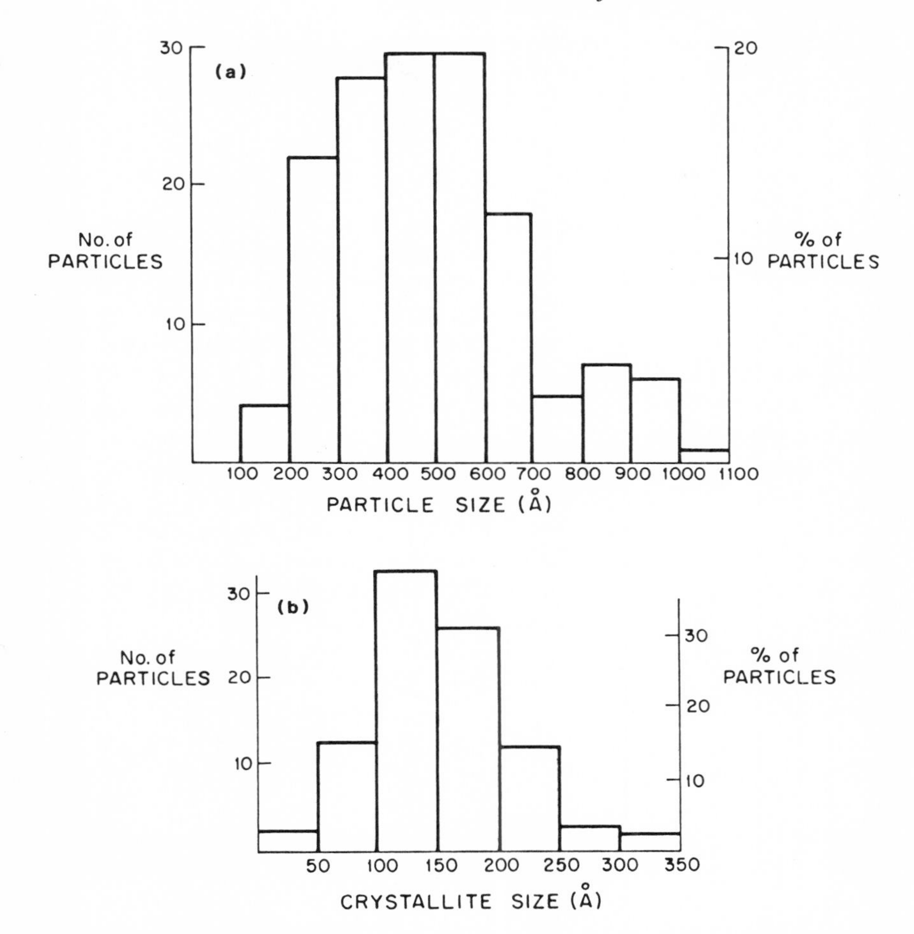

Figure 3.29. (a) Particle size distribution of silicon powders in lot 209S. (b) Crystallite size distribution within the same powders.

in particle diameters is smaller for these laser-synthesized powders than for powders prepared by conventional processing techniques.

The physical characteristics of the different lots of Si powders are compared in Table 3.8. These results clearly indicate that the laser process is capable of yielding powders with similar physical characteristics from one synthesis run to the next.

3.2.2.2a.2. Chemical characteristics. The chemistry of these powders is as important as their physical properties. Table 3.9 shows the chemical analysis of powder 209S. It is 99% Si with trace amounts of N and H and larger amounts of O. (It should be noted that these Si powders do not oxidize pyrophorically with exposure to air.)

Table 3.8. Physical Characteristics of Si *Powders Produced with Reference Conditions*

Run number	209S	213S	214S
SiH_4 flow (cm^3/min)	20	20	11
Surface area (m^2/g), BET	59	48	57
Size (Å), BET	436	543	458
Mean size (Å ± s.d.), TEM	490 ± 194	630 ± 275	470 ± 270
Size range, TEM	100–1000	100–1500	100–500

The 2.9 wt% O is comparable to Si powders produced by other techniques (Danforth, 1978). Table 3.9 also shows the O contents of 213S and 214S. The wt% O_2 is not as uniform as the other powder characteristics.

There are several possible sources of O_2: leaks in the synthesis or storage systems, O_2 in the reactant or inlet gases, and inadvertent exposure to the ambient atmosphere. While at times there may be very minor system leaks, calculated cell leak rates do not suggest sufficient O_2 to account for this level of contamination of the Si powders. We have measured 10–15 ppm O_2 in the argon buffer gases used in the system. If all of this O_2 (argon flow 1 l/min) were to react with the hot Si powders, it would amount to 0.1 wt% O. The indications are that this is not the primary source of O in the synthesis process. The postproduction handling of these Si powder is done in an inert atmosphere glove box with < 10 ppm O_2, which may account for some O, but it is doubtful that this is the major source.

The contamination very likely occurs during processing and is not intrinsic. Oxygen contamination can be improved by further reducing the O_2 content of the argon, by improving monitoring and gettering, by reduced system leak rates, and by reducing the O_2 content of the glove box.

3.2.2.2a.3. Crystalline characteristics. All of the Si powders we have produced are crystalline to both x-ray and electron diffraction. Debye–

Table 3.9. Chemical Characteristics of Si *Powders Produced with Reference Conditions*

Run number	209S	213S	214S
Si (wt%)	99	—	—
N (wt%)	0.02	—	—
H (wt%)	0.15	—	—
O (wt%)	2.92	1.0	3.4
Free Si (wt%)	100	100	100
Conversion (%)	70	85	72

Scherrer x-ray patterns revealed 11 peaks that indexed from Si(111) to Si(533). It remains to be seen whether they are fully crystalline or whether the powder is a mixture of crystalline and amorphous particles.

These x-ray patterns were also used to estimate the particle size from diffraction peak broadening using the expression (Cullity, 1956)

$$d = \frac{0.9\lambda}{B(p)\cos\theta_B} \tag{3.23}$$

where d is the crystallite diameter, λ is the wavelength, $B(p)$ is the peak breadth at half-intensity, and θ_B is the Bragg angle in radians. The particle size broadening $B(p)$ can be separated from the machine broadening by $B(g)$ by

$$B^2(p) = B^2(h) - B^2(g) \tag{3.24}$$

where $B(h)$ is the measured peak breadth at half-height. A > 50-μm particle size Si standard was used to measure the machine broadening $B(g)$. The peak breadths were taken as the difference between the two extreme angles at zero intensity (Warren, 1969).

The results of the x-ray analysis indicate a crystallite size of 154 Å for powder lot 209S, approximately one-third the TEM and BET particle sizes of 490 and 436 Å, respectively. Most of the Si particles consisted of several smaller crystallites with random orientations in dark field TEM. Figure 3.30 shows a dark field image where the particles that have a (111) Bragg orientation appear white. Figure 3.29b shows the size distribution of these microcrystallites measured from dark field images. The average crystallite size is 156 Å, which is in close agreement with the 159 Å value obtained from x-ray line broadening. The Si powders 213S and 214S are crystalline to x-ray and electron diffraction, and they also exhibit microcrystalline grain sizes that range from about one-fourth to one-half the particle diameters.

The structure of these Si powders must be accounted for in any modeling of their formation process; e.g., the particles may crystallize on cooling from an amorphous solid, or may grow as clusters of crystalline solid particles, or may crystallize on cooling from a liquid droplet. Until such time as the reaction zone temperatures are known, it will be difficult to describe the formation mechanism. It is apparent from the agreement between the diameters measured by BET and direct observation techniques that there is no void space between the crystallites in the particles that are accessible to the absorbate gas. It is, therefore, unlikely that the particles were formed from the agglomeration of individual small particles.

The synthesis of Si from SiH_4 has a conversion efficiency from 65 to 85% based on the amount of SiH_4 passing through the laser beam. We are

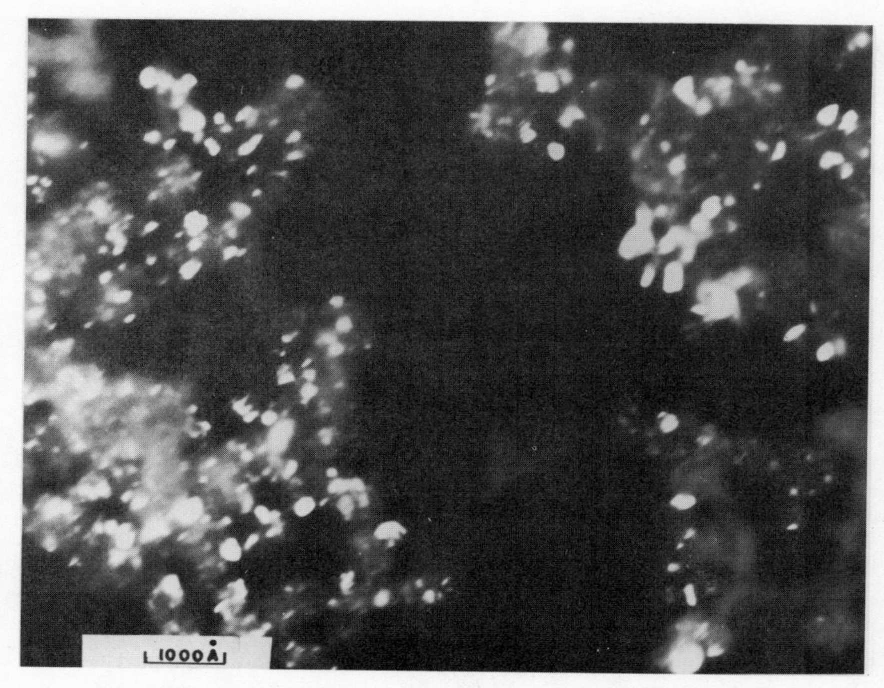

Figure 3.30. Dark field TEM photomicrograph of Si powders from lot 209S.

confident that the conversion efficiency can be improved to the 95% level observed with Si_3N_4 by increasing the optical path and/or preheating the SiH_4.

3.2.2.2b. Si_3N_4 *Powders.* The laser synthesis of Si_3N_4 proceeds as

$$3SiH_4(g) + 4NH_3(g) \xrightarrow{h\nu} Si_3N_4(s) + 12H_2(g)$$

The reaction between SiH_4 and NH_3 has been investigated to provide powders for sintering studies and to determine the effect of process variables on powder characteristics. Combined, they will enable accurate modeling of the synthesis process and provide a basis for manipulating variables to yield optimized powder characteristics.

3.2.2.2b.1. Physical characteristics. The Si_3N_4 powders all have the same general physical characteristics. Si_3N_4 powder lot 603SN made under reference conditions will be described in detail. The effects on powder characteristics of departures from reference conditions will then be presented.

The Si_3N_4 powder 603SN (Figure 3.31 and Table 3.10) consists of spherical particles that are uniform in size and, unlike the Si particles, lack any internal structure. They have a surface area of 117 m^2/g corresponding

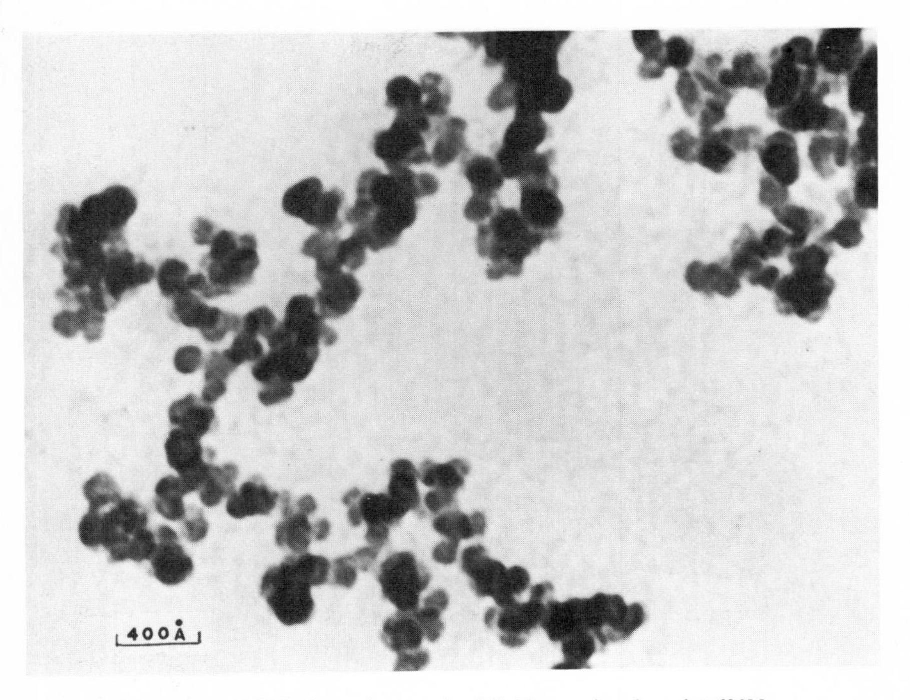

Figure 3.31. TEM photomicrograph of Si_3N_4 powders from lot 603N.

to an equivalent spherical diameter (corrected for composition) of 176 Å. The TEM measured average particle size is 168 ± 39 Å, and the particles range in size from 100 to 250 Å. Figure 3.32 shows the particle size distribution of the Si_3N_4 powders. The Si_3N_4 powders have a much smaller average size and narrower particle size distribution than the Si powders described above. The powders from several runs were combined to yield sufficient powder for density measurement by He displacement pycnometry.‡ The measured particle density of a 3-g powder sample was

Table 3.10. Physical Characteristics of Si_3N_4 *Powder (603SN) Produced with Reference Conditions*

Surface area (m^2/g), BET	117
Size (Å), BET	176
Mean size (Å ± s.d.), TEM	168 ± 39
Size range (Å), TEM	100–250

‡ Micrometrics, Norcross, Ga.

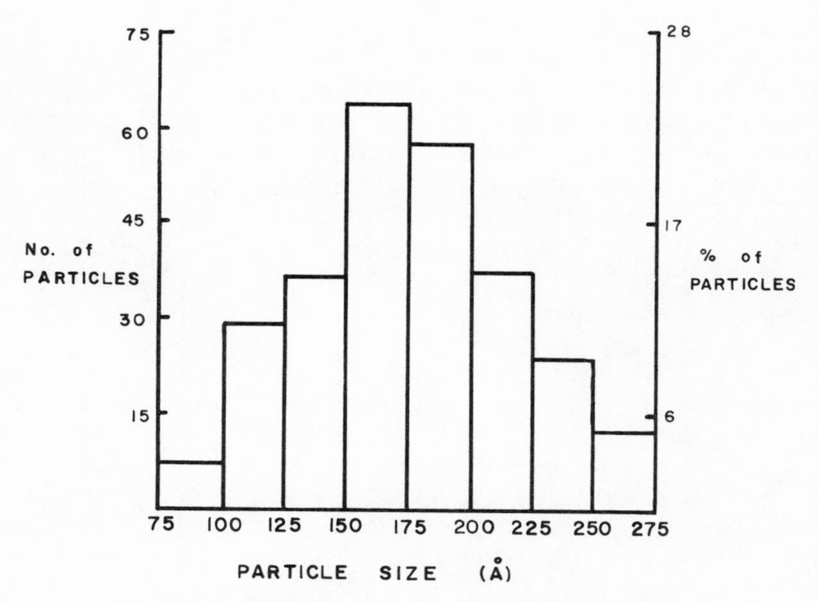

Figure 3.32. Particle size distribution of 603SN Si_3N_4 powder.

2.75 g/cm^3. The calculated density of these powders was 2.88 g/cm^3, based on a typical chemistry of 65% Si_3N_4 (3.18 g/cm^3) and 35% Si (2.33 g/cm^3). These results indicate that the individual particles have essentially theoretical density (within the accuracy of the test) and contain no pores that are not accessible to the surface. The close agreement between particle diameters measured by TEM and BET techniques is further confirmation that there is no porosity. If the powders contained internal porosity, which is open to the surface, the BET equivalent spherical diameter would be consistently smaller than that measured by TEM. This has not been the case. In addition, TEM examination has not revealed any indication that these Si_3N_4 (or any other) powders contain any features that could be interpreted as porosity.

3.2.2.2b.2. Chemical characteristics. Table 3.11 summarizes the chemical analyses of 603SN Si_3N_4 powder produced under reference conditions.

The chemical analyses reveal that these powders contain 72 wt% Si and 26 wt% N. Stoichiometric Si_3N_4 contains 60 wt% Si and 40 wt% N, indicating 35 wt% free Si for Si_3N_4 powders produced under reference conditions. These powders contain very small amounts of $H < 0.5$ wt% and an O content of 1.25 wt%. The small hydrogen content is because of the free H from the dissociation reactions of SiH_4 and NH_3.

It is estimated that with a surface area of 117 m^2/g, an adsorbed monolayer of SiO_2 would account for 4 wt% O. The O content of Si_3N_4 was observed to increase from 1.52 to 2.48 wt% after exposure to air for 24 hr. Although the increase was substantial, again the indication is that the majority of the O content of the powders results during the synthesis process, not afterward. It was anticipated that minor changes in the synthesis apparatus and handling procedures would result in reduced O content of the Si_3N_4 powders to acceptable levels. These efforts have resulted in a decrease in the O content from typical values of 1–3 wt% for early runs to 0.3–0.5 wt% for more recent runs. The only other detected impurity was 20 ppm of Cu (emission spectrographic analysis), no doubt from the brass and Cu system employed.

3.2.2.2b.3. Crystalline characteristics. The crystal structure of the synthesized Si_3N_4 powders was studied using Debye–Scherrer and electron diffraction techniques. The x-ray diffraction patterns reveal one very broad peak that corresponds to the most intense Si_3N_4 peaks [α(100), (210), and (102) and β(200) and (201)] as well as sharper peaks that correspond to Si. The broad Si_3N_4 peak could correspond to a crystallite size of approximately 20 Å or could result from the overlapping of broadened peaks from larger crystallites. Alternatively, it could result from a highly distorted or amorphous structure. We observed no evidence of crystalline characteristics in Si_3N_4 powders using TEM techniques. So, we feel that the Si_3N_4 powders are amorphous. The x-ray line-broadening analysis of the free Si peaks in the Si_3N_4 powder indicates a crystallite size of 95 Å. This size is close to that determined for Si powder produced from SiH_4 gas. Those results indicate that, under reference conditions, the Si_3N_4 powder consists of a mixture of amorphous Si_3N_4 particles and crystalline Si particles.

3.2.2.2b.4. Departures from reference conditions. To understand more fully the effects of processing variables on the synthesis of these powders, systematic departures were made from the reference conditions. These experiments were designed to show empirically how the departures affected

Table 3.11. Chemical Characteristics of Si_3N_4 *Powder (603SN) Produced with Reference Conditions*

Si (wt%)	72
N (wt%)	26
H (wt%)	< 0.5
O (wt%)	1.25
Free Si (wt%)	35
Conversion (%)	92

the powder characteristics and also to yield a basis to test the accuracy of the synthesis modeling. It should be noted that, in some experiments, it was not possible to alter one variable independent of all others (refer to the specific process parameters in Table 3.6).

The process variables studied were the laser intensity, the reaction zone pressure, the reactant gas velocity, and the effect of diluting the reactant gases with NH_3 and argon. The influence of these variables on powder size, chemistry, and reaction temperature was determined. The next section discusses these changes in relation to process models.

Laser intensity. Of the variables that were examined, variation in the laser intensity had by far the greatest effect on the powder synthesis process and powder characteristics. This observation is largely true because the range examined, a factor of 10^2, was much larger than could be made with any other variables. Table 3.12 shows the variation of powder size, reaction temperature, and stoichiometry as the laser power density is increased from 760 to 2×10^4 and then 1×10^5 W/cm^2 by focusing the beam with a 13-cm focal length lens.

The particle size decreases progressively as the intensity increases. Earlier experiments revealed that for $I = 10^9$ W/cm^2 (pulselength $= 10^{-9}$ sec) the particle size was 25–50 Å. These results indicate that a variation in particle size by a factor of 2–4 can be achieved by manipulation of the intensity over many decades.

Laser intensity has a major effect on the powder chemistry. Powders produced under reference conditions were brown to tan in color and contained 35 wt% free Si. Powders produced with 2×10^4 W/cm^2 were a light tan–white in color, showed no evidence of Si by x-ray, and had 2 wt% free Si by wet chemical analyses. At 10^5 W/cm^2, the powders were pure white and showed no Si by x-ray. There was insufficient powder to perform chemical analysis on the 10^5 W/cm^2 powder; but, based on color, they appear to be stoichiometric.

Although the laser intensity had a strong influence on the particle size and powder stochiometry, there was no change in the crystal structure. All Si_3N_4 powders were amorphous to both x-ray and electron diffraction,

Table 3.12. The Influence of Laser Intensity on the Particle Size, Chemistry, and Reaction Temperature for Si_3N_4 *Power Synthesis*

Run number	603SN	609SN	621SN
Laser power density (W/cm^2)	760	2×10^4	1×10^5
Particle size (Å), BET	176	114	98
Free Si (wt%)	35	2	Very low
Reaction temperature (°C)	867	985–1020	> 1020
Crystallinity	Amorphous	Amorphous	Amorphous

Table 3.13. The Influence of Pressure on the Particle Size, Chemistry, and Reaction Temperature for Si_3N_4 *Powder Synthesis*

Run number	611SN	603SN	610SN	023SN	021SN
Pressure (atm)	0.08	0.2	0.5	0.75	0.75
Particle size (Å), BET	167	176	211	220	221
Free Si (wt%)	60	35	14	23	15.6
Reaction temperature (°C)	—	867	1060–1080	1050–1125	—
Crystallinity	Amorphous	Amorphous	Amorphous	Amorphous	Amorphous

despite an apparent higher reaction temperature. When free Si is present in these powders, it is crystalline.

Cell pressure. The effects of variations in the cell pressure are shown in Table 3.13. As the pressure was raised, the particle size increased steadily. Despite the slightly higher power intensities of runs 021SN and 023SN at 0.75 atm, these had a larger particle size than powder lot 610SN, which was produced at 0.5 atm. Although the reaction temperature is not absolutely consistent, it shows a definite trend toward increased reaction zone temperature with increased pressure. It can also be seen in Table 3.13 that the Si_3N_4 powders become more stoichiometric as the pressure is raised from 0.08 to 0.5 atm. There was no variation in the crystallinity of the Si_3N_4 powders with increased pressure.

Gas velocity. The third process variable studied was the reactant gas velocity. Table 3.14 shows two pairs of comparable experiments where the velocity was altered. In powder lot 603SN and 605SN, the NH_3/SiH_4 ratio was 10/1, while in lots 020SN and 614SN it was 5/1. In both sets of experiments the particle size decreases, and the powders become less stoichiometric with increasing velocity. There was no detectable change in the crystal structure of these powders.

Reactant gas dilution. The influence of diluting the reactant gas with either NH_3 or argon was also investigated. Table 3.15 shows the particle

Table 3.14. The Influence of Reactant Gas Velocity on the Particle Size, Chemistry and Reaction Temperature for Si_3N_4 *Powder Synthesis*

Run number	603SN	605SN	020SN	614SN
NH_3/SiH_4 ratio	10/1	10/1	5/1	5/1
Reactant gas velocity, avg/max (cm/sec) at laser beam	474/948	825/1650	191/382	477/954
Particle size (Å), BET	176	151	210	155
Free Si (wt%)	35	—	18.5	35.6
Powder color	Light brown-tan	Dark brown	Light brown	Brown
Reaction temperature (°C)	867	—	—	1025
Crystallinity	Amorphous	Amorphous	Amorphous	Amorphous

Table 3.15. The Influence of Dilution of Reactant Gases with NH_3 *and Argon on the* Si_3N_4 *Powder Characteristics*

Run number	614SN	603SN	615SN	607S[illegible]
NH_3/SiH_4 ratio	5/1	10/1	20/1	[illegible]
% Argon dilution	0	0	0	[illegible]
Particle size (Å), BET	155	176	155	[illegible]
Free Si (wt%)	35.6	35	38.1	[illegible]
Powder color	Brown	Light brown–tan	Brown	[illegible]
Crystallinity	Amorphous	Amorphous	Amorphous[a]	Amo[illegible]
Temperature (°C)	1025	867	800	990

[a] No crystalline Si present.

size, free Si wt%, and reaction zone temperature for NH_3/SiH_4 ratios of 5/1, 10/1, and 20/1 under otherwise reference conditions. The results reveal that there is no apparent trend in particle size with reaction zone temperature or dilution. It is also not clear whether there is a significant change in stoichiometry for dilution with NH_3.

A similar experiment was performed by diluting the reactant gas stream with argon 607SN. Table 3.15 shows that diluting the gas stream with 45% argon had virtually no influence on the particle size (176–173 Å) or chemistry (35–36 wt% free Si) but caused an increase in the reaction temperature.

3.2.2.2c. SiC Powders. Synthesis experiments have been carried out to determine the possibility of producing SiC powders from laser-heated reactant gases. Silane was used as the silicon source as in synthesis experiments with other powders. Methane (CH_4) and ethylene (C_2H_4) were used as carbon sources. The proposed reactants for these experiments were

$$SiH_4(g) + CH_4(g) \xrightarrow{h\nu} SiC(s) + 4H_2(g)$$

and

$$2SiH_4(g) + C_2H_4(g) \xrightarrow{h\nu} 2SiC(s) + 6H_2(g)$$

Methane was expected to react more readily than ethylene because of its simple molecular structure, even though it has no absorption bands in the vicinity of 10.6 μm. More complete information suggests that methane pyrolyzes via an ethylene intermediate (Powell, 1966) so the initial hypothesis may not, in fact, be valid. Ethylene was selected because it was reported (Patty *et al.*, 1974) to exhibit high absorptivity to the *P*(20) line of 00°1–10°0 band in a CO_2 laser. Patty *et al.* reported an absorptivity of 1.64 cm^{-1} atm^{-1} for dilute concentrations of ethylene in air. This absorptivity level is intermediate between measured levels of SiH_4 and NH_3 and is a very high absorptivity by any comparison with other gases.

The process conditions used in these synthesis experiments are summarized in Table 3.16 along with the characteristics of resulting powders. In general, the process conditions were nearly the same as the reference conditions used for Si and Si_3N_4 powders. The gas mixtures had nearly stochiometric silicon-to-carbon ratios.

The physical characteristics of resulting powders are similar to those observed with Si and Si_3N_4 powders. TEM analysis (Figure 3.33) showed them to be small, uniform, loosely agglomerated spheres. The particles ranged from 180 to 260 Å with an average size of 210 Å. BET equivalent diameters and directly measured diameters were essentially equal to each other, again, indicating an absence of internal porosity that was open to the surface of the particles.

Chemical analysis showed that the powders formed with CH_4 contained virtually no carbon. The methane apparently did not react with exposure to the laser beam. Those formed with C_2H_4 did contain 14% by weight carbon compared with 33% for stoichiometric SiC. IR spectral analysis indicated that carbon combined as SiC in these powders, although we have not determined whether all of the carbon was combined in this manner.

3.2.2.2c.1. Crystalline characteristics. The powders made from SiH_4 and CH_4 (613SC) were crystalline to both x-ray and electron diffraction. The diffraction patterns were identical to those for Si powder made from SiH_4. The particles ranged from 180 to 260 Å with an average size of 210 Å. The

Table 3.16. Summary of SiC *Process Conditions and Powder Characteristics*

Gases	$SiH_4 + CH_4$	$SiH_4 + C_2H_4$
Cell pressure (atm)	0.2	0.2
Laser intensity (W/cm^2)	760	860
Measured reaction temperature (°C)	710	865
Powder chemistry (wt%)		
Si	99	80
C	0.2	14
O	0.3	1.4
Powder size		
BET equivalent (Å)	305	274
TEM	—	230
Powder crystallinity		
x-ray	Crystalline Si	Poorly crystalline
Electron diffraction	Crystalline Si	Poorly crystalline

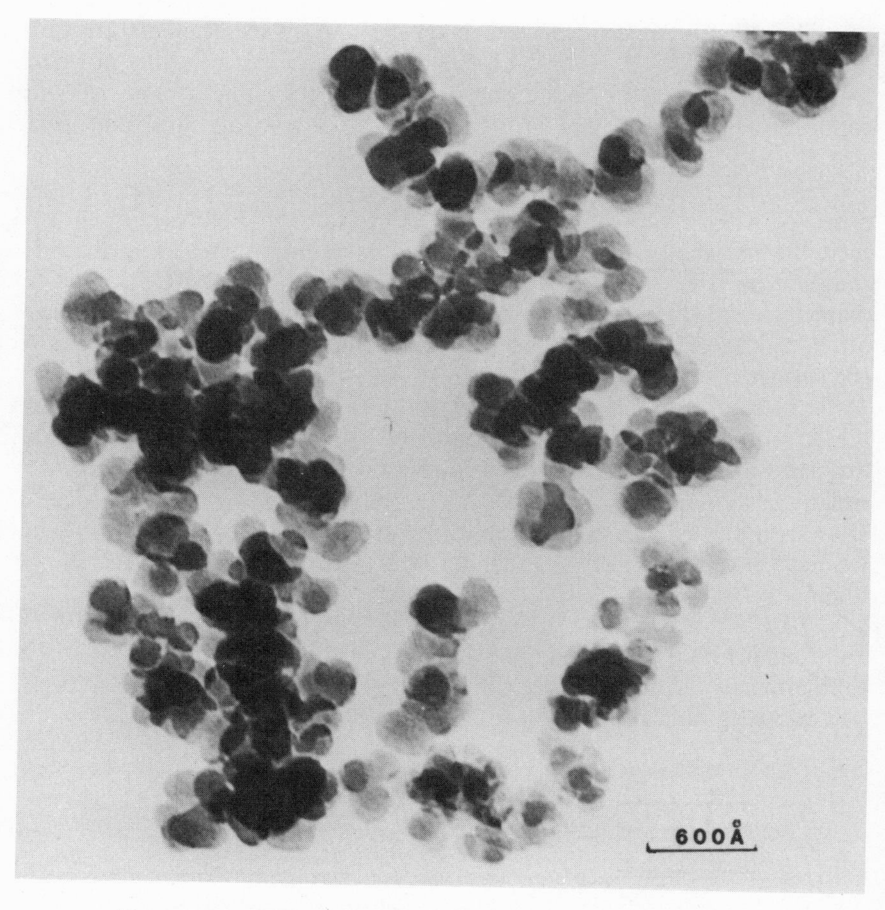

Figure 3.33. TEM photomicrograph of SiC powders from lot 022SC.

powders from run 022SC (SiH_4 and C_2H_4) showed some evidence of crystallinity in both electron and x-ray diffraction. The peaks were broader than 613SC indicating that these crystallite powders are either smaller or more poorly crystalline in nature.

It is not known whether the higher reactivity of the C_2H_4 gas resulted primarily because of its higher optical absorptivity, an intrinsically high reactivity with silane, or some combination of factors. It is interesting to speculate whether resonance effects between the photons and the gas molecules enhanced the reaction. It is also apparent that its absorptivity improved the coupling efficiency relative to the CH_4/SiH_4 gas mixture in which the CH_4 acted as an optical diluent. As shown in Table 3.16, the measured reaction temperature was higher with C_2H_4 than with CH_4.

Although the absolute temperature levels probably are not accurate, the relative levels are likely correct. So the cause for the higher reactivity of C_2H_4 may simply result from a higher temperature at the point of reaction, attributable to the higher optical absorptivity. If this is so, higher laser intensities should cause CH_4 to react with SiH_4. Higher reactant gas pressures should also improve the stoichiometry of the powders.

The question of possible resonance effects on reactions is fundamentally important for our understanding, controlling, and exploiting this laser synthesis process. If they are important, unusual reaction paths and kinetics will be possible. It may also be possible to cause reactions in the presence of unwanted impurities that will not be incorporated proportionally if they do not couple to the laser light. If resonance effects do not occur, it will be correct to describe the reaction as a very fast "thermal domain" reaction as we are now approaching the problem.

3.2.2.2d. Particle Agglomeration. The laser synthesized Si, Si_3N_4, and SiC powders exhibit most of the stated characteristics for an ideal powder. These powders appear to fall short of "ideal" in their apparent agglomeration. All of the TEM analyses have revealed two-dimensional networks of powders as shown in Figures 3.28, 3.31, and 3.33. The particles are usually in contact with one or more particles, yet there is no evidence of actual neck formation or other high-strength bonding between the particles.

Possibly the observed loose agglomeration results from the sample preparation techniques that were employed. Typically, powders examined by TEM had been captured in a filter and were probably subjected to shear when removed.

To examine this question, a TEM grid was placed inside the reaction cell and powders were produced using reference gas flow and pressure conditions with 100-msec pulses of the CO_2-laser beam. These powders again revealed loose two-dimensional agglomeration. This preliminary observation suggests that agglomeration occurs prior to capture in the filter assembly.

A simple calculation has been made to determine the frequency of collision and thus assess the likelihood of agglomeration before leaving the flame. Two possible solutions have been presented in the literature (Mason, 1977).

The perikinetic solution assumes particles of concentration N in a fluid medium of viscosity η move entirely by Brownian motion. For equisized noninteracting spheres, the frequency of collision of a single sphere with any other is

$$f = \frac{8}{3}\frac{NkT}{\eta} \tag{3.25}$$

where N is the number per unit volume; k is the Boltzmann constant; T is the temperature in degrees Kelvin; η is the viscosity.

The second is the orthokinetic solution that assumes particles in a flowing stream are brought together by the shearing action of flowing streams. For equisized noninteracting spheres the frequency of collision is

$$f = \tfrac{32}{3} b^3 N G \tag{3.26}$$

where b is the radius of the particles, and G is the gradient in the flow velocity. The following are values calculated for the reference conditions and for synthesizing silicon nitride: $N = 8.2 \times 10^{11}$ cm^{-3}; $\eta = 3 \times 10^{-4}$ poise; $G_{avg} = 9.5 \times 10^3$ cm/sec cm; $T = 1100$°K; $b = 7.8 \times 10^{-7}$ cm.

For the perikinetic case $f = 1.1 \times 10^3$ sec^{-1} and for the orthokinetic case $f = 5.5 \times 10^{-2}$ sec^{-1}. Clearly collisions because of Brownian (perikinetic) motion dominate. The collision frequency will be approximately 1000/sec. The collision frequency will be even higher if Van der Waals' attractive forces are considered, but will be lower if all particles have identical charge, with the same sign and electrostatic forces considered. These calculations have not yet been made.

Under the reference conditions the average flow rate in the flame is estimated to be 200 cm/sec, and the dwell time of particles in the flame is 1.2×10^{-2} sec. On the average, a particle will undergo approximately 12 collisions while in the flame. On this basis, particles would be expected to undergo about 15 more collisions before reaching the filter.

Particles will not necessarily stick when they collide and may rebound. We have made no estimate as to the frequency of sticking but assume that it will depend on the velocity of the particle, the Van der Waals' attractive forces, and possibly the electrostatic forces. Once the particles do stick, they may be held together by surface forces that reduce the total surface energy of the particle by forming the contact area between them (Easterling and Thölen, 1972). Sintering may even occur in the flame because of the same forces, although we see no evidence of this from TEM observations.

To avoid collisions between particles within the flame, this analysis dictates that the particle number density be decreased, the temperature be decreased, and/or the viscosity be increased. Parametric changes in run conditions will affect the number density of particles and local temperature and, to a limited extent, the viscosity. The effects of varied process parameters on powder characteristics show that conditions necessary to produce other desirable powder properties may not necessarily be compatible with those that inhibit agglomeration. This is a fundamentally important issue that needs to be investigated.

3.2.2.3. Interaction between Process Variables and Powder Characteristics

We have discussed the laser-heated gas-phase reaction and the resulting powders separately. It is our ultimate objective to develop an understanding of the relationship between the two that is based on equilibrium thermodynamics and the kinetics of the gas-phase process. With a somewhat improved description of the process, it should be possible to rationalize observations in terms of classical homogeneous nucleation and growth theory.

The variables that have a direct effect on the nucleation and growth kinetics include reaction temperature, heating rate, partial pressure of reactants, total pressure, and dilution by inert gases. Most of these variables have been manipulated to demonstrate their empirical effect on Si_3N_4 powder characteristics. We have insufficient data to provide a basis for discussing Si and SiC synthesis processes.

We examined the following powder characteristics: size, size distribution, shape, stoichiometry, chemical impurities, and crystallinity. Process variables that were changed are laser intensity, cell pressures, gas velocities, and, to a lesser extent, gas mixture. Of the characteristics and process variables examined, only particle size and stoichiometry were influenced by laser intensity and gas pressure. No other variable had any significant effect on any other characteristics within the range of conditions examined.

Increased laser intensity caused Si_3N_4 particle size to decrease and caused their Si/N ratio to approach the stoichiometric composition.

We have shown that virtually all of the silane gas is consumed during particle growth and that the particle size is limited by impingement of overlapping volumes of depleted reactant gas. The final particle size is therefore controlled by the number of embryos that reach supercritical dimensions and grow until depleted volumes impinge on one another. The homogeneous nucleation rate is very small until either the degree of supersaturation approaches a critical value or, if kinetically limited, a critical temperature is reached. The rate changes many orders of magnitude with small changes beyond either of these critical values. The narrow particle size distribution indicates that there was not appreciable nucleation during the time period between the appearance of the initial nuclei and cessation of growth with impingement. This can occur for two reasons. The growth process itself may effectively terminate the nucleation process by, for instance, reducing the supersaturation level below the critical level. Or, at the critical supersaturation level where nucleation rates first become appreciable, the growth rates may be extremely fast. In this case, the reaction effectively goes to completion and terminates with the appearance of the first nuclei.

The decreased particle size resulting with increased laser intensity results from forcing the spontaneous reaction temperature to higher levels, and correspondingly higher degrees of supersaturation, with the higher heating rates. Heating rates at the reaction temperature ranged from 10^6 to 10^8 °C/sec for laser intensities ranging from 760 to 10^5 W/cm^2. The particle size is reduced because the nucleation rate is larger at higher temperatures, and it increases proportionally faster than the growth rate. This is the same conclusion reached for TiO_2 powders produced from $TiCl_4$ (Suyama and Kato, 1976).

The general deficiency of nitrogen in resulting particles could result either from slow reaction kinetics involving N_2-bearing species or from preferential vaporization of N_2 from hot particles. These results indicate that higher temperatures permit achievement of stoichiometric compositions for kinetic reasons. The effect of increased kinetic rates is self-assisting because the increasingly rapid exothermic reactions involved in the growth process drive the particle temperatures even higher.

Increased reactant pressures caused Si_3N_4 particle sizes to increase and caused their compositions to be more nearly stoichiometric.

Previous analyses of the process showed that heating rate was independent of the pressure of optically absorbing species because the absorptivity and heat capacity both change proportionally with pressure. Pressure, itself, has less effect on nucleation rate than does temperature, so we would not anticipate that pressure would have any appreciable effect on how far the temperature could be driven beyond the level where nucleation rates become appreciable. This critical temperature level and the corresponding nucleation rates are expected to be essentially independent of relatively small pressure changes.

The results of our depletion volume analyses support this conclusion and explain the particle size dependence on pressure. For the range of pressures examined, the distance between growing nuclei was shown to be constant. The observed particle volumes are directly proportional to the mass of reactant gas within the depletion volumes and so are proportional to pressure.

The reason for increasing nitrogen content with increasing pressure has not been determined. It could be simply that the increased partial pressure of N_2-bearing species causes the increasing nitrogen content in the particles. It is also probable that the particle growth rates will increase with increasing reactant particle pressures. Consequently, the rate at which exothermic heat is liberated at the particles will increase with pressure, causing the particle temperatures to be proportionally higher. As was already shown, higher reaction temperatures cause the particles to be more nearly stoichiometric. Pyrometric temperature measurements did indicate that the flame temperature rose with increasing pressure.

The gas velocity, the NH_3/SiH_4 ratio in the gas, and dilution with argon had negligible effects on particle characteristics. The absence of an apparent gas-velocity effect can be anticipated because the instantaneous heating rates at the reaction temperature change by only a factor of 2–3 with the range of gas velocities investigated. This variation would probably not influence nucleation rates. This same conclusion may apply to the gas composition and dilution effects although we do not have a sound basis for analyzing their effect at this point.

It is apparent that particle size is insensitive to process variables other than laser intensity and pressure for large ranges in conditions. Stoichiometry is manipulable in an orderly manner; powders can be made that range from mostly pure Si to stoichiometric compositions. All as-synthesized Si_3N_4 powders have been amorphous, but they crystallize rapidly at temperatures in excess of 1400°C. Silicon powders made from SiH_4 can be either amorphous or crystalline depending on laser intensity. It appears that we should be able to achieve simultaneously virtually all of the ideal powder characteristics that were sought while retaining some control over individual powder characteristics. Continuation of this work will concentrate on means of extending the achievable range of particle sizes while retaining the narrow particle size distributions.

3.3. Summary

The research program described in this chapter has investigated means of producing powders that will permit fabrication of ceramic bodies with superior characteristics. We have worked with Si_3N_4 and SiC because they are promising candidates for use in new generations of heat engines. Ideal powders will have particles that are small, uniform in size, equiaxed (tending toward spheres) in shape, composed of a specific phase(s), compositionally pure, and free of agglomerations. In principle, gas-phase synthesis processes can achieve these attributes; however, variations in time-temperature history throughout the conventionally heated reaction zones cause unacceptably large variations in particle characteristics. Also, the typically long exposure to elevated temperatures causes particles to bond to one another. We have elected to heat the reactant gases by absorbing IR light emitted from a laser. This unique means of transferring energy to the gas permits precise, uniform heating with unusually high heating rates and small reaction volumes. We anticipated that this laser-heated synthesis process would overcome the deficiencies of conventional gas-phase synthesis processes while retaining their advantages.

In this process, optically absorbing gases are passed through a laser beam to cause a reaction in a definite volume within the region where the two interact. We have investigated process geometries in which the gas stream and the laser beam intersect orthogonally and also where they intersect coaxially from opposite directions (counterflow). A CO_2 laser was used as the heat source to drive reactions in gases containing active components, such as SiH_4, NH_3, CH_4, and C_2H_4, as well as various inert dilutants. The resulting Si, Si_3N_4, and SiC powders were collected and characterized as a function of process variables.

The results of this research demonstrate that this laser-heated synthesis process produces powders with virtually all of the desired characteristics. The resulting particles are small, uniform in size, round, and pure. The particles appear to be attached to one another in chainlike agglomerations; however, direct examination by TEM revealed no neck formation between particles. We anticipate that they can be dispersed. Besides producing powders with ideal characteristics, this laser-heated synthesis process is extremely efficient. Approximately 95% of SiH_4 is converted to Si_3N_4 powder in a single pass through the laser beam. Also, Si, Si_3N_4, and SiC powders can be produced from these reactants with as little as 2 kWhr of energy per 1 kg of powder. It is likely that this process can produce both a superior and a lower cost powder than conventional gas-phase or solid-phase synthesis processes.

Much of our efforts have focused on developing an analytical description of the laser-heated synthesis process. To develop a model, many fundamental property measurements were required, such as detailed absorptivity measurements for reactant gases as a function of pressure and emitted wavelength. Emissions from the reaction have been studied to identify reaction species and to estimate the reaction temperature. Computer analyses of the gas flow were used to predict gas stream dimensions and velocities. Combined with direct observations of both the reaction positions relative to the laser beam and the reaction temperatures, these analyses and characterizations have been used to describe the time-temperature history of the reactant gases throughout the course of the reaction.

With the process conditions used for the majority of these synthesis experiments (laser intensity = 765 W/cm^2, pressure = 0.2 atm, gas velocity = 500 cm/sec), reaction products were evident within 3–5 mm penetration into the laser beam. Heating rates to the reaction temperature (approximately 800°C) were approximately 10^6 °C/sec. The reaction was initiated in approximately 10^{-3} sec and was completed in, at most, 7.5×10^{-3} sec. The individual particles grew from and depleted a volume of gas equal to approximately 1×10^{-4} cm in diameter.

Most process variables were manipulated to determine their effect on particle characteristics. These interactions were interpreted in terms of changes in the process. Laser intensities up to 10^5 W/cm^2 produced heating rates in excess of 10^8 °C/sec. Variations in heating rates within different gas streamlines were analyzed in terms of the Gaussian intensity in the laser beam and the parabolic velocity profile in the gas stream. The effects of nonabsorbing gases were also considered. Other than laser intensity, most process variables had very little effect on particle size. The gas depletion volume remained essentially constant. Increased laser intensity caused the particle size to decrease progressively and raised the reaction temperature. Qualitatively, the results follow expectations based on nucleation and growth theory. The general characteristics of the powders of the three materials are similar to one another, but they differ in detail. Silicon powders were crystalline under all but the lowest laser intensity conditions. Individual crystalline particles consisted of multiple grains that were approximately 150 Å in diameter. Mean Si particle diameters were in the range of 500–600 Å with a standard deviation of 35–50%. Silicon nitride powders were always amorphous. Mean particle sizes ranged from 25 to 220 Å, depending primarily on laser intensity. These powders were more uniform than the silicon powders. The standard deviations were approximately 25%, and the ratio between the largest to the smallest observed particle was less than 2.5. Their stoichiometry varied with processing conditions. High laser intensities yielded stoichiometric powders; lower intensities produced powders that were rich in silicon. The silicon carbide powders are similar to the Si_3N_4 powders, amorphous and tending to be rich in silicon, with a mean particle size of approximately 230–250 Å. Powders of all three materials have spherical particles whose BET equivalent diameters equaled the directly observed mean diameters. This observation and direct density measurements indicate that individual particles contain no porosity. All of the materials were free of contaminants. The oxygen impurity level decreased progressively as handling procedures improved. In later powder batches, the O_2 content was less than 0.2% by weight. With the exception of Cu at a concentration of 20 ppm, no other impurities were detected.

This laser-heated powder synthesis process may be among the few known at this writing in which a desired product is obtained in better yield and at lower cost with a laser than by more conventional means. Powders that result from the laser-heated, gas-phase synthesis process have most of the characteristics presumed to be ideal for ceramic powders. The process also appears capable of reducing the cost of these powders because it requires very little energy per 1 kg of powder and utilizes feed materials very efficiently. The process should be developed further to eliminate the tendency of forming chainlike agglomerates and to increase the mean particle size. For many materials, it is desirable to have mean particle sizes of approximately 1000 Å rather than 250 Å typically produced by this process. Achieving a larger particle size will require an improved understanding of the nucleation and growth processes. It is also important to apply the process to other materials. There are a number of processes involving electronic, magnetic, and optical ceramics that would benefit from using powders with the characteristics displayed by the materials we have been able to produce using the laser synthesis method.

ACKNOWLEDGMENTS

The authors gratefully acknowledge the financial support for this research program and the contributions by other researchers. This work has largely been funded by the Department of Defense under ARPA order No. 3449 and was monitored by the Office of Naval Research under contract No. N00014-77-C-0581. Several M.I.T. staff and students participated in this interdisciplinary program. Like the two authors, Dr. S. C. Danforth who worked on all aspects of the program is associated with the Energy Laboratory as well as the Department of Materials Science and Engineering. Mr. R. A. Marra, who has developed gas-flow and heating-rate models, is a Research Assistant in the Department of Materials Science and Engineering. Professor C. F. Dewey and Mr. J. H. Flint, who conducted the absorptivity and threshold experiments, are members of the Mechanical Engineering Department staff. Mr. C. Reiser, who investigated the feasibility of inducing unimolecular reactions, is a Research Assistant in the Chemistry Department.

References

Benson, S. W., 1960, *The Foundations of Chemical Kinetics*, McGraw-Hill, New York, Chapter 9.

Braker, W., and Mossman, A. L., 1971, *Matheson Gas Data Book*, 5th Edition, Matheson Gas Products, E. Rutherford, N.J., p. 506.

Coble, R. L., 1960, Proc. of the 4th Intl. Conf. on Reactivity of Solids, Amsterdam.

Cochet, G., Mellottee, H., and Delbougo, R., 1975, Proc. 5th International Conference on Chem. Vapour Deposition (eds. J. M. Bloucher and T. E. Hintermann), Fuler Res. Int., pp. 43–55.

Cox, D. M., and Horsley, J. A., 1980, *J. Chem. Phys.* **72**:864.

Cullity, B. D., 1956, *Elements of X-Ray Diffraction*, Addison-Wesley, Reading, Mass., p. 261.

Danforth, S. C., June, 1978, Effective Control of the Microstructure of Reaction Bonded Si_3N_4 as Related to Improved Mechanical Properties, Ph.D. Thesis, Brown University, Providence, RI.

Deutsch, T. F., 1979, *J. Chem. Phys.* **70**:1187.

Dressler, K., and Ramsay, D. A., 1959, *Phil. Trans. Roy. Soc.* **A,251**:553.

Dubois, I., 1968, *Can. J. Phys.* **46**:2485.

Easterling, K. E., and Thölen, A. R., 1972, *Acta Met.* **20**:1001–8.

Freund, S. M., and Danen, W. C., October 1977, Purification of Materials using Infrared Lasers: Removal of Diborane from Silane, Proc. 9th Annual Electro-Optics/Laser Conference and Exhibition, Anaheim, Calif., p. 609.

Freund, S. M., and Danen, W. C., 1979, *Inorg. Nucl. Chem. Lett.* **15**:45.

Galasso, F. S., Veltri, R. D., and Croft, W. J., 1978, *Bull. Am. Ceram. Soc.* **57**:453.

Garing, J. S., Nielsen, H. H., and Rao, K. N., 1959, *J. Mol. Spectrosc.* **3**:496–527.

Gaydon, A. G., and Wolfhard, H. G., 1970, *Flames: Their Structure, Radiation, and Temperature*, Chapman and Hall, London.

Greskovich, C. D., Prochazka, S., and Rosolowskii, J. H., November 1976, *Basic Research on*

Technology Development for Sintered Ceramics, Report AFML-TR-179, G.E. Research and Development Laboratory under contract with the AF Materials Laboratory.

Haas, C. H., and Ring, M. A., 1975, *Inorg. Chem.* **14**:2253.

Haggerty, J. S., and Cannon, W. R., October 1978, *Sinterable Powders from Laser Driven Reactions*, Report on Contract N00014-77-C-0581, M.I.T., Cambridge, Mass.

JANAF Thermochemical Tables, 1971, NSRDS-NBS37, Washington, D.C.

Jensen, C., Steinfeld, J. I., and Levine, R. D., 1978, *J. Chem. Phys.* **69**:1432.

Lin, S.-S., 1977, *J. Electrochem. Soc.* **124**:1945.

Lin, S. T., and Ronn, A. M., 1978, *Chem. Phys. Lett.* **56**:414.

Lyman, J. L., Quigley, G. P., and Judd, O. P., 1980, Single-Infrared-Frequency Studies of Multiple-Photon Excitation and Dissociation of Polyatomic Molecules, in *Multiple-Photon Excitation and Dissociation of Polyatomic Molecules* (ed. C. Cantrell), Springer-Verlag, Heidelberg and Berlin.

Mason, S. C., 1977, *J. Colloid and Interface Sci.* **58**:275–285.

Patanker, S. V., and Spalding, D. B., 1970, *Heat and Mass Transfer in Boundary Layers*, 2nd Edition, Intertext Books, London.

Patty, R. R., Russwarm, G. M., and Morgan, D. R., 1974, *Appl. Optics* **13**:2850–4.

Pearse, R. W. B., and Gaydon, A. G., 1976, *The Identification of Molecular Spectra*, 4th edition, Chapman and Hall, London.

Powell, C. F., 1966, Chemically Deposited Nonmetals in Vapor Deposition (eds. C. F. Powell, J. H. Oxley, and J. M. Blocher), John Wiley, New York, p. 343.

Purnell, J. H., and Walsh, R., 1966, *Proc. Roy. Soc.* (*London*), **A293**:543.

Robinson, P. J., and Holbrook, K. A., 1972, *Unimolecular Reactions*, Wiley–Interscience, New York.

Ronn, A. M., and Earl, B. L., 1977, *Chem. Phys. Lett.* **45**:556.

Spalding, D. B., 1977, *Genmix: A General Computer Program for Two Dimensional Parabolic Phenomena*, Pergamon Press, Oxford.

Steinfeld, J. I., Burak, I., Sutton, D. G., and Nowak, A. V., 1970, *J. Chem. Phys.* **52**:5421.

Suyama, Y., and Kato, A., 1976, *J. Am. Chem. Soc.* **59**:146–49.

Tindal, C. H., Straley, J. W., and Nielsen, H. H., 1942, *Phys. Rev.* **62**:151–59.

Vande Hulst, H. C., 1957, *Light Scattering of Small Particles*, John Wiley, New York.

Warren, B. E., 1969, *X-Ray Diffraction*, Addison-Wesley, Reading, Mass., p. 258.

Woodin, R. L., Bomse, D. S., and Beauchamp, J. E., 1978, *J. Am. Chem. Soc.* **100**:3248.

4

Laser-Induced Chemical Reactions: Survey of the Literature, 1965–1979

J. I. Steinfeld

4.1. Introduction

One of the basic tools of the practising chemist is an awareness of the body of reported work pertaining to a particular class of reactions or compounds. For this purpose, a number of compilations of such reaction data have been established and are now a standard part of the chemical literature. Such a compilation is not yet available for that class of reactions that are induced by laser excitation; while many specialized review articles and introductory surveys in this field have appeared, there is still no convenient tabulation of data for individual reactions. The present survey attempts to fill this need. In addition to a listing of reactions, tabulated by reactant, product, excitation source, and literature citation, which should be of use to active researchers, we also include a bibliography of the aforementioned review articles for the benefit of those who wish to begin learning about this field. Cited literature references are given, arranged alphabetically by author.

In order to make this task a manageable one, we have not included references to the use of lasers for purely spectroscopic or energy-transfer investigations or to their application as diagnostic devices for product-state analysis in thermal or molecular-beam reactions. These are entire subfields

J. I. Steinfeld · Department of Chemistry, Massachusetts Institute of Technology, Cambridge, Massachusetts 02139

in themselves and are periodically reviewed elsewhere. Parametric studies of infrared multiphoton dissociation (mostly of SF_6) are not listed, except for a few representative papers, since they are thoroughly covered in the chapter by Galbraith and Ackerhalt. Also, the work summarized is almost purely experimental in nature, dealing with specific reaction systems; most of the papers presenting theoretical models, or dealing in speculations about "possible" experiments, are not included. Somewhat more arbitrarily, we have chosen to omit many of the conventional single-photon photochemistry and flash-photolysis studies that happen to use a laser as a light source. These omissions still leave a large and rapidly growing body of experimental results in which an infrared or ultraviolet laser is used to drive a chemical reaction at conditions far removed from thermal equilibrium. While the attempt has been made to include references to as wide a variety of laser-induced reactions as possible, we do not claim complete coverage of all published literature in the field, and certainly not for unpublished technical reports and conference papers.

The tables are organized according to the type of excitation source employed. Table 4.1A lists those reactions that have been induced by infrared laser excitation. It is divided (again somewhat arbitrarily) into Table 4.1A, which lists reactions that are apparently directly driven by vibrational excitation, and Table 4.1B, which lists reactions that clearly take place following thermalization or that involve the use of nonreacting sensitizer, such as SF_6 or BCl_3. Since most of the excitation involves a CO_2 laser we have used a convenient shorthand for specifying the transition used: $P_1(J)$ or $R_1(J)$ denotes transitions in the 00°1–10°1, or "10.6-μm" band; $P_2(J)$ or $R_2(J)$, those in the 00°1–02°0, or "9.6-μm" band. Isotopic enrichment factors are given when they are stated in the original papers.

Table 4.2 lists reactions induced by visible or ultraviolet laser excitation, including both single- and multiple-photon processes. Finally, a variety of miscellaneous effects, including reactions in solid matrices or at catalytic surfaces, luminescence, ionization, etc., are listed in Table 4.3.

Table 4.1A. Reactions Directly Induced by Single- and Multiple-Infrared-Photon Absorption

Reactant	Added reagents	Products	Laser			Conditions, remarks	References
			Type	λ (μm)	Power/energy		
BCl_3	$B(CH_3)_3$	CH_3BCl_2	CO_2 $P_1(44)$–$R_1(34)$	10.86–10.16	6–12 W CW		Bachmann *et al.* (1977a)
BCl_3	H_2	$HBCl_2$	CO_2	10.6	Pulsed	80 Torr total pressure	Rockwood and Hudson (1975)
BCl_3	H_2S	Not collected	CO_2 $P_1(16)$	10.55	0.1 J/pulse	^{11}B depletion	Freund and Ritter (1975)
BCl_3	CH_4	$HBCl_2$	CO_2 $P_1(22)$, $R_1(24)$	10.6, 10.2		CH_4 pyrolysis at high pressure	Schramm (1979)
$B(CH_3)_3$	HBr	$B(CH_3)_2Br$, BCH_3Br_2	CO_2 $R_1(12)$	10.30	4.5 W CW	600 Torr total pressure	Bachmann *et al.* (1975)
B_2H_6		$B_{20}H_{16}$	CO_2 $R_1(16)$	10.28	1.5 W CW	50–400 Torr pressure	Bachmann *et al.* (1974)
B_2H_6		Dissociated	CO_2			SiH_4 not decomposed	Freund and Danen (1979)
B_2H_6	H_2S, D_2S	$HB(SH)_2$, HSB_2H_5	CO_2 $R_1(16)$	10.28	6–7 W CW	0.5 atm pressure	Bachmann *et al.* (1976)
H_3BPF_3		B_2H_6, PF_3	CO_2 $P_1(12)$–$P_1(34)$	10.51–10.74	0.3 W/cm^2 CW		Lory *et al.* (1975)
CH_2D_2	Cl	CHD_2Cl	CO_2 $R_2(22)$	9.26	CW	$\beta[D/H] \simeq 1.72$ at 190°K	Hsu and Manuccia (1978)
CH_3Br	Cl_2	CH_2BrCl	CO_2 $P_1(10)$, $P_1(14)$	10.5	160 W/cm^2 CW	$\beta[^{79}Br/^{81}Br] \simeq 1.04$	Manuccia *et al.* (1978)
CH_3Br	NH_3	None	CO_2 $P_1(24)$	10.63	8 J/cm^2	No reaction observed	Hwang *et al.* (1979)
CH_3F	Cl_2	CH_2FCl, CH_3Cl	CO_2		Pulsed		Earl and Ronn (1976)
CH_3F	Br	CH_2FBr	CO_2 $P_2(20)$, $P_2(32)$	9.55, 9.65	14–18 W CW	^{12}C, ^{13}C enrichment	Molin *et al.* (1978)
CH_3F	NH_3	None	CO_2 $P_2(22)$	9.57	8 J/cm^2	No reaction with ammonia	Hwang *et al.* (1979)
CH_3F	BCl_3	BCl_2F, CH_3Cl, C_2H_2	CO_2 $P_2(22)$	9.57	8 J/cm^2	Surface reaction on window	Hwang *et al.* (1979)
CH_2F_2	Cl_2	CH_2FCl, CF_2Cl_2, $C_2F_4Cl_2$	CO_2 $R_2(18)$	9.28			Lin and Ronn (1977)
CHF_2Cl		CF_2	CO_2 $R_2(40)$	9.17		CF_2 detected by LIF (laser-induced fluorescence)	Duperrex and van den Bergh (1979)
CHF_2Cl		Dissociation	CO_2 $R_2(40)$	9.17	1–3 J/cm^2	Magnetic field increases yield	Duperrex and van den Bergh (1980)
CHF_2Cl		$CF_2 + HCl$	CO_2			IR energy input measured	Braun *et al.* (1978)
CHF_2Cl		$CF_2 + HCl$	CO_2 $R_2(32)$	9.21	6–40 MW/cm^2		King and Stephenson (1979)
CHF_2Cl		$CF_2 + HCl$	CO_2 $R_2(34)$	9.20		60 Torr pressure	Slater and Parks (1979)
CHF_2Cl		$CF_2 + HCl$	CO_2 $R_2(32)$	9.21	10–1000 MW/cm^2	Yield increases w. added Ar	Stephenson *et al.* (1979)
CHF_2Cl		$CF_2 + HCl$	CO_2 $R_2(26)$	9.24	20–30 J/cm^2	Molecular beam	Sudbø *et al.* (1978a, b)
$CHCl_2F$		$CFCl + HCl$	CO_2 $P_2(10)$	9.47		Molecular beam	Sudbø *et al.* (1978a, b)
CDF_3		$CF_2 + DF$	CO_2 $R_1(26) + R_1(28)$	10.2		Yield increases w. added Ar	Herman and Marling (1980a,b)
CDF_3		$CF_2 + DF$	CO_2 $R_1(14)$	10.29		$\beta[D/H] \approx 5000$	Tuccio and Hartford (1979)
CDF_3		C_2F_4	CO_2 $R_1(6)$–$R_1(36)$	10.14–10.35	30 J/cm^2	$\beta[D/H] \gtrsim 20{,}000$	Herman and Marling (1980)
CF_2Cl_2		C_2^* luminescence	CO_2				Ambartzumyan *et al.* (1975b)
CF_2Cl_2		Dissociated	CO_2 $P_1(20)$	10.6		Threshold ≈ 5 J/cm^2	Gower and Billman (1977)
CF_2Cl_2		COF_2	CO_2				Fettweis and Nève de Mévergnies (1979)

continued overleaf

Table 4.1A. (continued)

Reactant	Added reagents	Products	Laser: Type	Laser: λ (μm)	Laser: Power/energy	Conditions, remarks	References
CF_2Cl_2		$CF_2Cl + Cl$	CO_2 $R_2(26)$	9.24			Hudgens (1978)
CF_2Cl_2		Cl_2, C_2F_4	CO_2 $P_1(18)$–$P_1(34)$	10.6, also 9.6		^{35}Cl, ^{37}Cl enrichment	Huie *et al.* (1978)
CF_2Cl_2, CF_2Br_2	$(CH_3)_2C{=}CH_2$ $CH_3C{=}CH_2$ $CH_2{=}CH_2$	$(CH_3)_2\overline{CCH_2CF_2}$, C_2F_4, $CF_2{=}CH_2$	CO_2 $P_1(36)$, $R_2(18)$–$R_2(26)$	10.76, 9.28–9.24	0.2–0.3 J	^{13}C, ^{12}C enrichment	Ritter (1978)
CF_2Cl_2		$CFCl_3$	CO_2	10–40 W CW			Slezak *et al.* (1978)
CF_2Cl_2		$CF_2Cl + Cl$	CO_2 $R_2(38)$	9.18	3–10 J/cm^2	Molecular beam	Sudbø *et al.* (1978a, 1979)
CF_2Br_2		$CF_2Br + Br$	CO_2 $R_2(36)$	9.19	3–10 J/cm^2	Molecular beam	Sudbø *et al.* (1978a, 1979)
CF_2Cl_2		$CF_2 + 2Cl$	CO_2 $P_1(30)$–$P_1(40)$, $P_2(10)$–$P_2(34)$, $R_2(10)$–$R_2(34)$	10.7–10.8, 9.2–9.7			King and Stephenson (1978)
CF_3Cl, CF_3Br, CF_3I		C_2F_6	CO_2 $R_2(4)$–$R_2(32)$, $P_2(8)$–$P_2(26)$, $P_1(16)$–$P_1(44)$, $R_1(30)$–$R_1(38)$	9.2–9.6, 10.5–10.8, 10.1		^{13}C enrichment	Drouin *et al.* (1979)
CF_3Cl, CF_3Br, CF_3I		CF_3 + halogen	CO_2 $R_2(12)$, $R_2(20)$, $R_2(40)$	9.3, 9.17		Molecular beam	Sudbø *et al.* (1978a, 1979)
CF_3Br		C_2F_6, CF_4, CF_2Br_2, Br_2	CO_2 $R_2(12)$–$R_2(34)$	9.25–9.32	0.2 J/pulse	2–100 Torr pressure	Jalenak and Nogar (1979)
CF_3Br	HI	CF_3, CF_2, Br, F; HBr, HF	CO_2 $R_2(26)$, $R_2(28)$	9.23	2 J/cm^2		Würzberg *et al.* (1978)
CF_3I		C_2F_6, I_2	CO_2 $R_2(14)$	9.30	0.5–1.5 J/cm^2	$\beta[^{13}C/^{12}C] \approx 600$	Bittenson and Houston (1977)
CF_3I		C_2F_6, I_2	CO_2 $R_2(14)$ + XeCl, XeF	9.30 + 0.31, 0.35		^{12}C enrichment	Knyazev *et al.* (1978)
CF_3I		C_2F_6, I_2	CO_2 $R_2(14)$	9.30	0.3–3 J/cm^2		Bagratashvili *et al.* (1979)
CF_3I	Br_2, NOCl, O_3, NO_2	CF_3Br, etc.	CO_2 $R_2(16)$	9.29	5 J/pulse	Measured CF_3 rate constants	Rossi *et al.* (1979a, b)
$CFCl_3$		$CFCl{=}CFCl$, $CCl_2{=}CF_2$	CO_2 $R_2(20)$	9.27	0.1–0.4 J/cm^2	Pressure $\approx$ 60 Torr	Dever and Grunwald (1976)
$CFCl_3$		$CFCl_2 + Cl$	CO_2 $R_2(20)$	9.27			Hudgens (1978)
$CFCl_3$		$CFCl_2 + Cl$	CO_2 $R_2(14)$, $R_2(20)$	9.27, 9.30	5–10 J/cm^2	Molecular beam	Sudbø *et al.* (1978a, 1979)
CCl_4		Dissociation	NH_3 (pumped by CO_2 $R_2(16)$)	12.8	~ 35 mJ/pulse	^{13}C enrichment	Ambartzumyan *et al.* (1978a, b)
$H_2CO(HDCO)$		$HD + CO$	CO_2 $P_1(20)$	10.6	4 J/pulse	$\beta[D/H] \approx 40$	Koren *et al.* (1976)
H_2CO		$H_2 + CO$	DF $P_1(5)$–$P_2(6)$	3.58–3.73	20–200 mJ/pulse		Evans *et al.* (1979)
HCOOH		$CO + H_2O$	HF $P_1(8)$	2.78	70–170 mJ/pulse		Corkum *et al.* (1977)
HCOOH		$CO + H_2$	HF $P_2(5)$	2.79			Evans *et al.* (1978)
HCOOH		$CO_2 + H_2/CO + H_2O$	CO_2 $R_2(20)$	9.6	10 W CW	At Pt catalytic surface	Umstead and Lin (1979)
CH_3NH_2		$CH_3 + NH_2$	CO_2			NH_2 detected by LIF	Ashfold *et al.* (1979)
CH_3NH_2		NH_2	CO_2 $P_1(10)$–$P_1(30)$	10.5–10.7		NH_2 detected by LIF	Filseth *et al.* (1979a)

CH_3NO_2		CH_2O, NO, N_2O, CH_3OH	CO_2	9.4–10.9			Hartford, Jr. (1979)
CH_3NO_2		Ionization current	CO_2	9.2–10.75			Avouris *et al.* (1979a)
CH_3NO_2	H_2	HCN, H_2O; CN*	CO_2 $R(22)$	9.26		Above 20 Torr pressure	Nève de Mévergnies and Fettweis (1979)
CH_3OH		H_2, CO, CH_4, C_2H_6, C_2H_4, C_2H_2, H_2O	CO_2 $P_2(34)$ *or* HF	9.67 or 2.7			Bhatnagar *et al.* (1979)
CH_3OH		CH, OH	CO_2 $P_2(20)$	9.55			Bialkowski and Guillory (1977, 1978)
CH_3OH		H_2, CO, CH_4	HF $P_1(6)$, DF $P_2(4)$–$P_2(8)$	2.7 3.7		$\beta[H/D] \approx 60$ with DF excitation	McAlpine *et al.* (1979)
CH_3OH	Br	Not collected	HF $P_1(5)$–$P_1(7)$	2.7		Apparent D enrichment	Mayer *et al.* (1970)
CH_3OH, CH_3NO_2, CH_3CN		C_2*, CH*, OH*, CN* luminescence	CO_2		Pulsed		Ambartzumyan *et al.* (1975b)
CH_3NC		CH_3CN	CO_2 $P_1(22)$	10.61	$\gtrsim 20$ mJ/pulse	Induced thermal explosion	Bethune *et al.* (1978)
CH_3NC		CH_3CN	CO_2 $P_1(34)$	10.74	0.5 J/pulse	$\beta[^{15}N/^{14}N] \approx 1.34$	Hartford and Tuccio (1979)
CH_3NC		CH, CN, C_2; C_2H_2, HCN, CH_3CN	CO_2 $P_1(30)$	10.69			Hicks *et al.* (1979)
CH_3CN		CN, CH, C_2	CO_2 $P_2(20)$, $P_1(32)$	9.55 10.72	0.22 J/pulse 0.18 J/pulse		Lesiecki and Guillory (1977, 1978)
C_2H_4	O	No reaction	CO_2		3–6 W CW	No increase over thermal rate	Manning *et al.* (1976)
C_2H_4		C_2*$(a^3\Pi_u)$	CO_2		Pulsed	C_2 detected by LIF	Chekalin *et al.* (1977)
C_2H_4		C_2*	CO_2 $P_1(16)$	10.55	to 150 J/cm^2		Hall *et al.* (1978)
C_2H_4		C_2*, CH*	CO_2		Pulsed		Ambartzumyan *et al.* (1975b)
C_2H_4		C_2*, CH_2*	CO_2	9.2–10.6	10–100 J/cm^2	Luminescence observed	Filseth *et al.* (1979c)
$(C_2H_4)_2$		C_2H_4	CO_2	9–11	Pulsed	Pulsed molecular beam	Hoffbauer *et al.* (1979)
$CH_2{=}CHCl$		C_2H_2 + HCl	CO_2	9.2–10.8	0.1–1.0 J/pulse		Lussier and Steinfeld (1977)
$CH_2{=}CHCl$		C_2H_2 + HCl	CO_2	9.2–10.8	0.1–1.0 J/pulse		Reiser *et al.* (1980)
$CH_2{=}CCl_2$		$ClC{\equiv}CH$ + HCl					
$CHCl{=}CHCl$		*Trans* → *cis* isomerization	CO_2 $P_1(30)$	10.69			Karny and Zare (1977)
$CH_2{=}CHCl$		C_2HCl_2 + Cl	CO_2 $P_1(36)$	10.76		Molecular beam	Sudbø *et al.* (1978a, b)
$CF_2{=}CHCl$		C_2F_2 + HCl	CO_2 $R_1(8)$	10.33	5–10 J/cm^2	Molecular beam	Sudbø *et al.* (1978a, b)
$CF_2{=}CHCl$	D_2S	C_2F_2HD, $C_2F_2D_2$	CO_2 $R_1(18)$	10.26	Pulsed		Reiser and Steinfeld (1980)
$CH_2{=}CF_2$		$FC{\equiv}CH$ + HF	CO_2 $P_1(20)$	10.59	40 J/cm^2	HF IR luminescence observed	Quick and Wittig (1978a, b); Levy *et al.* (1980)
$CH_2{=}CHF$		C_2H_2 + HF					
$CH_2{=}CF_2$		$FC{\equiv}CH$ + HF	CO_2 $P_1(20)$	10.59	30 J/cm^2	HF IR luminescence observed	Quick *et al.* (1979)
C_2F_3Cl		C_2F_4, $ClFC{=}CFCl$	CO_2 multiline	9.6	1 J/pulse		Nagai *et al.* (1979)
C_2F_3Cl		CF_2, CFCl	CO_2 $P_2(24)$	9.59	2–20 J/cm^2	Detected by LIF	Bialkowski *et al.* (1979); Stephenson *et al.* (1980)
CH_3CH_2F		C_2H_4 + HF	CO_2 $R(16)$	10.27	0.1–0.5 J/cm^2	Nonthermal below 2 Torr	Richardson and Setser (1977)
CH_3CF_3		$CH_2{=}CF_2$ + HF	CO_2 $P_2(18)$	9.54	0.1–0.5 J/cm^2	Nonthermal below 2 Torr	Richardson and Setser (1977)
CH_3CHF_2		$CH_2{=}CHF$ + HF	CO_2 $P_1(20)$	10.59	40 J/cm^2	HF IR luminescence obs.	Quick *et al.* (1978b)
CH_3CH_2F		C_2H_4 + HF	CO_2 $P_2(20)$	9.55	40 J/cm^2	HF IR luminescence obs.	Quick *et al.* (1978b)
CF_3CHF_2		C_2F_4, C_2F_6	CO_2 $R_2(20)$	9.27	0.4–2 J/pulse		Hackett *et al.* (1980)
CH_3CH_2F		C_2H_4 + HF	CO_2 $P_2(20)$	9.55	1–3 J/cm^2	Pressure increases yield	Jang and Setser (1979)

continued overleaf

Table 4.1A. (continued)

Reactant	Added reagents	Products	Laser: Type	Laser: λ (μm)	Laser: Power/energy	Conditions, remarks	References
CH_3CF_3		$CH_2{=}CF_2 + HF$	CO_2 $R_1(16)$	10.27	1–3 J/cm^2	Pressure increases yield	Jang and Setser (1979)
CH_3CF_3		$CH_2{=}CF_2 + HF$	CO_2 $R_1(16)$	10.27		HF IR luminescence obs.	West *et al.* (1978)
CH_3CF_2Cl		$CH_2{=}CFCl + HF$; $CH_2{=}CF_2 + HCl$	CO_2 $P_1(18)$	10.57		HF, HCl luminescence obs.	West *et al.* (1978)
C_2H_5Cl		$C_2H_4 + HCl$	CO_2 $P_1(18)$	10.57		HCl luminescence obs.	West *et al.* (1978)
C_2H_5Cl		$C_2H_4 + HCl$	OPO (optical parametric oscillator)	3.3–3.5; 1.64–1.75	2.7–3.5 mJ/pulse		Dai *et al.* (1979)
CH_3CF_2Cl		$CH_2{=}CF_2 + HCl$	CO_2 $P_1(12)$–$P_1(24)$ $R_1(4)$–$R_1(12)$	10.5–10.6, 10.35–10.30	5–25 W CW		Zitter and Koster (1978); Zitter *et al.* (1979)
CH_3CF_2Cl		$CH_2{=}CF_2 + HCl$	CO_2 $P_1(6)$	10.46		Molecular beam	Sudbø *et al.* (1978a, b, 1979)
CH_3CCl_3		$CH_2CCl_2 + HCl$	CO_2 $R_2(12)$	9.32		Molecular beam	Sudbø *et al.* (1978a, b, 1979)
CF_3CF_2Cl		$CF_3CF_2 + Cl$	CO_2 $R_1(24)$	10.22		Molecular beam	Sudbø *et al.* (1978a, b, 1979)
CF_3CHCl_2		$CF_2{=}CFH$	CO_2 $P_1(26)$, $P_1(28)$	10.65	10 J/cm^2	$\beta[D/H] \approx 1400$	Marling and Herman (1979); Marling *et al.* (1980)
C_2F_6	H_2, C_6H_{14}	CHF_3, C_2F_4	CO_2 $R_2(36)$	9.19	6 J/cm^2		Fisk (1978)
C_2H_5I			CO_2	10.53	20 W CW	Thermal explosion	Bellows and Fong (1975)
C_2H_5OH		CH_4, C_2H_4, C_2H_6	HF $P_1(7)$	2.74			Selwyn *et al.* (1978)
C_2H_5OH		$C_2H_4 + H_2O$	CO_2 $P_2(28)$	9.62	3 J/cm^2		Danen (1979)
$CH_2{=}C{=}O$		H_2, CO	CO_2 $R_2(22)$	9.26	0.2 J/pulse	Collisionally dominated	Jalenak *et al.* (1979)
$CH_2{=}C{=}O$		CH_2	CO_2 $R_2(20)$	9.27	5–50 J/cm^2	Detected by LIF	Grimley and Stephenson (1980)
$CH_2{=}C{=}O$		CH_2^*, C_2^*	CO_2	9.3, 10.5	1–100 J/cm^2	Luminescence observed	Filseth *et al.* (1979c)
$HC{\equiv}CCHO$		CO, C_2H_2	CO_2 $P_1(14)$–$P_1(20)$	10.5	2.5 J/pulse	also two-step UV + IR	Stafast *et al.* (1979)
C_2H_3CN		C_2	CO_2 $P_1(20)$	10.59			Filseth *et al.* (1979a)
C_2H_3CN	NO	$CN^*(A^2\Pi, B^2\Sigma^+)$	CO_2 $P_1(12)$	10.51	2.5 J/pulse	$C_2(a^3\Pi) + NO \rightarrow CN^* + CO$	Reisler *et al.* (1979)
C_2H_5OH, $C_2H_5NH_2$, C_2H_5CN		$C_2^*(a^3\Pi_u)$	CO_2 $R_2(18)$	9.28	0.4 J/pulse		Campbell *et al.* (1978); Levy *et al.* (1980)
C_2H_5CN		Luminescence	CO_2 $P_1(20)$	10.59			Yu *et al.* (1979)
C_2H_5NC		C_2H_5CN	CO_2 $R_1(34)$	10.16	0.4 J/pulse	^{15}N enrichment	Hartford and Tuccio (1979)
C_3H_4 (allene)		C_3^*	CO_2 $P_2(20)$	9.55	0.5 J/pulse	Luminescence observed	Lesiecki *et al.* (1980)
C_3H_6 (cyclopropane)		Propylene	CO_2 $P_2(20)$	9.55			Karny and Zare (1977)
(propylene)		CH_4, C_2H_2, C_2H_4, C_2H_6	CO_2 $P_1(34)$	10.74			Karny and Zare (1977)
Cyclopropane		Propylene	OPO	3.22	2 J/cm^2	Studied vs. Ar pressure	Hall and Kaldor (1979)
Cyclopropane		CH_4, C_2H_2, C_2H_6, C_3H_6	CO_2 $P_2(14)$	9.50	10–60 J/cm^2	Studied vs. Ar pressure	Hall and Kaldor (1979)
Cyclopropane		CH_4, C_2H_2, C_2H_4; $(C_2^+)^*$	CO_2 $P_2(20)$	9.55	> 1 GW/cm^2		Lesiecki and Guillory (1979)
$C_3F_6^+$		$C_2F_4^+$, CF_2	CO_2	9.2–10.2	30 W/cm^2 CW	In ICR (Ion Cyclotron Resonance)	Woodin *et al.* (1979)
C_3F_6	NO, O_2	C_2F_4, C_2F_6	CO_2 $P_2(4)$–$P_2(40)$	9.4–9.8	1.4 J/cm^2		Nip *et al.* (1980)
$C_2F_4S_2$		$F_2C{=}S$	CO_2 $R_2(16)$	9.3	0.12 J/cm^2		Plum and Houston (1980)
CD_3COCD_3		C_2D_6, CD_4	CO_2 $P_2(34)$	9.68	0.8 J/pulse		McNesby and Scanland (1979)

CF_3COCF_3		C_2F_6, CO	CO_2 $P_1(6)$–$P_1(18)$, $R_1(4)$–$R_1(18)$	10.46–10.57, 10.26–10.36	3–9 J/cm^2	^{18}O, ^{13}C enrichment	Hackett *et al.* (1979b)
CF_3COCF_3		C_2F_6, CO	CO_2 $R_1(12)$, $P_2(36)$	9.7, 10.3	0.01–10 J/cm^2		Hackett *et al.* (1978, 1979a, b)
CF_3COCF_3		C_2F_6	CO_2 $P_1(12)$–$P_1(24)$, $R_1(12)$–$R_1(28)$	10.3, 10.6	> 1 J/cm^2		Avatkov *et al.* (1979)
CF_3COCF_3		C_2F_6, CO	CO_2 $P_1(18)$–$P_1(30)$, $R_1(6)$–$R_1(28)$	10.3, 10.6	~ 5 J/cm^2		Fuss *et al.* (1979)
CF_3OOCF_3		F_2CO, CF_3OF, C_2F_6	CO_2 $P_2(30)$–$R_2(40)$	9.1–9.6	0.01–0.5 J/pulse		Francisco *et al.* (1980)
C_4H_8 (2-butene)		CH_4, C_2H_4, C_3H_6	CO_2 $P_1(20)$	10.59	1 J/pulse		Yogev and Loewenstein-Benmair (1973)
i-C_4H_9Cl, Br, I		Numerous fragments	CO_2 $P_1(36)$	10.76	0.3 J/pulse		Braun and Tsang (1976)
$(CH_3)_2C{=}CHCl$		$CH_3C{\equiv}CCH_3$, $CH_2{=}CHCH{=}CH_2$	CO_2 $P_2(16)$, $P_2(18)$	9.5	Pulsed		Reiser and Steinfeld (1980)
cyclo-C_4F_6		$CF_2{=}CFCF{=}CF_2$	CO_2 P_1, R_1	10.13–10.50	0.15–0.40 J/pulse		Yogev and Benmair (1977, 1979)
cyclo-C_4F_8		C_2F_4	CO_2 $P_1(14)$	10.55	0.75 J/pulse		Preses *et al.* (1977)
$(CH_3CO)_2O$		CH_2	CO_2 $R_1(20)$	10.25	4.5 J/pulse	CH_2 detected by LIF	Ashfold *et al.* (1979b)
$(CH_3CO)_2O$		CH_2*	CO_2	9.3–10.2	1–30 J/cm^2	Luminescence observed	Filseth *et al.* (1979c)
$CH_3COOCH_2CH_3$		CH_3COOH, C_2H_4	CO_2	9.57	2–4 J/cm^2		Knott and Pryor (1979)
$CH_3COOCH_2CH_3$		CH_3COOH, C_2H_4	CO_2 $P_2(16)$–$P_2(32)$	9.6	0.7–0.8 J/cm^2	Nonthermal below 3 Torr	Danen *et al.* (1977)
$CH_3COOCH_2CH_3$		CH_3COOH, C_2H_4	CO_2		Pulsed	IR energy input measured	Braun *et al.* (1978)
$C_2H_5OCH{=}CH_2$		CH_3CHO, C_2H_4; $CH_2{=}C{=}O$, C_3H_7CHO, C_4H_{10}	CO_2 $P_2(26)$	9.6		Studied vs. IR peak power	Brenner (1978)
C_5H_8, ▷═/		Various C_5H_8 isomers	CO_2 $P_1(24)$	10.63	5.5 J/cm^2		Farneth *et al.* (1979)
C_6H_{10}		*cis* → *trans* isomerization	CO_2 $R_1(18)$, $P_1(38)$	10.79, 10.26	0.9–1.6 J/pulse		Buechele *et al.* (1979)
$C_6H_{12}O_2$ (tetramethyl-dioxetane)		$(CH_3)_2CO$; luminescence	CO_2 $R_1(24)$, $R_1(26)$	10.2	0.5 J/pulse		Haas and Yahav (1977); Yahav and Haas (1978); Yahav *et al.* (1980)
$C_4H_9OCH{=}CH_2$		C_2H_2, C_2H_4, CH_3CHO, C_4H_8, CH_2CO, C_4H_{10}	CO_2 $P_1(20)$, $P_2(20)$	9.55, 10.59	3 J/pulse		Hofmann *et al.* (1980)
$((C_2H_5)_2O)_2H^+$		$(C_2H_5)_2OH^+$, $(C_2H_5)_2O$	CO_2 $P_1(20)$	10.59	1–10 W/cm^2 CW	In ICR	Woodin *et al.* (1978)
$C_{10}H_{12}$(tetralin)		C_8H_8, $\phi CH{=}CH_2$, etc.	CO_2 $P_1(18)$	10.57	0.8 J/pulse		Berman *et al.* (1980)
$C_{10}H_{16}$ (limonene)		C_2H_2, C_4H_4, C_4H_6, C_6H_6, etc.	CO_2 $P_1(20)$	10.59	5 W CW	Apparently thermal	Yogev *et al.* (1972)
$Fe(CO)_4$	CO, Xe, CH_4	$Fe(CO)_5$, etc.	CO	5.26	CW	In CO, etc., matrix	McNeish *et al.* (1976); Poliakoff *et al.* (1978)
$Fe(CO)_4$	CO, etc.	$Fe(CO)_5$	Spin–flip Raman	5.32	CW	In CO matrix	Poliakoff *et al.* (1979)
HBr	NO	Br_2	HBr P_3–P_1 + doubled dye	4.1–4.3	13 mJ/pulse	$\beta[^{81}Br/^{79}Br] \sim 1.5$	Zittel and Little (1979)
HCl	O	OH, Cl	HCl($1 \rightarrow 0$)	3.6	Pulsed		Kneba and Wolfrum (1978)
HCl	K	KCl	HCl($1 \rightarrow 0$)	3.6	Pulsed	$[\sigma(v = 1)/\sigma(v = 0)] \approx 100$	Odiorne *et al.* (1971)
HCl	K	KCl	HCl($1 \rightarrow 0$; J)	3.6	Pulsed	Rotational state dependence studied	Dispert *et al.* (1979)

continued overleaf

Table 4.1A. (continued)

Reactant	Added reagents	Products	Laser: Type	Laser: λ (μm)	Laser: Power/energy	Conditions, remarks	References
MoF_6		Dissociation	CO_2 $P_1(10)$–$P_1(22)$	10.5	Pulsed	^{92}Mo–^{100}Mo enrichment	Freund and Lyman (1978)
NH_3			CO_2 $P_1(16)$ + u.v.	10.55	Pulsed	Apparent ^{15}N enrichment	Ambartzumyan *et al.* (1972, 1973)
NH_3		NH_2	CO_2 $P_1(22)$, $P_1(32)$	10.6	3 J/pulse		Campbell *et al.* (1976)
NH_3	NO	HNO, NH_2	CO_2 $P_1(32)$	10.7	0.1–0.4 J/pulse	Yield increases with pressure	Avouris *et al.* (1979b)
HN_3, DN_3		ND_2*, N_2	CO_2 $P_1(18)$	10.57			Hartford (1978)
N_2F_4		NF_2	CO_2		Pulsed		Lyman and Jensen (1972)
N_2F_4		NF_2	CO_2 $R_1(20)$	10.25	1–10 J/cm^2	Molecular beam	Sudbø *et al.* (1979)
N_2H_4		NH_2	CO_2 P_1	10.6			Filseth *et al.* (1979b)
NO	O_3	NO_2*	CO $P_9(13)$	5.31	3 W CW		Stephenson and Freund (1976)
$(N_2O)_2$		N_2O	Diode	4.48		In supersonic beam	Gough *et al.* (1978)
NO_2		NO	Nd:YAG SHG + CO_2 $P_1(14)$, $P_2(20)$	0.532 + 9.55, 10.55	5 mJ/pulse 1 J/pulse	NO observed by LIF	Feldmann *et al.* (1979)
O_3		Dissociation	CO_2 $P_2(30)$	9.64	> 50 J/cm^2	First reported dissoc. of triatomic	Proch and Schröder (1979)
O_3	NO	NO_2*	CO_2 $P_2(12)$	9.49	Multipulse, 200 W	$[k(\text{exc.})/k(0)] \approx 20$	Gordon and Lin (1973)
O_3	NO	NO_2*	CO_2		CW		Braun *et al.* (1974)
O_3	NO	NO_2*	CO_2 $P_2(30)$	9.64	Pulsed, 5 mJ/cm^2		Freund and Stephenson (1976)
O_3	NO	NO_2*	CO_2		40 mJ/cm^2	Temp. dependence studied	Bar-Ziv *et al.* (1978)
O_3	SO	SO_2*	CO_2		CW	$[k(\text{exc.})/k(0)] \approx 2.5$	Kaldor *et al.* (1974)
OCS		Dissociation	CO_2 $P_2(24)$	9.59	> 85 J/cm^2	First reported dissoc. of triatomic	Proch and Schröder (1979)
OCS	O	No reaction	CO_2		3–6 W CW	No increase over thermal rate	Manning *et al.* (1976)
OsO_4		Dissociation, luminescence	CO_2 P_1	10.4–10.7	65 MW/cm^2		Ambartzumyan *et al.* (1977)

OsO_4		Dissociation, luminescence	CO_2			"Two-color" IR	Ambartzumyan *et al.* (1978c)
SF_6		Dissociation	CO_2 $P_1(16)$, $P_1(40)$	10.55, 10.8		First reported IR multiphoton dissoc.; ^{32}S, ^{34}S enrichment	Ambartzumyan *et al.* (1975a)
SF_6		Dissociation	CO_2 $P_1(16)$–$P_1(32)$	10.6		^{34}S enrichment	Lyman *et al.* (1975)
SF_6		SOF_2	CO_2	9.6 + 10.6		"Two-color" IR	Ambartzumyan *et al.* (1976a)
SF_6		Dissociation	CO_2			"Two-color" IR	Gower and Gustafson (1977)
SF_6		Dissociation	CO_2 $P_1(20)$	10.59		Threshold ≈ 5 J/cm^2	Gower and Billman (1977)
SF_6		Dissociation	CO_2 $P_1(16)$	10.55	10 J/cm^2	0.25 g SF_6 (50% ^{34}S) produced per hour with 150 Hz laser	Bagratashvili *et al.* (1977)
SF_6	H_2	Dissociation	CO_2 $P_1(20)$	10.59	2 J/pulse	Collisionally dominated	Lin *et al.* (1978)
SF_5Cl		SF_5 + Cl	CO_2 $P_1(58?)$	11.0	6 J/cm^2		Karl and Lyman (1978); Leary *et al.* (1978)
SF_5NF_2		SF_4, S_2F_{10}, N_2F_4, NF_3	CO_2 $P_1(4)$–$P_1(56)$	10.4–10.9	1.2 J/pulse		Lyman *et al.* (1979)
S_2F_{10}		SF_5	CO_2 $P_1(20)$–$P_1(44)$	10.6–10.8	0.01–0.6 J/cm^2		Lyman and Leary (1978)
$(SF_6)_n$		SF_6	CO_2		10–15 W CW	In supersonic jet	Kim *et al.* (1979)
SeF_6		Dissociation	NH_3 [pumped by CO_2 $R_1(16)$]	12.8	0.75 J/pulse	Se isotope enrichment	Tiee and Wittig (1978a, b)
SiF_4		Luminescence	CO_2	10.6			Merchant (1978)
SiH_4		SiH_2, H_2; luminescence	CO_2 $P_1(20)$	10.6	0.03–4.0 J/cm^2		Deutsch (1979)
UF_6	SiH_4	SiH_3F, UF_5, UF_4	Diode laser	~ 16 μm	25 mW/cm^2	In SiH_4 matrix	Catalano *et al.* (1979)
UF_6	H_2	HF, $(UF_5)_n$ solid	CF_4 [pumped by CO_2 $R_2(12)$]	~ 16 μm			Rabinowitz *et al.* (1978)
UF_6	H_2	HF	CF_4 [pumped by CO_2 $R_2(18)$] + CO_2	(9.28 μm)		"Two-color" IR	Tiee and Wittig (1978c)
UO_2—$(hfacac)_2$—THF		$UO_2(hfacac)_2$, THF	CO_2 $P_1(8)$, $R_2(8)$	9.34, 10.47		^{18}O enrichment	Cox *et al.* (1979)
UO_2—$(hfacac)_2$—THF		$UO_2(hfacac)_2$, HF	CO_2 $P_1(6)$–$P_1(26)$, $R_1(4)$–$R_1(12)$	10.6, 10.3	0.001–0.3 J/cm^2		Cox and Horsley (1980)

Table 4.1B. *Thermal and Photosensitized Infrared Laser Induced Reactions*

Reactant	Sensitizer	Added reagents	Products	Laser Type	Laser λ, μm	Laser Power/energy	Conditions, remarks	References
BCl_3			Dissociation, luminescence	CO_2		100 W CW	High pressure	Karlov *et al.* (1970)
BCl_3		H_2		CO_2		600 W CW	Thermal explosion	Karlov *et al.* (1971)
B_2H_6	SF_6	N_2O		CO_2			Thermal explosion	Bauer *et al.* (1978)
B_2H_6	SF_6		B_5H_9, B_5H_{11}, $B_{10}H_{14}$	CO_2 P_1	10.5–10.7	CW		Riley and Shatas (1979)
CF_2Cl_2	(SF_6)		CF_3Cl	CO_2 $P_1(20)$	10.59	300 W CW		Freeman *et al.* (1974)
$COCl_2$	BCl_3		Decomposition	CO_2		100 W CW		Merritt and Robertson (1977)
$COCl_2$	BCl_3		Decomposition	CO_2		1–5 W CW		Bachmann *et al.* (1979)
C_2Cl_4	BCl_3		C_6Cl_6	CO_2 $P_1(24)$	10.6	6 W CW	200 Torr	Bachmann *et al.* (1977b)
CH_4	SF_6	O_2	Combustion	CO_2		10 J/pulse	75–200 Torr	Hill and Laguna (1980)
C_2H_6	SF_6		CH_4, C_2H_4	CO_2	10.6	350 W CW		Tardieu de Maleissye *et al.* (1976a, b);
C_2H_4	SF_6		Decomposition	CO_2	10.6	660 W CW		
C_3H_8	SF_6		C_2H_4, CH_4	CO_2	10.6	350 W CW		Pazendeh *et al.* (1979)
C_2H_4, C_3H_6, C_3H_4, etc.			H_2, C_2H_2, CH_4, etc.; luminescence	CO_2	10.6	100 W CW	120–200 Torr	Cohen *et al.* (1967)
C_4H_{10} (2-butene)	SF_6		*cis* → *trans*	CO_2 $P_1(28)$	10.67	10 W/cm^2 CW	Also decomposed CCl_2F_2, CHF_3, cyclo-C_4F_8, C_2H_6, C_2H_4, C_2H_2, $Fe(CO)_5$, B_2H_6	Shaub and Bauer (1975)
C_3H_6 (cyclopropane)	SF_6		Propylene	CO_2 $P_1(28)$	10.67	10 W/cm^2 CW		Shaub and Bauer (1975)

$CHCl{=}CHCl$	SF_6		*cis* → *trans*	CO_2 $P_1(28)$	10.67	10 W/cm^2 CW		Shaub and Bauer (1975)
C_3H_7Cl	SF_6		$C_3H_6 + HCl$	CO_2 $P_1(28)$	10.67	10 W/cm^2 CW		Shaub and Bauer (1975)
cyclo-C_6H_8			$C_6H_6 + H_2$	CO_2 $P_1(28)$	10.67	10 W/cm^2 CW		Shaub and Bauer (1975)
C_4H_6O (cyclobutanone)	NH_3		$C_3H_6 + CO$; $C_2H_4 + CH_2{=}C{=}O$	CO_2 $R_2(16)$	9.29			Steel *et al.* (1979)
$C_7H_7NO_2$ (*o*-nitrotoluene)	SF_6	C_6H_6	$(C_6H_5)_2$, $C_6H_5C_6H_4CH_3$	CO_2 $P_1(18)$	10.57	0.5 J/pulse		Lewis *et al.* (1980)
C_7H_8	SiF_4		$C_5H_6 + C_2H_2$	CO_2 $P_2(42)$	9.75	0.3 J/pulse	Retro-Diels–Alder reaction	Garcia and Keehn (1978)
C_6H_8	SiF_4		$C_4H_4 + C_2H_4$	CO_2 $P_2(42)$	9.75			
$C_{10}H_{12}$ (tetralin)	SiF_4		$C_{10}H_8$, $\phi CH{=}CH_2$, etc.	CO_2 $P_2(40)$	9.73			Berman *et al.* (1980)
$C_{10}H_{12}$ (dicyclopentadiene)			C_5H_6, C_2H_4, C_2H_2, H_2	CO_2 $P_1(20)$	10.59	0.3 J/pulse	Thermal and/or breakdown	Hoffmann and Ahrens-Botzong (1980)
H_2	HF	D	HD	HF	2.8		$k(1)/k(0) \sim 10^4$	Kneba *et al.* (1979)
NH_3			Luminescence	CO_2		CW	shown to be thermal by Bauer and Buchwald	Bordé *et al.* (1966a, b)
NH_3		O_2	N_2, H_2O, N_2O	CO_2 $P_1(32)$	10.72		Single-pulse explosion	Lin and Bertran (1978)
$Pb(CH_3)_4$	SF_6	N_2O		CO_2			Single-pulse explosion	Bauer *et al.* (1978)
SF_6		H_2	S(solid), HF	CO_2		1 J/pulse	Laser-initiated breakdown	Ronn (1976)
SF_6			SF_4, SOF_4, SO_2F_2, S(solid)	CO_2 $P_1(20)$	10.59	1–2 J/pulse	20–160 Torr; thermal breakdown	Lin and Ronn (1978)
SiH_4	SF_6		S_2^*, etc.	CO_2		4 J/pulse	Explosive chain reaction	Bauer and Habermann (1978)
$Sn(CH_3)_4$	SF_6	N_2O		CO_2			Single-pulse explosion	Bauer *et al.* (1978)
UF_6		H_2	$(UF_5)_n$(solid), HF	CO_2 $P_1(20)$	10.59	1–2 J/pulse	Nonresonant breakdown	Ronn and Earl (1977)

Table 4.2. Reactions Induced by Visible and Ultraviolet Laser Excitation

Reactant	Added reagents	Products	Laser Type	Laser λ, nm	Laser Power/energy	Conditions, remarks	References
$Al(CH_3)_3$		Al (solid)	Ar^+ (frequency doubled)	257		Deposited Al film	Deutsch *et al.* (1979)
AgCl (solid)		Ag (solid, colloidal)	Ruby	694.3		Two-photon	Rousseau *et al.* (1965)
AsH_3		Decomposed	ArF	193		Removed from SiH_4	Clark and Anderson (1978)
B_2H_6		Decomposed	ArF	193		Removed from SiH_4	Clark and Anderson (1978)
Br_2	HI	HBr	Nd:YAG (doubled)	558, 532		^{79}Br, ^{81}Br selectivity	Leone and Moore (1974)
Br_2	C_4F_8	$C_4F_8Br_2$	Ruby	693.4–694.3			Tiffany *et al.* (1967); Tiffany (1968)
$Cd(CH_3)_2$		Cd	Ar^+ (frequency doubled)	257		Deposited Cd film	Deutsch *et al.* (1979)
Cl_2	C_2Cl_4	C_2Cl_6	Ar^+	488		^{37}Cl enrichment	Suzuki *et al.* (1978)
CO		C^*, C_2^*	ArF	193			Bokor *et al.* (1980)
Cs	D_2	CsD (solid)	Dye	601.05		Laser "snow"	Tam *et al.* (1977)
H_2CO		H_2, CO	Ruby (doubled)	347.2	8 MW/pulse	$\beta[D/H] \approx 6$; first published laser isotope separation	Yeung and Moore (1972)
H_2CO		H_2, CO	He–Cd	325		D enrichment	Marling (1975)
H_2CO		H_2, CO	Ne^+	337.8, 332.4, 325.0		D, ^{13}C, ^{17}O, ^{18}O enrichment	Marling (1977)
H_2CO		H_2, CO	Dye	303.2		^{12}C enrichment	Clark *et al.* (1975)
H_2CO		H, HCO	Dye (doubled)	294		HCO obs. by intracavity absorption	Clark *et al.* (1978)
H_2CO		CO, CO_2	He–Cd	325		In liq. Xe soln; D enrichment	Freund *et al.* (1978)
H_2CO		H_2, CO	Dye	345.61		$\beta[D/H] \sim 254$	Mannik *et al.* (1979)
CH_3Br		CH^*	ArF	193		Multiphoton	Baronavski and McDonald (1978)
CHI_3		CHI_2, I	Nd:glass (tripled)	353	2 mJ/pulse	Solution-phase	van den Ende *et al.* (1973)
CCl_4, $CFCl_3$		CCl^*, CCl_2^*, Cl_2^*	ArF	193	20 mJ/pulse	Multiphoton	Tiee *et al.* (1980)
CF_2Br_2		CF_2^*	KrF, ArF	249, 193			Wampler *et al.* (1979)
CBr_4		Br_2^*	KrF	249		Multiphoton	Sam and Yardley (1979)
CF_2Br_2		CF_2^*	KrF	249		Multiphoton	Sam and Yardley (1979)
CH_3NC		CH_3CN	Dye		(Intracavity)	C–H overtone excitation	Reddy and Berry (1979a)
C_2H_2	C_2H_3, C_2H_2, O_2, C_2H_6, CH_4	$C_2^*(a^3\Pi_u)$	ArF	193		Multiphoton	Donnelly and Pasternack (1979)
C_2H_2		C_2^*, CH^*	ArF	193		Multiphoton	McDonald *et al.* (1978)
C_2H_2	HBr, HCl	None	Dye			Overtone C–H exc.; negative result	Herman and Marling (1979a)
C_2N_2		CN	ArF	195		Two-photon	Jackson and Halpern (1979)
C_2N_2, ClCN, C_2H_2, C_2H_4, H_2O, CH_3OH		CN^*, OH^*, CH^*	ArF	193		Multiphoton	Jackson *et al.* (1978)
C_2F_5I		C_2F_5, I^*	Dye (doubled)	295			Comes and Pionteck (1976)
$C_2H_2N_4$ (s-tetrazine)		HCN, N_2	Dye	551.5		^{13}C, ^{15}N enrichment	Karl and Innes (1975)

$C_2H_2N_4$		HCN, N_2	Dye			In low-temperature matrix	Hochstrasser and King (1975, 1976)
$C_2H_2N_4$		HCN, N_2	Dye			Vapor phase; $\beta[^{13}C/^{12}C] > 1000$	Boesl *et al.* (1979)
$C_2H_2N_4$		HCN, N_2	Dye	492.3		Obs HCN ($v_3 = 1$) emission	Coulter *et al.* (1978)
$C_6H_4O_2$ (*p*-benzoquinone)	C_8H_8	$C_{14}H_{12}O_4$ adduct	Ar^+	514.5		Oxidative addition; CCl_4 soln.	Gardner *et al.* (1973)
$CH_2{=}CHCH_2NC$		$CH_2{=}CHCH_2CN$	Dye	524–746	(Intracavity)	C–H overtone excitation	Reddy and Berry (1979b)
C_7H_8(cycloheptatriene, CHT); 7-MeCHT, 7-Et CHT, (7, 7-diEt)CHT		Isomerization	Nd:YAG (quadrupled)	265	150 mJ/pulse		Hippler *et al.* (1978)
$C_{10}H_8{}^+$		$C_8H_6{}^+$, C_2H_2	Dye	< 700		In ICR (ion cyclotron resonance)	Kim and Dunbar (1980)
Free-base phthalocyanine		Dissociation	Ruby	694.3	Q switched	Multiphoton	Porter and Steinfeld (1966)
Styrene		Polymerization	Ruby	694.3		Free-radical chain initiation	Pao and Rentzepis (1965)
Thymine, adenine		Photobleaching	Nd:YAG (quadrupled)	266	30 mJ/pulse		Kryukov *et al.* (1979)
Eu^{3+}(aq)	SO_4^{2-}	$EuSO_4$	ArF	195		Aqueous solution	Donohue (1977)
Ce^{3+}(aq)	$IO_3{}^-$	$Ce(IO_4)_4$	KrF	249		Aqueous solution	Donohue (1977)
$Fe(CO)_5$		Fe*	KrF	249		Multiphoton	Karny *et al.* (1979a)
$Fe(C_2O_4)_3$		$Fe(C_2O_4)_2{}^{2-}$, CO_2	Ruby	694.3	20–70 MW/cm^2	Two-photon	Zipin and Speiser (1975)
HCl	C_2Cl_4, C_4H_6, C_6H_8	None	Dye			Overtone C–H excitation; negative result	Herman and Marling (1979a)
HD		None	Dye	603.5	15 W CW	Overtone H–D excitation; negative result	Herman (1980)
I_2	C_2H_2	$C_2H_2I_2$	Dye	604.0	100 MW	^{129}I enrichment	Kushawaha (1980)
I_2	1-Hexene	I_2 removal	Ar^+	514.5, 501.7		Claimed "ortho/para" separation	Balykin *et al.* (1976)
ICl	BrCH=CHBr	ClCH=CHCl	Dye	605.4		^{37}Cl enrichment	Datta *et al.* (1975)
ICl	H_2	HI, HCl	Dye	588–667			Harris (1977)
Li	H_2O	LiOH	Dye	670.8			Muller *et al.* (1978)
$Mn(CO)_5$		Mn*	ArF	193		Multiphoton	Karny *et al.* (1979a)
NO_2	CO	NO, CO_2	Ar^+	454.5–514.5	4 W		Herman *et al.* (1976, 1978)
			Dye	575–620	0.5 W		
NO_2	C_2H_4	CO_2	Ar^+	488	1 W	At Pt catalytic surface	Umstead *et al.* (1980)
PH_3		PH_2*, PH*	ArF	193		Multiphoton	Sam and Yardley (1978)
PH_3		Decomposed	ArF	193		Removed from SiH_4	Clark and Anderson (1978)
$Pb(C_2H_5)_4$		Pb*	ArF	193		Multiphoton	Karny *et al.* (1979a)
S (solid)	C_2H_4	$\overline{CH_2CH_2S}$	KrF	249			Betteridge and Yardley (1979)
S_2Cl_2	H_2S	HCl, HS	Dye (doubled)	300		Cl chain reaction initiated	Braithwaite and Leone (1978)
$S_2O_6F_2$		SO_3F*	Ar^+	488, 496		Also thermal initiation	Warren (1978)
UF_6		UF_5*	KrF	249		Multiphoton	Rice *et al.* (1980)
UO_2F_2	HF	UF_4	Dye	448, 455		In CH_3OH soln.; ^{18}O enrichment	Rofer-de Poorter and de Poorter (1979)
$Zn(CH_3)_2$		Zn*	KrF	249			Karny *et al.* (1979a)

Table 4.3. Miscellaneous Laser-Induced Effects (Surface Reactions, Ionization, Luminescence, etc.)

Reactant	Added reagents	Products or results	Laser			Conditions, remarks	References
			Type	λ, μm	Power/energy		
BCl_3 (matrix)		Reported ^{10}B, ^{11}B selective desorption	CO_2	10.6		Result shown to be spurious (Davies *et al.*, 1979)	Karlov *et al.* (1976)
BCl_3	H_2; Ti, Pb catalyst	$B_2H_2Cl_4$, B_2Cl_4	CO_2 $P_1(16)$	10.55			Lin and Atvars (1978)
CH_4	Rh catalyst	None	He–Ne	3.39	38 mW	No rate enhancement	Brass *et al.* (1979)
CH_4, C_4H_8, C_5H_{12}, C_3H_6, C_4H_6		C_2H_2	CO_2	10.6	10 MW/pulse	Acetylene produced under breakdown conditions	Yogev and Loewenstein–Benmair (1976)
CH_3OH, CH_3NO_2, $CH_2{=}CHCN$, C_2H_4, C_4H_6O, NH_3, $C_2H_5OCH{=}CH_2$		Ionization	CO_2	9.2–10.7	0.4 J/pulse		Avouris *et al.* (1980)
$CH_2{=}CHCHO$ (acrolein)		Luminescence	CO_2 $P_1(20)$	10.6	6 J/pulse		Blinov *et al.* (1978)
F_2CO		Luminescence	CO_2 $P_1(18)$	10.57	50 J/cm^2	Internal V → E	Hudgens *et al.* (1979)
$(CH_3CO)_2$		Luminescence	CO_2 $P_1(20)$	10.59		Internal V → E	Burak *et al.* (1979)
Gasoline	Air	Ignition	CO_2	10.6	1 J/pulse	Laser-ignited internal combustion engine	Dale *et al.* (1977)
Saccharine, coumarin, KCl, NaCl		Luminescence	Nd:glass	1.06	200 MW/cm^2	Induced luminescence in solids	Hardy *et al.* (1979)
CrO_2Cl_2		Luminescence	CO_2 $R_1(26)$–$R_1(30)$	10.2		Internal V → E	Karny *et al.* (1979b); Nieman and Ronn (1980)
NaN_3 (solid)		Decomposition	CO_2	10.6	5–10 W CW *or* 1 J/pulse	Presumably thermal	Chiu *et al.* (1979)
OsO_4		Luminescence	CO_2 $P_1(16)$ + $P_1(38)$	10.55 + 10.79		Internal V → E	Ambartzumyan *et al.* (1979)
SF_6 (matrix)		Reported ^{32}S, ^{34}S selective desorption	CO_2 $P_1(22)$, $P_1(24)$	10.6		Result shown to be spurious (Davies *et al.*, 1979)	Ambartzumyan *et al.* (1976b)
SF_6, N_2F_4, NF_3, OCS, O_3 (matrix)		None	CO_2			No dissociation in low-temperature matrix	Crocombe *et al.* (1978)
SF_6 (Ar matrix, 10°K)		None	CO_2 $P_1(24)$	10.6	0.12 J/cm^2	No dissociation in low-temperature matrix	Jones *et al.* (1979)
SF_6 (matrix)			CO_2			Surface ablation artifact	Davies *et al.* (1979)
SF_6	CH_4, C_2H_2, C_2H_4, C_2H_6	CH*, C_2* luminescence	CO_2 $P_1(20)$	10.6	0.8 MW/pulse		Orr and Keentok (1976)
SF_6	Hydrocarbon	Ionization	CO_2 $P_1(20)$	10.6	0.5 J/pulse		Crim *et al.* (1977)
SO_2		Luminescence	CO_2 $R_2(30)$	9.22			Bialkowski and Guillory (1979)
SiH_4		Si (deposited)	CO_2 $R_1(8)$	10.33	5–50 W CW	Vapor deposition of Si	Christensen and Lakin (1978)

4.2. References for Tables 4.1–4.3

Ambartzumyan, R. V., Letokhov, V. S., Makarov, G. N., and Puretzkii, A. A., 1972, *JETP Lett.* **15**:501.

Ambartzumyan, R. V., Letokhov, V. S., Makarov, G. N., and Puretzkii, A. A., 1973, *JETP Lett.* **17**:63.

Ambartzumyan, R. V., Gorokhov, Yu. A., Letokhov, V. S., and Makarov, G. N. 1975a, *JETP Lett.* **21**:375.

Ambartzumyan, R. V., Chekalin, N. V., Letokhov, V. S., and Ryabov, E. A., 1975b, *Chem. Phys. Lett.* **36**:301.

Ambartzumyan, R. V., Furzikov, N. P., Gorokhov, Yu. A., Letokhov, V. S., Makarov, G. N., and Puretzkii, A. A., 1976a, *Optics Commun.* **18**:517.

Ambartzumyan, R. V., Gorokhov, Yu. A., Makarov, G. N., Puretzkii, A. A., and Furzikov, N. P., 1976b, *JETP Lett.* **24**:256.

Ambartzumyan, R. V., Gorokhov, Yu. A., Makarov, G. N., Puretzkii, A. A., and Furzikov, N. P., 1977, *Chem. Phys. Lett.* **45**:231.

Ambartzumyan, R. V., Vasil'ev, B. I., Grazyuk, A. Z., Dyad'kin, A. P., Letokhov, V. S., and Furzikov, N. P., 1978a, *Sov. J. Quantum Electron.* **8**:1015.

Ambartzumyan, R. V., Furzikov, N. P., Letokhov, V. S., Dyad'kin, A. P., Grazyuk, A. Z., and Vasil'ev, B. I., 1978b, *Appl. Phys.* **15**:27.

Ambartzumyan, R. V., Letokhov, V. S., Makarov, G. N., and Puretzkii, A. A., 1978c, *Optics Commun.* **25**:69.

Ambartzumyan, R. V., Makarov, G. N., and Puretzkii, A. A., 1979, *JETP Lett.* **28**:647.

Ashfold, M. N. R., Hancock, G., and Ketley, G., 1979, *Disc. Faraday Soc.* **67**:204.

Ashfold, M. N. R., Hancock, G., Ketley, G., and Minshull-Beech, J. P., 1980, *J. Photochem.* **12**:75.

Avatkov, O. N., Aslanidi, E. B., Bakhtadzhe, A. B., Zainulin, R. I., and Turischchev, Yu. A., 1979, *Sov. J. Quantum Electron.* **6**:388.

Avouris, P., Chan, I. Y., and Loy, M. M. T., 1979a, *J. Chem. Phys.* **70**:5315.

Avouris, P., Loy, M. M. T., and Chan, I. Y., 1979b, *Chem. Phys. Lett.* **63**:624.

Avouris, P., Chan, I. Y., and Loy, M. M. T., 1980, *J. Chem. Phys.* **72**:3522.

Bachmann, H. R., Nöth, H., Rinck, R., and Kompa, K. L., 1974, *Chem. Phys. Lett.* **29**:627.

Bachmann, H. R., Nöth, H., Rinck, R., and Kompa, K. L., 1975, *Chem. Phys. Lett.* **33**:261.

Bachmann, H. R., Bachmann, F., Kompa, K. L., Nöth, H., and Rinck, R., 1976, *Chem. Ber.* **109**:3331.

Bachmann, F., Nöth, H., Rinck, R., Fuss, W., and Kompa, K. L., 1977a, *Ber. Bunsengesellschaft Phys. Chem.* **81**:313.

Bachmann, H. R., Rinck, R., Nöth, H., and Kompa, K. L., 1977b, *Chem. Phys. Lett.* **45**:169.

Bachmann, H. R., Nöth, H., Rinck, R., and Kompa, K. L., 1979, *J. Photochem.* **10**:433.

Bagratashvili, V. N., Kolomiiski, Yu. R., Letokhov, V. S. Ryabov, E. A., Baranov, V. Yu., Kazakov, S. A., Nizjev, V. G., Pismenny, V. D., Staradubtsev, A. I., and Velikhov, E. P., 1977, *Appl. Phys.* **14**:217.

Bagratashvili, V. N., Dolzhikov, V. S., Letokhov, V. S., and Ryabov, E. A., 1979, in *Laser-Induced Processes in Molecules* (eds. K. L. Kompa and S. D. Smith), Springer-Verlag, pp. 179–185, *Ser. Chem. Phys.* No. 6, Berlin, 1979.

Balykin, V. I., Letokhov, V. S., Mishin, V. I., and Semchishen, V. A., 1976, *Chem. Phys.* **17**:111.

Baronavski, A. P., and McDonald, J. R., *Chem. Phys. Lett.* **56**:369.

Bar-Ziv, E., Moy, J., and Gordon, R. J., 1978, *J. Chem. Phys.* **68**:1013.

Bauer, S. H., and Haberman, J. A., 1978, *I.E.E.E. J. Quantum Electronics* **QE-14**:233.

Bauer, S. H., Bar-Ziv, E., and Haberman, J. A., 1978, *I.E.E.E. J. Quantum Electronics* **QE-14**:237.

Bellows, J. C., and Fong, F. K., 1975, *J. Chem. Phys.* **63**:3035.

Berman, M. R., Comita, P. B., Moore, C. B., and Bergman, R. G., 1980, in press.
Bethune, D. S., Lankard, J. R., Loy, M. M. T., Ors, J., and Sorokin, P. P., 1978, *Chem. Phys. Lett.* **57**:479.
Betteridge, D. R., and Yardley, J. T., 1979, *Chem. Phys. Lett.* **62**:570.
Bhatnagar, R., Dyer, P. E., and Oldershaw, G. A., 1979, *Chem. Phys. Lett.* **61**:339.
Bialkowski, S. E., and Guillory, W. A., 1977, *J. Chem. Phys.* **67**:2061.
Bialkowski, S. E., and Guillory, W. A., 1978, *J. Chem. Phys.* **68**:3339.
Bialkowski, S. E., and Guillory, W. A., 1979, *Chem. Phys. Lett.* **60**:429.
Bialkowski, S. E., King, D. S., and Stephenson, J. C., 1979, *J. Chem. Phys.* **71**:4010.
Bittenson, S., and Houston, P. L., 1977, *J. Chem. Phys.* **67**:4819.
Blinov, S. I., Zalesskaya, G. A., and Kotov, A. A., 1978, *Izvest. Akad. Nauk SSSR, Ser. Fiz.* **42**:383.
Boesl, U., Neusser, H. J., and Schlag, E. W., 1979, *Chem. Phys. Lett.* **61**:57, 62.
Bokor, J., Zavelovich, J., and Rhodes, C. K., 1980, *J. Chem. Phys.* **72**:965.
Bordé, Ch., Henry, A., and Henry, L., 1966a, *Compt. Rend. Acad. Sci.* (*Paris*) **B262**:1389.
Bordé, Ch., Henry, A., and Henry, L., 1966b, *Compt. Rend. Acad. Sci.* (*Paris*) **B263**:619.
Braithwaite, M., and Leonard, S. R., 1978, *J. Chem. Phys.* **69**:389.
Brass, S. G., Reed, D. A., and Ehrlich, G., 1979, *J. Chem. Phys.* **70**:5245.
Braun, W., and Tsang, W., 1976, *Chem. Phys. Lett.* **44**:354.
Braun, W., Kurylo, M. J., Kaldor, A., and Wayne, R. P., 1974, *J. Chem. Phys.* **61**:461.
Braun, W., Herron, J. T., Tsang, W., and Churney, K., 1978, *Chem. Phys. Lett.* **59**:492.
Brenner, D. M., 1978, *Chem. Phys. Lett.* **57**:357.
Buechele, J. L., Weitz, E., and Lewis, F. D., 1979, *J. Am. Chem. Soc.* **101**:370.
Burak, I., Quelly, T. J., and Steinfeld, J. I., 1979, *J. Chem. Phys.* **70**:334.
Campbell, J. D., Hancock, G., Halpern, J. B., and Welge, K. H., 1976, *Chem. Phys. Lett.* **44**:404.
Campbell, J. D., Yu, M. H., and Wittig, C., 1978, *Appl. Phys. Lett.* **32**:413.
Catalano, E., Barletta, R. E., and Pearson, R. K., 1979, *J. Chem. Phys.* **70**:3291.
Chekalin, N. V., Dolzhikov, V. S., Letokhov, V. S., Lokhman, V. N., and Shibanov, A. N., 1977, *Appl. Phys.* **12**:191.
Chiu, H. Y., Somers, R. M., and Benson, R. C., 1979, *Chem. Phys. Lett.* **61**:203.
Christensen, C.P., and Lakin, K. M., 1978, *Appl. Phys. Lett.* **32**:254.
Clark, J. H., and Anderson, R. G., 1978, *Appl. Phys. Lett.* **32**:46.
Clark, J. H., Haas, Y., Houston, P. L., and Moore, C. B., 1975, *Chem. Phys. Lett.* **35**:82.
Clark, J. H., Moore, C. B., and Reilly, J. P., 1978, *Internat. J. Chem. Kinetics* **10**:427.
Cohen, C., Borde, Ch., and Henry, L., 1967, *Compt. Rend. Acad. Sci.* (*Paris*), **B625**:267.
Comes, F. J., and Pionteck, S., 1976, *Chem. Phys. Lett.* **42**:558.
Corkum, R., Willis, C., and Back, R. A., 1977, *Chem. Phys.* **24**:13.
Coulter, D., Dows, D., Reisler, H., and Wittig, C., 1978, *Chem. Phys.* **32**:429.
Cox, D. M., Hall, R. B., Horsley, J. A., Kramer, G. M., Rabinowitz, P., and Kaldor, A., 1979, *Science* **205**:390.
Cox, D. M., and Horsley, J. A., 1980, *J. Chem. Phys.* **72**:864.
Crim, F. F., Kwei, G. H., and Kinsey, J. L., 1977, *Chem. Phys. Lett.* **49**:526.
Crocombe, R. A., Smurl, N. R., and Mamantov, G., 1978, *J. Am. Chem. Soc.* **100**:6526.
Dai, H.-L., Kung, A. H., and Moore, C. B., 1979, *Phys. Rev. Lett.* **43**:761.
Dale, J. D., Smy, P. R., Way-Nee, D., and Clements, R. M., 1977, *Combustion and Flame* **30**:319.
Danen, W. C., 1979, *J. Am. Chem. Soc.* **101**:1187.
Danen, W. C., Munslow, W. D., and Setser, D. W., 1977, *J. Am. Chem. Soc.* **99**:6961.
Datta, S., Anderson, R. W., and Zare, R. N., 1975, *J. Chem. Phys.* **63**:5503.
Davies, B., Poliakoff, M., Smith, K. P., and Turner, J. J., 1979, *Chem. Phys. Lett.* **58**:28.
Deutsch, T. F., 1979, *J. Chem. Phys.* **70**:1187.

Deutsch, T. F., Ehrlich, D. J., and Osgood, R. M., Jr., 1979, *Appl. Phys. Lett* **35**:175.

Dever, D. F., and Grunwald, E., 1976, *J. Am. Chem. Soc.* **98**:5055.

Dispert, H. H., Geis, M. E., and Brooks, P. R., 1979, *J. Chem. Phys.* **70**:5317.

Donnelly, V. M., and Pasternack, L., 1979, *Chem. Phys.* **39**:427.

Donohue, T., 1977, *J. Chem. Phys.* **67**:5402.

Drouin, M., Gauthier, M., Pilou, R., Hackett, P. A., and Willis, C., 1979, *Chem. Phys. Lett.* **60**:16.

Duperrex, R., and van den Bergh, H., 1979, *J. Chem. Phys.* **71**:3613.

Duperrex, R., and van den Bergh, H., 1980, *J. Chem. Phys.* **73**:585.

Earl, B. L., and Ronn, A. M., 1976, *Chem. Phys. Lett.* **41**:29.

Evans, D. K., McAlpine, R. D., and McCluskey, F. K., 1978, *Chem. Phys.* **32**:81.

Evans, D. K., McAlpine, R. D., and McCluskey, E. K., 1979, *Chem. Phys. Lett.* **65**:226.

Farneth, W. A., Thomsen, M. W., and Berg, M. A., 1979, *J. Amer. Chem. Soc.* **101**:6468.

Feldmann, D., Zacharias, H., and Welge, K. H., 1980, *Chem. Phys. Lett.* **69**:466.

Fettweis, P., and Nève de Mévergnies, M., 1979, *J. Appl. Phys.* **49**:5699.

Filseth, S. V., Hancock, G., Fournier, J., and Meier, K., 1979 a, *Chem. Phys. Lett.* **61**:288.

Filseth, S. V., Danon, J., Feldmann, D., Campbell, J. D., and Welge, K. H., 1979b, *Chem. Phys. Lett.* **63**:615.

Filseth, S. V., Danon, J., Feldmann, D., Campbell, J. D., and Welge, K. H., 1979c, *Chem. Phys. Lett.* **66**:329.

Fisk, G. A., 1978, *Chem. Phys. Lett.* **60**:11.

Francisco, J., Findeis, M., and Steinfeld, J. I., 1980, *Intl. J. Chem. Kinetics*, in press.

Freeman, M. P., Travis, D. N., and Goodman, M. F., 1974, *J. Chem. Phys.* **60**:231.

Freund, S. M., and Danen, W. C., 1979, *Inorg. Nucl. Chem.* **15**:45.

Freund, S. M., and Lyman, J. L., 1978, *Chem. Phys. Lett.* **55**:435.

Freund, S. M., and Ritter, J. J., 1975, *Chem. Phys. Lett.* **32**:255.

Freund, S. M., and Stephenson, J. C., 1976, *Chem. Phys. Lett.* **41**:157.

Freund, S. M., Maier, W. B., II, Holland, R. F., and Beattie, W. H., 1978, *J. Chem. Phys.* **69**:1961.

Fuss, W., Kompa, K. L., and Tablas, F. M. G., 1979, *Disc. Faraday Soc.* **67**:180.

Garcia, D., and Keehn, P. M., 1978, *J. Am. Chem. Soc.* **100**:6111.

Gardner, E. J., Squire, R. H., Elder, R. C., and Wilson, R. M., 1973, *J. Am. Chem. Soc.* **95**:1693.

Gordon, R. J., and Lin, M. C., 1973, *Chem. Phys. Lett.* **22**:262.

Gough, T. E., Miller, R. E., and Scoles, G., 1978, *J. Chem. Phys.* **69**:1588.

Gower, M. C., and Gustafson, T. K., 1977, *Optics Commun.* **23**:69.

Gower, M. C., and Billman, K. W., 1977, *Appl. Phys. Lett.* **30**:514.

Grimley, A. J., and Stephenson, J. C., 1980, in press.

Haas, Y., and Yahav, G., 1977, *Chem. Phys. Lett.* **48**:63.

Hackett, P. A., Gauthier, M., and Willis, C., 1978, *J. Chem. Phys.* **69**:2924.

Hackett, P. A., Gauthier, M., and Willis, C., 1979a, *J. Chem. Phys.* **71**:546.

Hackett, P. A., Willis, C., and Gauthier, M., 1979b, *J, Chem. Phys.* **71**:2682.

Hackett, P. A., Willis, C., Drouin, M., and Weinberg, E., 1980, *J. Phys. Chem.* in press.

Hall, J. H., Jr., Lesiecki, M. L., and Guillory, W. A., 1978, *J. Chem. Phys.* **68**:2247.

Hall, R. B., and Kaldor, A., 1979, *J. Chem. Phys.* **70**:4027.

Hardy, G. E., Chandra, B. P., Zink, J. I., Adamson, A. W., Fukuda, R. C., and Walters, R. T., 1979, *J. Am. Chem. Soc.* **101**:2787.

Harris, S. J., 1977, *J. Am. Chem. Soc.* **99**:5798.

Hartford, A., Jr., 1978, *Chem. Phys. Lett.* **57**:352.

Hartford, A., Jr., 1979, *Chem. Phys. Lett.* **63**:503.

Hartford, A., Jr., and Tuccio, S. A., 1979, *Chem. Phys. Lett.* **60**:431.

Herman, I. P., 1980, *J. Chem. Phys.* **72**:5777.

Herman, I. P., and Marling, J., 1979a, *J. Chem. Phys.* **71**:643.
Herman, I. P., and Marling, J., 1979b, Infrared photolysis of CDF_3, Lawrence Livermore Laboratory Preprint UCRL-82341 (Feb. 14, 1979).
Herman, I. P., and Marling, J., 1980, *J. Chem. Phys.* **72**:516.
Herman, I. P., Mariella, R. P., and Javan, A., 1976, *J. Chem. Phys.* **65**:3792.
Herman, I. P., Mariella, R. P., and Javan, A., 1978, *J. Chem. Phys.* **68**:1070.
Herman, I. P., Marling, J., and Thomas, S. J., 1980, *J. Chem. Phys.* **72**:5603.
Hicks, K. W., Lesiecki, M. L., Riseman, S. M., and Guillory, W. A., 1979, *J. Phys. Chem.* **83**:1936.
Hill, R. A., and Laguna, G. A., 1980, *Optics Commun.* **32**:435.
Hippler, H., Luther, K., Troe, J., and Walsh, R., 1978, *J. Chem. Phys.* **68**:323.
Hochstrasser, R. M., and King, D. S., 1975, *J. Am. Chem. Soc.* **97**:4766.
Hochstrasser, R. M., and King, D. S., 1976, *J. Am. Chem. Soc.* **98**:5443.
Hoffbauer, M. A., Gentry, W. R., and Giese, C. F., 1979, Pulsed molecular beam study of ethylene dimer dissociation with a CO_2 laser, in *Laser Induced Processes in Molecules* (eds. K. L. Kompa and S. D. Smith), Springer-Verlag *Ser. Chem. Phys.* No. 6, Berlin.
Hoffmann, G., and Ahrens-Botzong, R., 1980, *Chem. Phys. Lett.* **71**:83.
Hoffmann, H., Klopffer, W., and Schafer, G., 1980, *Chem. Phys. Lett.*, in press.
Hsu, D. S. Y., and Manuccia, T. J., 1978, *Appl. Phys. Lett.* **33**:915.
Hudgens, J. W., 1978, *J. Chem. Phys.* **68**:777.
Hudgens, J. W., Durant, J. L., Jr., Bogan, D. J., and Coveleskie, R. A., 1979, *J. Chem. Phys.* **70**:5906.
Huie, R., Herron, J. T., Braun, W., and Tsang, W., 1978, *Chem. Phys. Lett.* **56**:193.
Hwang, W. C., Herm, R. R., Kalsch, J. F., and Gust, G. R., 1979, Multiple-photon chemistry induced by a pulsed CO_2 laser at moderate fluences, Aerospace Corp. Tech. Report ATR-79(8420)-1 (May 15, 1979).
Jackson, W. M., and Halpern, J. B., 1979, *J. Chem. Phys.* **70**:2373.
Jackson, W. M., Halpern, J. B., and Lin, C.-S., 1978, *Chem. Phys. Lett.* **55**:254.
Jalenak, W. A., and Nogar, N. S., 1979, *Chem. Phys.* **41**:407.
Jalenak, W. A., Schultz, D., Fisher, M., and Nogar, N.S., 1979, *Chem. Phys. Lett.* **64**:457.
Jang, J. C., and Setser, D. W., 1979, *J. Phys. Chem.* **83**:2809.
Jones, L. H., Ekberg, S., and Asprey, L., 1979, *J. Chem. Phys.* **70**:1566.
Kaldor, A., Braun, W., and Kurylo, M. J., 1974, *J. Chem. Phys.* **61**:2496.
Karl, R. R., Jr., and Innes, K. K., 1975, *Chem. Phys. Lett.* **36**:275.
Karl, R. R., Jr., and Lyman, J. L., 1978, *J. Chem. Phys.* **69**:1196.
Karlov, N. V., Petrov, Yu. N., Prokhorov, A. M., and Stel'makh, O. M., 1970, *JETP Lett.* **11**:135.
Karlov, N. V., Karpov, N. A., Petrov, Yu. N., Prokhorov, A. M., and Stel'makh, O. M., 1971, *JETP Lett.* **14**:140.
Karlov, N. V., Petrov, R. P., Petrov, Yu. N., and Prokhorov, A. M., 1976, *JETP Lett.* **24**:258.
Karny, Z., and Zare, R. N., 1977, *Chem. Phys.* **23**:321.
Karny, Z., Naaman, R., and Zare, R. N., 1979a, *Chem. Phys. Lett.* **59**:33.
Karny, Z., Gupta, A., Zare, R. N., Lin, S. T., Nieman, J., and Ronn, A. M., 1979b, *Chem. Phys.* **37**:15.
Kim, K. C., Filip, H., and Person, W. B., 1979, *Chem. Phys. Lett.* **54**:253.
Kim, M. S., and Dunbar, R. C., 1980, *J. Chem. Phys.* **72**:4405.
King, D. S., and Stephenson, J. C., 1978, *J. Am. Chem. Soc.* **100**:7151.
King, D. S., and Stephenson, J. C., 1979, *Chem. Phys. Lett.* **66**:33.
Kneba, M., and Wolfrum, J., 1978, *Proc. 17th Internat. Combustion Symp.*, Leeds, pp. 497–504.
Kneba, M., Wellhausen, U., and Wolfrum, J., 1979, *Ber. Bunsengesellschaft Phys. Chem.* **83**:940.
Knott, R. B., and Pryor, A. W., 1979, *J. Chem. Phys.* **71**:2946.

Knyazev, I. N., Kudryavtsev, Yu. A., Kuzima, N. P., Letokhov, V. S., and Sarkisian, A. A., 1978, *Appl. Phys.* **17**:427.
Koren, G., Oppenheim, U. P., Tal, D., Okon, M., and Weil, R., 1976, *Appl. Phys. Lett.* **29**:40.
Kryukov, P. G., Letokhov, V. S., Nikogosyan, P. N., Borodavkin, A. V., Budowsky, E. I., and Simukova, N. A., 1979, *Chem. Phys. Lett.* **61**:375.
Kushawaha, V. S., 1980, *Opt. and Quantum Electronics* **12**:269.
Leary, K. M., Lyman, J. L., Asprey, L. B., and Freund, S. M., 1978, *J. Chem. Phys.* **68**:1671.
Leone, S. R., and Moore, C. B., 1974, *Phys. Rev. Lett.* **33**:269.
Lesiecki, M. L., and Guillory, W. A., 1977, *Chem. Phys. Lett.* **49**:92.
Lesiecki, M. L., and Guillory, W. A., 1978, *J. Chem. Phys.* **69**:4572.
Lesiecki, M. L., and Guillory, W. A., 1979, *J. Chem. Phys.* **70**:4317.
Lesiecki, M. L., Hicks, K. W., Orenstein, A., and Guillory, W. A., 1980, *Chem. Phys. Lett.* **71**:72.
Lewis, K. E., McMillen, D. F., and Golden, D. M., 1980, *J. Phys. Chem.* **84**:227.
Levy, M. R., Reisler, H., Mangir, M. S., and Wittig, C., 1980, *Opt. Eng.* **19**:29.
Lin, C. T., and Atvars, T. D. Z., 1978, *J. Chem. Phys.* **68**:4233.
Lin, C. T., and Bertran, C. A., 1978, *J. Phys. Chem.* **82**:2299.
Lin, S. T., and Ronn, A. M., 1977, *Chem. Phys. Lett.* **49**:255.
Lin, S. T., and Ronn, A. M., 1978, *Chem. Phys. Lett.* **56**:414.
Lin, S. T., Lee, S. M., and Ronn, A. M., 1978, *Chem. Phys. Lett.* **53**:260.
Lory, E. R., Manuccia, T., and Bauer, S. H., 1975, *J. Phys. Chem.* **79**:545.
Lussier, F. M., and Steinfeld, J. I., 1977, *Chem. Phys. Lett.* **50**:175.
Lyman, J. L., and Jensen, R. J., 1972, *Chem. Phys. Lett.* **13**:421.
Lyman, J. L., and Leary, K. M., 1978, *J. Chem. Phys.* **69**:1858.
Lyman, J. L., Jensen, R. J., Rinck, J., Robinson, C. P., and Rockwood, S. D., 1975, *Appl. Phys. Lett.* **27**:87.
Lyman, J. L., Danen, W. C., Nilsson, A. C., and Nowak, A. V., 1979, *J. Chem. Phys.* **71**:1206.
Mannik, L., Keyser, G. M., and Woodall, K. B., 1979, *Chem. Phys. Lett.* **65**:231.
Manning, R. G., Braun, W., and Kurylo, M. J., 1976, *J. Chem. Phys.* **65**:2609.
Manuccia, T. J., Clark, M. D., and Lory, E. R., 1978, *J. Chem. Phys.* **68**:2271.
Marling, J. B., 1975, *Chem. Phys. Lett.* **34**:84.
Marling, J. B., 1977, *J. Chem. Phys.* **66**:4200.
Marling, J. B., and Herman, I. P., 1979, *Appl. Phys. Lett.* **34**:439.
Marling, J. B., Herman, I. P., and Thomas, S. J., 1980, *J. Chem. Phys.* **72**:5603.
Mayer, S. W., Kwok, M. A., Gross, R. W. F., and Spencer, D. J., 1970, *Appl. Phys. Lett.* **17**:516.
McAlpine, R. D., Evans, D. K., and McClusky, F. K., 1979, *Chem. Phys.* **39**:263.
McDonald, J. R., Baronavski, A. P., and Donnelly, V. M., 1978, *Chem. Phys.* **33**:161.
McNeish, A., Poliakoff, M., Smith, K. P., and Turner, J. J., 1976, *J. Chem. Soc. Chem. Comms.* 859.
McNesby, J. R., and Scanland, C., 1979, *Chem. Phys. Lett.* **66**:303.
Merchant, V. E., 1978, *Optics Commun.* **25**:259.
Merritt, J. A., and Robertson, L. C., 1977, *J. Chem. Phys.* **67**:3545.
Molin, Yu. N., Panfilov, V. N., and Strunin, V. P., 1978, *Chem. Phys. Lett.* **56**:557.
Muller, C. H., III, Schofield, K., and Steinberg, M., 1978, in Proc. Natl. Bur. Stds. 10th Materials Res. Symp. on Characterization of High Temp. Vapors and Gases, Gaithersburg, Md., 1978.
Nagai, K., Katayama, M., Mikuni, H., and Takahashi, M., 1979, *Chem. Phys. Lett.* **62**:499.
Nève de Mevergnies, M., and Fettweis, P., 1979, in *Laser-Induced Processes in Molecules* (eds., K. L. Kompa and S. S. Smith), Springer Verlag, pp. 205–208, *Ser. Chem. Phys.* No. 6, Berlin, 1979.
Nieman, J., and Ronn, A. M., 1980, *Opt. Eng.* **19**:39.

Nip, W. S., Drouin, M., Hackett, P. A., and Willis, C., 1980, *J. Phys. Chem.* **84**:932.
Odiorne, T. J., Brooks, P. R., and Kasper, J. V. V., 1971, *J. Chem. Phys.* **55**:1980.
Orr, B. J., and Keentok, M. V., 1976, *Chem. Phys. Lett.* **41**:68.
Pao, Y.-H., and Rentzepis, P. M., 1965, *Appl. Phys. Lett.* **6**:93.
Pazendeh, H., Marsal, C., Lempereur, F., and Tardieu de Maleissye, J., 1979, *Internat. J. Chem. Kinetics* **11**:595.
Plum, C. N., and Houston, P. L., 1980, *Chem. Phys.* **45**:159.
Poliakoff, M., Davies, B., McNeish, A., Tranquille, M., and Turner, J. J., 1978, *Ber. Bunsengesellschaft Phys. Chem.* **82**:121.
Poliakoff, M., Breedon, N., Davies, B., McNeish, A., and Turner, J. J., 1979, *Chem. Phys. Lett.* **56**:474.
Porter, G., and Steinfeld, J. I., 1966, *J. Chem. Phys.* **45**:3456.
Preses, J. M., Weston, R. E., Jr., and Flynn, G. W., 1977, *Chem. Phys. Lett.* **46**:69.
Proch, D., and Schröder, H., 1979, *Chem. Phys. Lett.* **61**:426.
Quick, C. R., Jr., and Wittig, C., 1978a, *Chem. Phys.* **32**:75.
Quick, C. R., Jr., and Wittig, C., 1978b, *J. Chem. Phys.* **69**:4201.
Quick, C. R., Jr., Tiee, J. J., Fischer, T. A., and Wittig, C., 1979, *Chem. Phys. Lett.* **62**:435.
Rabinowitz, P., Stein, A., and Kaldor, A., 1978, *Optics Commun.* **27**:381.
Reddy, K. V., and Berry, M., 1979a, *Disc. Faraday Soc.* **67**.
Reddy, K. V., and Berry, M., 1979b, *Chem. Phys. Lett.* **66**:223.
Reiser, C., Lussier, F. M., Jensen, C. C., and Steinfeld, J. I., 1979, *J. Am. Chem. Soc.* **101**:350.
Reiser, C., and Steinfeld, J. I., 1980, *J. Phys. Chem.* **84**:680.
Reisler, H., Mangir, M., and Wittig, C., 1979, *J. Chem. Phys.* **71**:2109.
Richardson, C., and Shatas, R., 1979, *J. Phys. Chem.* **83**:1679.
Richardson, J. H., and Setser, D. W., 1977, *J. Phys. Chem.* **81**:2301.
Rice, W. W., Wampler, F. B., Oldenborg, R. C., Lewis, W. B., Tiee, J. J., and Park, R. T., 1980, *J. Chem. Phys.* **72**:2948.
Riley, C., and Shatas, R., 1979, *J. Phys. Chem.* **83**:1679.
Ritter, J. J., 1978, *J. Am. Chem. Soc.* **100**:2441.
Rockwood, S. D., and Hudson, J. W., 1975, *Chem. Phys. Lett.* **34**:542.
Rofer-de Poorter, C. K., and de Poorter, G. L., 1979, *Chem. Phys. Lett.* **61**:605.
Ronn, A. M., 1976, *Chem. Phys. Lett.* **42**:202.
Ronn, A. M., and Earl, B. L., 1977, *Chem. Phys. Lett.* **45**:556.
Rossi, M. J., Barker, J. R., and Golden, D. M., 1979a, *Chem. Phys. Lett.* **65**:523.
Rossi, M. J., Barker, J. R., and Golden, D. M., 1979b, *J. Chem. Phys.* **71**:3722.
Rousseau, D. L., Leroi, G. E., and Link, G. L., 1965, *J. Chem. Phys.* **42**:4048.
Sam, C. L., and Yardley, J. T., 1978, *J. Chem. Phys.* **69**:4621.
Sam, C. L., and Yardley, J. T., 1979, *Chem. Phys. Lett.* **61**:509.
Schramm, B., 1979, in *Laser induced processes in molecules*, (eds., K. L. Kompa and S. D. Smith), Springer-Verlag, pp. 274–276, *Ser. Chem. Phys.* No. 6, Berlin, 1979.
Schulz, P. A., Sudbø, Aa. S., Krajnovich, D. J., Kwok, H. S., Shen, Y. R., and Lee, Y. T., 1979, *Ann. Rev. Phys. Chem.* **30**:379.
Selwyn, L., Back, R. A., and Willis, C., 1978, *Chem. Phys.* **32**:323.
Shaub, W. M., and Bauer, S. H., 1975, *Internat. J. Chem. Kinetics* **7**:509.
Slater, R. C., and Parks, J. H., 1979, *Chem. Phys. Lett.* **60**:275.
Slezak, V., Caballero, J., Burgos, A., and Quel, E., 1978, *Chem. Phys. Lett.* **54**:105.
Stafast, H., Opitz, J., and Huber, J. R., 1979, in *Laser Induced Processes in Molecules* (eds., K. L. Kompa and S. D. Smith), Springer-Verlag, pp. 280–282, *Ser. Chem. Phys.* No. 6, Berlin, 1979.
Steel, C., Starov, V., Leo, R., John, P., and Harrison, R. G., 1979, *Chem. Phys. Lett.* **62**:121.
Stephenson, J. C., and Freund, S. M., 1976, *J. Chem. Phys.* **65**:1893, 4303.

Stephenson, J. C., King, D. S., Goodman, M. F., and Stone, J., 1979, *J. Chem. Phys.* **70**:4496.
Stephenson, J. C., Bialkowski, S. E., and King, D. S., 1980, *J. Chem. Phys.* **72**:1161.
Sudbø, Aa. S., Schulz, P. A., Grant, E. R., Shen, Y. R., and Lee, Y. T., 1978a, *J. Chem. Phys.* **68**:1306.
Sudbø, Aa. S., Schulz, P. A., Shen, Y. R., and Lee, Y. T., 1978*b*, *J. Chem. Phys.* **69**:2312.
Sudbø, Aa. S., Schulz, P. A., Grant, E. R., Shen, Y. R., and Lee, Y. T., 1979, *J. Chem. Phys.* **70**:912.
Suzuki, K., Kim, P. H., and Namba, S., 1978, *Appl. Phys. Lett.* **33**:52.
Tam, A. C., Happer, W., and Siano, D., 1977, *Chem. Phys. Lett.* **49**:320.
Tardieu de Maleissye, J., Lempereur, F., Marsal, C., and Ben-Aim, R. I., 1976, *Chem. Phys. Lett.* **42**:46, 472.
Tiee, J. J., and Wittig, C., 1978a, *Appl. Phys. Lett.* **32**:236.
Tiee, J. J., and Wittig, C., 1978b, *J. Chem. Phys.* **69**:4756.
Tiee, J. J., and Wittig, C., 1978c, *Optics Commun.* **27**:377.
Tiee, J. J., Wampler, F. B., and Rice, W. W., 1980, *J. Chem. Phys.* **72**:2925.
Tiffany, W. B., 1968, *J. Chem. Phys.* **48**:3019.
Tiffany, W. B., Moos, H. W., and Schawlow, A. L., 1967, *Science* **157**:40.
Tuccio, S. A., and Hartford, A., Jr., 1979, *Chem. Phys. Lett.* **65**:235.
Umstead, M. E., and Lin, M. C., 1979, *J. Phys. Chem.* **82**:2047.
Umstead, M. E., Talley, L. D., Tevault, D. E., and Lin, M. C., 1980, *Opt. Eng.* **19**:94.
van den Ende, A., Kimel, S., and Speiser, S., 1973, *Chem. Phys. Lett.* **21**:133.
Wampler, F. B., Tiee, J. J., Rice, W. W., and Oldenborg, R. C., 1979, *J. Chem. Phys.* **71**:3926.
Warren, C. H., 1978, *Chem. Phys. Lett.* **53**:509.
West, G. A., Weston, R. E., Jr., and Flynn, G. W., 1978, *Chem. Phys.* **35**:275.
Woodin, R. L., Bomse, D. S., and Beauchamp, J. L., 1978, *J. Am. Chem. Soc.* **100**:3248.
Woodin, R. L., Bomse, D. S., and Beauchamp, J. L., 1979, *Chem. Phys. Lett.* **63**:630.
Würzberg, E., Kovalenko, L. J., and Houston, P. L., 1978, *Chem. Phys.* **35**:317.
Yahav, G., and Haas, Y., 1978, *Chem. Phys.* **35**:141.
Yahav, G., Haas, Y., Carmeli, B., and Nitzan, A., 1980, *J. Chem. Phys.* **72**:3410.
Yeung, E. S., and Moore, C. B., 1972, *Appl. Phys. Lett.* **21**:109.
Yogev, A., and Loewenstein-Benmair, R. M. J., 1973, *J. Am. Chem. Soc.* **95**:8487.
Yogev, A., and Loewenstein-Benmair, R. M. J., 1976, British Patent No. 1,429,426.
Yogev, A., and Benmair, R. M. J., 1977, *Chem. Phys. Lett.* **46**:290.
Yogev, A., Loewenstein, R. M. J., and Amar, D., 1972, *J. Am. Chem. Soc.* **94**:1091.
Yu, M. H., Reisler, H., Mangir, M., and Wittig, C., 1979, *Chem. Phys. Lett.* **62**:439.
Zipin, H., and Speiser, S., 1975, *Chem. Phys. Lett.* **31**:102.
Zittel, P. F., and Little, D. D., 1979, *J. Chem. Phys.* **71**:2748.
Zitter, R. N., and Koster, D. F., 1978, *J. Am. Chem. Soc.* **100**:2259.
Zitter, R. N., Koster, D. F., Cantoni, and Pleil, J., 1979, in *Laser-induced Processes in molecules* (eds., K. L. Kompa and S. D. Smith), Springer-Verlag, pp. 277–279, *Ser. Chem. Phys.* No. 6, Berlin, 1979.

4.3. Selected Review Articles, Monographs, and References to Theory and Diagnostic Techniques

Akhmanov, A. S., Baranov, V. Yu., Pismenny, V. D., Bagratashvili, V. N., Kolomiisky, Yu. R., Letokhov, V. S., and Ryabov, E. A., 1977, *Opt. Commun.* **23**:357. Optoacoustic energy deposition measurements in SF_6, C_2H_4, CH_3F.

Basov, N. G., Belenov, E. M., Markin, E. P., Ora'evskii, A. N., and Pankratov, A. V., 1973, *Soc. Phys. JETP* **37**:247. A model of laser heating is described, along with specific reactions of $N_2F_4 + NO$, N_2O, H_2, CH_4, and BCl_3; $SiH_4 + BCl_3$ and SF_6; and $HNF_2 + CH_4$, SiH_4, and H_2.

Bauer, S. H., 1978, How energy accumulation and disposal affect the rates of reactions, *Chem. Revs.* **78**:147. Role of rotational and vibrational excitation in controlling the rate of chemical reactions; 313 references.

Berry, M. J., 1975, Laser studies of gas phase chemical reaction dynamics, *Ann. Rev. Phys. Chem.* **26**:259. Includes laser diagnostics as well as laser-induced chemical reactions; 288 references.

Birely, J. H., and Lyman, J. L., 1975, Effect of reagent vibrational energy on measured reaction rate constants, *J. Photochem.* **4**:269. Unsuccessful attempt at correlating rate enhancement data with macroscopic observables.

Braun, W., Kurylo, M. J., and Kaldor, A., 1974, *Chem. Phys. Lett.* **28**:440. Measurement of energy flow in SiF_4–O_3–O_2 mixtures.

Bylinsky, G., 1977, Laser alchemy is just around the corner, *Fortune Magazine*, Sept. **1977**:186–190. Popularized account.

Cantrell, C. D., Freund, S. M., and Lyman, J. L., 1980, Laser induced chemical reactions and isotope separation, in *Laser Handbook*, Vol. III (b) (ed., M. Stitch), North-Holland, Amsterdam.

Fuss, W., Kompa, K. L., Proch, D., and Schmid, W. E., 1977, High power infrared laser chemistry, in *Lasers in Chemistry* (ed., M. A. West), Elsevier, Amsterdam, pp. 235–244.

Gordiets, B. F., Osipov, A. I., and Panchenko, V. Ya., 1977, On the separation of isotopes in collisional chemical reactions with selective excitation of gas molecules vibrations by laser radiation, *Sov. Phys. Doklady* **22**:325. Model for CW excitation.

Grunwald, E., Dever, D. F., and Keehn, P. M., 1978, *Megawatt Infrared Laser Chemistry*, John Wiley, New York. Primarily discusses Brandeis work; 109 references, list of laser lines.

Happer, W., *et al.*, 1979, *Laser Induced Photochemistry*, JASON Report JSR-78-11, Stanford Research Institute (February, 1979). Discusses prospects of laser-induced chemistry for commercial applications.

Houston, P. L., 1979, Initiation of atom–molecule reactions by infrared multiphoton dissociation, *Adv. Chem. Phys.* in press.

Kaldor, A., and Hastie, J. W., 1972, Infrared laser modulated molecular beam mass spectrometer, *Chem. Phys. Lett.* **16**:328. Diagnostic technique.

Karlov, N. V., 1974, Laser induced chemical reactions, *Appl. Opt.* **13**:301. Forty-nine references.

Karlov, N. V., and Prokhorov, A. M., 1976, Laser separation of isotopes, *Soviet Physics Uspekhii* **118**:153. Sixty-seven references.

Karlov, N. V., and Prokhorov, A. M., 91977, *Sov. Phys. Uspekhii* **123**:57. Includes matrix and surface reactions; 59 references.

Kimel, S., and Speiser, S., 1977, Lasers and chemistry, *Chem. Revs.* **77**:437. Eight hundred forty-six references.

Kneba, M., and Wolfrum, J., 1980, Bimolecular reactions of vibrationally excited molecules, *Ann. Rev. Phys. Chem.* **31**:47. Two hundred fifty-four references.

Kompa, K. L., Fuss, W., Proch, D., and Schmid, W. E., 1978, Recent progress in the study of molecular dissociation and isotope separation by high power infrared lasers, in *Electronic and Atomic Collisions* (ed., G. Watel), North-Holland Publishing Co., Amsterdam, pp. 737–746.

Kompa, K. L., Fuss, W., Proch, D., Schmid, W. E., Smith, S. D., and Schröder, H., 1979, Towards an understanding of infrared multiphoton absorption and dissociation, in *Nonlinear Behavior of Molecules, Atoms, and Ions in Electric, Magnetic, or Electromagnetic Fields*, Elsevier Publishing Co., Amsterdam, pp. 55–63.

Lee, E. K. C., Laser photochemistry of selected vibrational and rotational states, *Accts. Chem. Research* **10**:319. Rotationally and vibrationally resolved ultraviolet photochemistry of cyclobutanone and formaldehyde.

Letokhov, V. S., 1972, On selective laser photochemical reactions by means of photopredissociation of molecules, *Chem. Phys. Lett.* **15**:221. Suggested experiment, published simultaneous with Yeung and Moore work on formaldehyde.

Letokhov, V. S., 1973, Use of lasers to control selective chemical reactions, *Science* **180**:451.

Letokhov, V. S., 1977, Laser separation of isotopes, *Ann. Rev. Phys. Chem.* **28**:133. One hundred thirteen references.

Lyman, J. L., Quigley, J. P., and Judd, O. P., 1980, Single-infrared-frequency studies of multiple-photon excitation and dissociation of polyatomic molecules, in *Multiple-Photon Excitation and Dissociation of Polyatomic Molecules* (ed., C. D. Cantrell), Springer-Verlag, Berlin, Heidelberg. Extensive tabular data on cross sections, pressure effects; 242 references.

McNair, R. E., Fulghum, S. F., Flynn, G. W., Feld, M. S., and Feldman, B. J., 1977, *Chem. Phys. Lett.* **48**:241. Non-Boltzmann energy distributions in $\nu_1, \ldots, \nu_6$ of CH_3F measured by infrared fluorescence.

Moore, C. B., 1973, Application of lasers to isotope separation, *Accts. Chem. Research* **6**:323. Emphasizes work on formaldehyde.

Moore, C. B., ed., 1977, *Chemical and Biochemical Applications of Lasers, Volume III*, Academic Press, New York. Includes the following two chapters: (I) V. S. Letokhov and C. B. Moore, Laser Isotope Separation, pp. 1–165 (367 references); (II) R. V. Ambartzumyan and V. S. Letokhov, Multiple Photon Infrared Laser Photochemistry, pp. 167–314 (112 references).

Ora'evsky, A. I., 1973, Chemical lasers and chemical reactions induced by lasers, in *Trends in Physics*, European Physical Society, Geneva, pp. 95–124. One-hundred twenty-nine references.

Ora'evskii, A. I., 1974, *Radiofizika* **17**:608. "Two-temperature" model.

Orel, A. E., and Miller, W. H., 1978, Infrared laser induced chemistry, *Chem. Phys. Lett.* **57**:362; also 1979, *J. Chem. Phys.* **70**:4393. Calculation of "effect" on $H + H_2$ cross section at 10^{12} W/cm^2.

Pert, G. J., 1973, *I.E.E.E. J. Quantum Electronics* **QE-9**:435. Calculation of multiphoton dissociation probability, two years before published experimental results.

Robinson, A. L., 1976, *Science* **193**:1230. Brief survey of status of current research.

Ronn, A. M., 1979, Laser Chemistry, *Scientific American*, **May 1979**:114–128. Popularized account.

Schultz, P. A., Sudbø, Aa. S., Krajnovich, D. A., Kwok, H. S., Shen, Y. R., and Lee, Y. T., 1979, Multiphoton dissociation of polyatomic molecules, *Ann. Rev. Phys. Chem.* **30**:379. One hundred forty references.

Sudbø, Aa. S., Krajnovich, D. J., Schulz, P. A., Shen, Y. R., and Lee, Y. T., 1979, Molecular beam studies of laser induced multiphoton dissociation, in *Multiple-Photon Excitation and Dissociation of Polyatomic Molecules* (ed., C. D. Cantrell), Springer-Verlag, Berlin. Review of molecular beam studies; 37 references.

Tablas, F. M. G., Schmid, W. E., and Kompa, K. L., 1976, Vibrational heating of ν_2 mode of NH_3 by a CO_2 laser, *Optics Commun.* **16**:136. Monitored $\nu_2(+)$, $\nu_2(-)$, $2\nu_2$ level populations by ultraviolet absorption.

Verdieck, J. F., 1969, Chemical reactions induced by laser radiation, *Nuclear Appls.* **6**:474. Earliest review of laser-induced chemistry; 29 references.

Wolfrum, J., 1977, Reactions of vibrationally excited molecules, *Ber. Bunsengesellschaft Phys. Chem.* **81**:114. One hundred and ten references.

Note Added in Proof

Supplementary Table of Laser-Induced Reactions

Reactant	Added reagents	Products	Laser: Type	Laser: λ (μm)	Laser: Power/energy	Conditions, remarks	References
CHF_2Cl		Dissociation	CO_2 $R_2(40)$	9.17	1–3 J/cm^2	CF_2 observed by UV absorption	Duperrex and van den Bergh (1980b)
CF_3Cl^+, CF_3Br^+, CF_3I^+		CF_3^+ + halogen	CO_2	10.5	0.5–1.0 W CW	$[CF_3X^+]$ in ion beam	Coggiola *et al.* (1980)
CF_3Br		C_2F_6	CO_2			^{13}C enrichment	Gauthier *et al.* (1980)
CF_3Br	SiO_2(solid)	Etching of SiO_2	CO_2	9.8	0.3 J/pulse	Surface etching by CF_3	Steinfeld *et al.* (1980)
CF_3I		C_2F_6, I_2	CO_2 $R_2(12)$	9.32	8 J/pulse	^{13}C enrichment	Fuss and Schmid (1979)
CH_3OHF^-		CH_3OH, F^-	CO_2	9.33, 10.65	0–6 J/cm^2	In ICR	Rosenfeld *et al.* (1980)
$(C_2H_4)_2$		C_2H_4	CO_2	10.2–10.7	1–12 W/cm^2 CW	Molecular beam	Casassa *et al.* (1980)
cyclo-C_4H_7Cl		C_2H_4, C_2H_3Cl, C_2H_2, C_4H_6, C_4H_8	CO_2 $P_2(34)$	9.67	0.1–0.4 J/pulse		Francisco and Steinfeld (1980)
SF_6	HCl, HBr, DBr, HI	HF	CO_2	10.6		Measured F + HX rate constants	Würzberg and Houston (1980)
SF_6		Dissociation	CO_2 $P_1(6)$–$P_1(36)$	10.4–10.8	1–8 J/cm^2	In molecular beam	Schulz *et al.* (1980)
SF_6		Dissociation	CO_2 $P_1(20)$	10.59	20 J/pulse	^{34}S enrichment	Fuss and Schmid (1979)
$[UO_2(hfacac)_2]_2$		$UO_2(hfacac)_2$	CO_2 $P_1(26)$–$P_1(32)$, $R_1(4)$–$R_1(6)$	10.7, 10.35	10^3 W/cm^2 CW	Isotope-selective	Cox and Maas (1980); Horsley *et al.* (1980)
Br_2	C_2H_2	*trans*-$C_2H_2Br_2$	Ar^+	0.488, 0.514	1–2 W		Kushawaha (1980b)

References to Supplementary Table

Casassa, M. P., Bomse, D. S., Beauchamp, J. L., and Janda, K. C., 1980, *J. Chem. Phys.* **72**:6805.
Coggiola, M. J., Cosby, D. C., and Peterson, J. R., 1980, *J. Chem. Phys.* **72**:6507.
Cox, D. M., and Maas, E. T., Jr., 1980, *Chem. Phys. Letts.* **71**:330.
Duperrex, R., and van den Bergh, H., 1980b, *J. Mol. Struct.* **61**:291.
Francisco, J., and Steinfeld, J. I., 1980, *Intl. J. Chem. Kinetics*, in press.
Fuss, W., and Schmid, W. E., 1979, *Ber. Bunsengesellschaft Phys. Chem.* **83**:1148.
Gauthier, M., Nip, W. S., Hackett, P. A., and Willis, C., 1980, *Chem. Phys. Letts.* **69**:372.
Horsley, J. A., Rabinowitz, P., Stein, A., Cox, D. M., Brickman, R. O., and Kaldor, A., 1980, *I.E.E.E. J. Quantum Electronics* **QE-16**:412.
Kushawaha, V. S., 1980b, *Phys. Scripta* **21**:179.
Rosenfeld, R. N., Jasinski, J. M., and Brauman, J. I., 1980, *Chem. Phys. Letts.* **71**:400.
Schulz, P. A., Sudbø, Aa. S., Grant, E. R., Shen, Y. R., and Lee, Y. T., 1980, *J. Chem. Phys.* **72**:4985.
Steinfeld, J. I., Anderson, T. G., Reiser, C., Denison, D. R., Hartsough, L. D., and Hollahan, J. R., 1980, *J. Electrochem. Soc.* **127**:514.
Würzberg, E., and Houston, P. L., 1980, *J. Chem. Phys.* **72**: 5915.

ACKNOWLEDGMENTS

Preparation of this survey was supported, in part, by contracts from the Air Force Office of Scientific Research and the Advanced Isotope Separation Technology Program, U.S. Department of Energy. The author would also like to thank Prof. Dr. Herbert Walther and Dr. Karl Kompa, directors of the Projektgruppe for Laserforschung of the Max-Planck-Gesellschaft, for their hospitality in Garching where a part of this review was prepared.

Author Index

Authors' names appearing only in the Tables and References of Chapter 4 are not listed separately.

Subject Index

Chemical compounds appearing only in the Tables on pp. 245–256 are not listed separately.